Origins and Successors of the Compact Disc

Philips Research

VOLUME 11

SCOPE TO THE '*PHILIPS RESEARCH BOOK SERIES*'

As one of the largest private sector research establishments in the world, Philips Research is shaping the future with innovative concepts that meet peoples' needs and desires. While the ultimate user benefits of these inventions end up on the high-street shelves, the often pioneering scientific and technological basis usually remains less visible.

This 'Philips Research Book Series' has been set up as a way for our researchers to contribute to the scientific community by publishing their comprehensive results and theories in a convenient book form.

Dr Peter Wierenga

Chief Executive Officer of Philips Research

For other titles published in this series, go to
www.springer.com/series/6437

Origins and Successors of the Compact Disc

Contributions of Philips to Optical Storage

By

Hans Peek
Fellow, IEEE;
formerly with Philips Research Laboratories, Eindhoven, The Netherlands

Jan Bergmans
Eindhoven University of Technology, Eindhoven, The Netherlands

Jos van Haaren
Philips Research Laboratories, Eindhoven, The Netherlands

Frank Toolenaar
Philips Research Laboratories, Eindhoven, The Netherlands

and

Sorin Stan
Philips Consumer Lifestyle, Eindhoven, The Netherlands

Springer

Authors
Hans Peek
Fellow, IEEE
formerly with Philips
Research Laboratories
Eindhoven
The Netherlands

Jan Bergmans
Eindhoven University of Technology
Eindhoven
The Netherlands
j.w.m.bergmans@tue.nl

Jos van Haaren
Philips Research Laboratories
Eindhoven
The Netherlands
j.a.m.m.van.haaren@philips.com

Frank Toolenaar
Philips Research Laboratories
Eindhoven
The Netherlands
frank.toolenaar@philips.com

Sorin Stan
Philips Consumer Lifestyle
Eindhoven
The Netherlands

ISBN 978-1-4020-9552-8 ISBN 978-1-4020-9553-5 (eBook)

DOI 10.1007/978-1-4020-9553-5

Library of Congress Control Number: 2008941734

Printed on acid-free paper

9 8 7 6 5 4 3 2 1

springer.com

Acknowledgements

The editors would like to thank the following persons for their help in providing historical information for Chapter 1 and Sections 2.1, 3.1 and 6.1: C.P.M. Baggen, H.F.M. Beckers, C. Buma, L.M.H. Driessen, T. Fujimoto, J.P.J. Heemskerk, H. Hoeve, R.J.van de Plassche, K. Schep, J.J.H.B. Schleipen, E.C. Schylander, J.P. Sinjou and G.J.J. Vos. They are further indebted to J.P.J. Heemskerk, V.P.J.M. Paulissen, R.J. van de Plassche, K. Schep, J.P. Sinjou, J. de Vries, F. Uittenbogaard and G.J.J. Vos for proofreading parts of the book. Thanks are also due to F.J.W. Schraven and J.C.M. de Valk-Roulaux for their typographic contributions.

Contents

Chapter 1

THE COMPACT DISC AS A HUMBLE MASTERPIECE

J.A.M.M. van Haaren

Philips Research Laboratories Eindhoven

Perhaps the simplest piece of art on display in the entire Museum of Modern Art in New York, is a 12 cm brightly reflecting, plastic disc with a small hole in its centre. This artefact, a Compact Disc with its rainbow-like colours, is exhibited in the museum's department of Architecture and Design. That department welcomes visitors with an explanation of the criteria the curators used for including contemporary design objects in their prestigious collection[1]. One of these criteria is *Innovation*. "Good designers transform the most momentous scientific and technological revolutions into objects that anybody can use." Other criteria, like *Cultural Impact*, are mentioned as well before arriving at the final criterion: Necessity. "Here is the ultimate litmus test: if this object had never been designed or produced, would the world miss it, even just a bit? As disarming as this question might seem, it really works. Try it at home."

The Compact Disc (CD) was inducted into the Museum of Modern Art (MoMA) in 2004 in an exhibition called *Humble Masterpieces*. At this exhibition, the CD occurred together with design highlights such as the paperclip and the tea bag. To be exhibited next to these common objects is an unexpected result and an honour for a development that started about 30 years earlier at Philips in Eindhoven, the Netherlands.

The seemingly simple disc represents a lot more than meets the eye. People may be intrigued by the rainbow interference colours from a Compact Disc. On a microscopic scale each CD contains millions of bits, coded and stored in a globally standardized form, and reproduced with unprecedented precision in millions of low-cost copies. The interference colours are a macroscopic manifestation of this.

This precious disc is useless without a player. The CD-player has evolved from a sophisticated laboratory set-up with state of the art contributions from many scientists and engineers with a wide variety of backgrounds, via advanced products that were costing a monthly salary, into a commodity product that sells for the price of two cinema tickets.

The CD stands for a new industry that created new formats for optical data storage and new applications in the decades that followed the launch of the CD. The CD arrived in a consumer electronics landscape of vinyl discs discs and magnetic tape that was dominated by analogue electronics. The CD-system was the first digital entertainment product brought to the consumer's home, and in this way it marks the change of a paradigm. The resulting benefit of robust, accurate, wearless play-back was clear from the onset.

The cultural impact of introducing digital technology for content storage, distribution and play-back proved to be even bigger than that. Digital content allows transfer to other media without loss of quality, even when the channel between the two is imperfect. This perfect-quality transfer leading to 'pure' sound was a technical ambition and inspiration to the experts in the seventies. By now it is a common notion for consumers, even for technical laymen. To some extent it has led to a separation of content (a song, a photo, a document) from medium (an optical disc, a memory card, a hard disk, or a file on a server). Even young children are aware that valuable content may be protected and preserved in its original form via digital copies, and in some cases it takes a considerable effort to explain that there was a time that this was not possible. In addition, the introduction of digital technology in mass markets for consumers has motivated an unprecedented, competitive race to more powerful and smarter devices at stunningly lower costs.

The CD was at the start of the digital entertainment era for consumers. The distribution of music turned out to be only a first step. Later, CD-ROM was standardized (1985) and also became popular. With the increasing popularity of personal computers (PCs), user-friendly and cost-effective ways of distribution of software and data became of crucial importance. This could initially only be done with magnetic media, the so-called floppies. In the early nineties, popular software releases encompassed typically a series of floppy discs. But then the software releases grew bigger: Windows 95, for example, was released on 13 floppy discs. This made the alternative offering of an even more complete CD-release of the same program very attractive. CD become a crucial enabler for the evolution of the PC industry in that period of time. In addition the CD also enabled the distribution of games. The extra storage space and the random access to content enabled games with increasing sophistication and with more appealing and realistic graphics.

Let's return to the show-case with the compact disc in the Museum of Modern Art in New York. Artifacts on display in the museum come with a small label that acknowledges the artists who created it. For the Compact Disc, the card-board label says: "Philips Research Laboratories, Dutch, est. 1891, and Sony Research Laboratories, Japanese, est. 1946". The label fails to specify the year of creation: it only says "1970s".

The reference to both Philips and Sony gives proper recognition of the excellent teamwork between these two established, global companies that was essential to the success of optical recording. There has been a much wider and crucial support of thousands of other companies that followed Philips and Sony. This wide industry support has led to globally accepted standards that have meant so much for consumers and for the industry.

The absence of a single creator, designer or inventor aligns well with the answer Philips consistently gives on a question often asked to them: Who invented the CD? Philips Research answers this question on its website[2] as follows. "The inventor of the CD does not exist. Nobody even invented one part of the technology alone. The CD was invented collectively by a large group of people working as a team. Emil Berliner, the founder of Deutsche Grammophon, might have been able to invent the gramophone record on his own in 1887, but the technology on which the CD is based is too complex for just one genius. "We needed all the skills that you would find in a large lab," says Piet Kramer, who at the time was head of the Optics group that made a significant contribution to the CD technology. "Electronics engineers, photographic experts, mechanical engineers, control engineers, you have to bring all of these experts together, and then look to see if it can be done." The pooling of creativity like this is typical of the way in which technological progress is made nowadays."

So what could be invented at the start of the CD, and what not?

When in the seventies, a skilled digital communication engineer or someone familiar with the recording of digital signals on magnetic tape or hard disk, looked at a track on Video Long Play (VLP) disc consisting of pits and lands, (s)he would have concluded that this new optical recording system was ideally suited to record and play back digital signals. In the worlds of patent and inventions, such a merging of two existing major technologies is called an evident step. And it cannot be patented. So there is not a single invention, nor are there inventors of the CD as such.

However, the successive digital signal processing operations and subsystems in the CD-system had to be adapted to the properties of the optical storage medium and the reading device. This required new ideas, with major inventive steps. Many of these inventions were done at Philips. The realization of each subsystem, taking into account the proper functioning of the other subsystems, required team work and this holds even stronger for the total CD system. When

Philips decided in the seventies to start the development of the CD player and the disc, they showed an exceptional vision. And when the CD system was unveiled to the public on March 8, 1979, it was the result of great team-work by experts and inventors from many different disciplines.

Scientists, engineers and business men and women have worked for years on making this happen. For some of them, this has been a single project. A step in their personal development and career. Others have built a life-long career in the optical disc industry. Some of the early contributors to the CD in the late 1970s had moved to senior positions in their companies around the year 2000. And the field of optical disc storage had grown into a global, mature industry. Trade fairs, supplier networks, specialized workshops, industrial roadmap committees and global conferences had become part of the routine in this industry.

Expertise fields had been introduced in the optical disc storage world as a topic of an individual scientist or engineer in Eindhoven or Tokyo, often with its roots in adjacent applications. Examples are lens design and manufacturing, solid state lasers and photodetectors, actuators and servo electronics, digital rights management, coding and signal processing for detection, materials for read-only, write once, and rewritable discs, disc mastering and replication. Around the year 2000, each of these fields had become specialisms with dedicated sessions at international optical-storage conferences, and in some cases dedicated supplier-companies of knowledge and tools.

Specialists of several companies and academia in Europe, Japan and the USA, but also in Korea, Taiwan, China and India met each other at these international events. They increased performance of their current products, for instance in the speed race for optical recording. At the same time they pushed down manufacturing costs. But they were also interested in inventive solutions for a next generation optical disc formats. In 1995 this resulted in the realization of a second generation optical disc standard for standard-definition video: the Digital Versatile Disc (DVD) with more than 7 times the storage capacity of CD. And while DVD was becoming a big market success, already at the turn of the century some of the people who started the Compact Disc worked intensively with younger generations to look even beyond DVD. They used their joint expertise for the creation of the new Blu-ray Disc (BD) format, boosting the storage capacity with another factor of 5 compared to DVD. This BD-format serves for the distribution of high-definition video content. Blu-ray Disc is now conquering the market as, perhaps, the ultimate optical disc format.

A lot of this has been facilitated by global standards. This has been started with CD-audio, but it was later followed by many more standards, on different modalities of optical storage (read-only, recordable, rewritable) and different applications (computers, audio, video). It may even be claimed that the worldwide recognition in the market has been a critical success factor that was enabled by the global standards. And this recognition could be identified both at consumer and supplier side. Consumers could be confident that CDs of different brands or different geographical origin would play in their appliances at home. And manufacturers could be confident that if they had met the specifications, their systems would find their place in the optical disc storage world.

The founders of CD have retired or are close to retirement now. It is appropriate to acknowledge and honour their contributions to this industry. Their heritage is a mass-market optical disc technology that has been pushed to its limits. Its specifications are far beyond the imagination of the original CD-workers in the 1970s. And its business impact and global proliferation have met only the most optimistic projections at its market introduction.

The optical data storage story has started with imagination and inventions in research laboratories. Its breakthrough success was, however, only possible because the ideas resonated in the business groups. People saw an opportunity and acted on that by creating appealing products. Right from the start, technical developers and business managers took a leading role in this process. This interplay of science, technology and business may be caught in a single term: Innovation.

In the end, the success of optical discs has been created in the markets. It is granted by our customers, and by our customers only. Since the invention of the compact disc, billions of discs have been sold and almost everybody on our planet uses them. They have enriched people's lives via the distribution and reproduction of music, and later also data, movies, software, and as a back-up medium of records ranging from digital pictures to tamper-free off-line back-up of mission-critical data.

This rich world may indeed be caught in a simple, shiny disc on display in an exhibition on contemporary design in a museum filled with masterpieces. The curators of the museum ask the *Necessity*-question as the litmus test for justification of its presence: "Would the world have missed it, if it had not been invented or produced?" We think the answer is yes.

About this book

The advent of the compact disc has been an important milestone for today's digital world. This book has been created at the occasion of the awarding of an IEEE Milestone in Electrical Engineering and Computing[3] to Philips to commemorate the first public announcement of the Compact Disc, at a press conference on March 8, 1979. The book provides a survey of the evolution of optical storage, with an emphasis on the contributions of Philips to this field. It covers 4 phases: (1) The work leading to the first prototype (Pinkeltje) and its public announcement, (2) The CD system as standardized by Philips and Sony, (3) the period following the market introduction of Compact Disc audio, with the proliferation of new formats, like CD-ROM, CD-I, and finally with DVD and its standards, and (4) the research leading to Blu-ray Disc, the highest capacity optical disc on the market today. For phases (1), (2) and (4) the book provides introductory historical perspectives, followed by reprints of seminal texts by Philips technical experts. For phase (3), it can be argued that the success of CD and DVD owes much to the development of worldwide standards for CD and DVD formats. For this reason the book covers phase (3) via a detailed account of these standards and formats.

While the editors have used their best efforts in preparing this book, they make no representation or warranties with respect to the accuracy of the contents.

References

[1] Paolo Antonelli and Christian Larsen, Department of Architecture and Design, New York Museum of Modern Art, New York, USA. See also www.moma.org.

[2] See www.research.philips.com.

[3] See http://www.ieee.org/web/aboutus/history_center/milestones_intro.html for a complete list of IEEE milestones.

Chapter 2

THE PHILIPS PROTOTYPE OF THE CD SYSTEM

2.1 Introduction to contributions on the Philips prototype of the Compact Disc digital audio system

J.B.H. Peek

On March 8, 1979, a prototype of the Compact Disc (CD) digital audio system was presented at Philips in Eindhoven, the Netherlands, to an audience of about 300 journalists. The system was presented and demonstrated by J.P. Sinjou, the head of the Compact Disc laboratory of Philips' main industry group Audio. The optical disc he showed had a diameter of 11.5 cm. The text of his presentation, together with the slides that he used, is reproduced in Sect. 2.2. Referring to this demonstration, R. Bernard noted in his paper ('Higher fi by digits', IEEE Spectrum, pp. 28-32, Dec. 1979) that "Demonstration systems have been impressive, and the total lack of background noise of any kind during pauses in musical passages is particularly dramatic". Since the prototype CD-player had such small dimensions, the engineers of the Compact Disc laboratory named it 'Pinkeltje' after a tiny dwarf who plays the central role in a Dutch fairy tale book. The text by J.P. Sinjou is followed by three papers that describe various subsystems used in the prototype player.

The demonstrated system was the conclusion of a successful merger of two major existing technologies. First, the optical read out, by using a laser, of information stored on a disc, and, second, the digital coding/decoding and digital processing of signals.

The optical playback of an analog color video signal by using a laser was introduced in 1973 by Philips with the VLP (Video Long Play) system. The development of the VLP system was the result of the combined effort of a team of specialists in very divergent fields. In 1974, the VLP player and the disc became available on the market. An introduction to the VLP system was presented by K. Compaan and P. Kramer in a paper (1973) that is reprinted here in Sect. 2.3. The experience that was obtained in developing the VLP system was crucial in the realization of the optical part of the CD prototype player. This is also true for the production, on a small scale, of CD discs for the prototype player.

In the VLP player there is no mechanical contact between the optical pick-up unit and the disc. The information on the VLP disc is present in the form of a spiral track that consists of a succession of pits and flat areas called lands. In the case of the VLP disc the length of a pit and also of a land is a continuous variable. This is in contrast to a CD disc where the length of a pit and a land is a discrete variable. The track is optically scanned by a laser beam that is focused by an objective lens on the information layer of the disc. Before the beam reaches the information layer it passes a transparent protective layer. When the spot of the beam falls on a land, the light is almost totally reflected. After that, the light is detected by a photodiode. However, when the spot falls on a pit, the depth of which is about a quarter of the wavelength of the laser light, interference and extinction occur which cause less light to be reflected and to reach the photo-diode. Hence, ideally the output signal of the photodiode is a fair representation of the originally recorded signal. Unfortunately, there are several sources of errors that can occur in or on an optical disc. First, small unwanted particles or air bubbles in the plastic material, or pit inaccuracies, may occur in the replication process. This can cause errors when the information is read out by a laser. Second, fingerprints or scratches may appear on the disc when handled. As a consequence of this, and of the small dimension of the pits, the errors mainly occur in bursts. A burst (dropout) implies that the signal pattern at the output of the photodiode differs for a long interval, encompassing many pits, from the originally recorded pattern.

There are two reasons why these errors do not seriously affect the picture quality in the VLP system. The first reason, generic to all optical storage systems, is that the diameter of the beam at the surface of the disc is much wider than the diameter of the spot at the information layer. As a result, local defects and imperfections at the disc surface effectively get blurred and de-emphasized in the readout signal. This effect is inherent in reading out a disc through a transparent substrate, and constitutes one of the key patents of the CD system (P. Kramer, "Reflective optical record carrier", U.S. patent 5,068,846). The second reason, specific to the VLP system, is that in a TV picture there is a high correlation between two successive lines. A dropout can be detected and rendered much less visible by replacing the affected line by the preceding line (U.S. patent 4,032,966). However, the correlation in a signal is not always present in a useful form to conceal errors. This was observed in 1975 with the failure of experiments made at Philips to play back high-fidelity analog audio signals recorded on an optical disc. In this case burst errors caused an unacceptable deterioration of the audio. It was at this time that it became clear to most people at Philips that the only solution to record high-fidelity audio signals was to go digital.

In the VLP player and also in the CD prototype player, three servo systems are used. The first servo system ensures that the light beam is kept on track. The second servo system keeps the spot focused on the information layer. The third servo system ensures that the beam scans the spiral track at a constant velocity. The function of the first servo system in the CD prototype can, if no precautions are taken, be disturbed by the digital recorded signal. To prevent this disturbance of the servo system in the CD prototype player, the digital signal is modulated before recording. By applying modulation prior to recording, the frequency spectrum of the recorded digital signal can be given a spectral null at zero frequency. As a consequence, the first servo system is only minimally disturbed. A modulation code called M3, invented by M.G. Carasso, W.J. Kleuters and J.J. Mons, was used in the CD prototype. Although this code was not described in a journal or conference paper, it is covered in a U.S. Patent (4,410,877) that was granted in 1983 to the three inventors.

When the stored signal is digital, a certain number of errors can be corrected by using error correcting codes. A high-fidelity analog audio signal can be digitized by using pulse code modulation (PCM). PCM was proposed by A. Reeves in 1937. An early, successful application of PCM was in the T1-carrier system developed by AT&T in 1962. In the DS1 version of the T1 system, 24 PCM speech signals (each with 8 bits) are transmitted over one twisted pair of copper wires. Each of the two audio signals(stereo) in the CD prototype system was PCM encoded using 14-bit uniform quantization.

A digital audio signal can be protected against errors by an error-correcting code that adds so-called parity bits before recording. The precise recipe for adding these parity bits depends on the mathematical properties of the applied error-correcting code. In 1950, R. Hamming gave a method for designing block codes that have a single error correction capability per block (R.W. Hamming, 'Error Detection and Error Correction Codes', Bell Syst. Techn. J., Vol. 29, pp. 147-160, 1950). With his work he started the discipline of error-correction coding that resulted in codes with greater error correcting capabilities per block.

Burst errors, which may exceed the capability of a given error-correcting code, may in general be corrected by an additional technique called interleaving. By using interleaving before recording, a burst of errors is, after de-interleaving, spread out in time. These dispersed errors can be corrected by a less powerful code that needs to correct only a few errors per block.

In the prototype CD system an interleaved convolutional error-correcting code was used. This code was chosen by L.B. Vries based on measured statistics of optical disc errors. His paper that describes the convolutional code is reprinted in Sect. 2.4. In the summary of his paper he wrote "Implementations made so far prove that a single-chip realization of a Philips Compact Disc Decoder is very well feasible". This is an important point, essential for realizing a Compact Disc player at an attractive price for the consumer. During the sixties and seventies, digital system engineers assumed that in the course of time, complex digital systems could be realized on one chip and that consequently the price of digital systems would go down. This assumption was based on Moore's law. In 1964 and 1975, G.E. Moore made predictions on the future growth of the transistor density in integrated circuits. He predicted in 1975 that the transistor density of integrated circuits would double every two years for the next decade. This prediction proved to be remarkably accurate and still holds after more than 40 years.

The presence of two monolithic 14-bit Digital-to-Analog (D/A) converters in the Philips prototype CD player shows the sophisticated and advanced level of IC technology at that time. The monolithic 14-bit D/A converter is described in a paper by R.J. van de Plassche and D. Goedhart that is reprinted in Sect. 2.5. In 1978, this D/A converter was the only one available on the market with that resolution. At that time more complex non-monolithic 12-bit D/A converters were priced between 250 and 500 US dollars. However, the availability of the Philips 14-bit D/A converter was an encouraging sign that in time all digital and mixed-signal subsystems needed in a CD player could be realized on just a few chips. A significant promise for future cost savings was also that the prototype CD player contained a solid state 'Aluminum Gallium Arsenide' laser.

Finally, it is important to note that the basic arrangement of the successive digital signal processing operations in the CD prototype system did not change when the CD system was standardized by Philips and Sony in June 1980. What changed, however, in the standardized CD system was that the successive digital signal processing operations became more effective and powerful.

2.2 Presentation of J.P. Sinjou on the public presentation of the Philips prototype of the CD system on March 8, 1979

J.P. Sinjou

Fig. 1. The presentation of the CD by J.P. Sinjou.

Ladies and gentlemen,

For the explanation of the technical specification of our new sound-reproduction system I like to describe:
- the disc and the player,
- the coding system,
- the optical read-out,
- track following,
- and the disc production.

After this you will hear classical music as well as popular music. The Compact Disc and its slip-case are shown in Fig. 2.

Fig. 2. From left to right: Compact Disc, prototype CD player and slip-case.

As you see it is a small disc, it is 115 mm in diameter, 1.1 mm thick and it is made of transparent plastic.
The recording takes the form of a helical track of etched pits commencing at the centre of the disc. A Compact Disc of this size can carry a stereo recording of 60 minutes. This is due to the track to track distance of 1.66 microns, as shown in Fig. 3.

Disc	
Diameter	: 115 mm
Thickness	: 1.1 mm
Track pitch	: 1.66 micron
Recording time	: 60 min. stereo 1 side recorded
Material	: Polyvinyl chloride

Fig. 3. Key physical parameters of the disc.

The disc is recorded on one side only and is covered by a metallic layer embedded beneath a transparent protective coating. It is light and in all respects more convenient than the conventional long play. The Compact Disc bears certain similarities to present day gramophone records, however, with regard to sound quality the similarity ceases to exist. This is due to the breakthrough achieved in storing the music information on the disc digitally and reading it out optically.

As a result of disc size, the Compact Disc player chassis need be no larger than a compact cassette tape-deck. The pick-up head is an optical device employing a miniature laser and a compact optical system. The light reflected back from the metallic layer in the disc contains all the signal information in digital form, with which to reproduce the original music information. The location of the optical pick-up unit determines the speed rotation of the disc and this changes inverse-linearly with the radius from 500 r.p.m. in the centre to 215 r.p.m. at the outer edge. Since there is no physical contact between the optical pick-up head and the disc, the optical pick-up unit generates signals, which indicate whether the disc is in focus and whether the spot is correctly following the track in the radial direction. The optical pick-up unit is mounted at the end of a moveable arm, which is driven by a linear motor.

The player can directly be connected to all existing Hifi-chains, e.g. amplifiers and loudspeakers. Operating the Compact Disc player amounts to no more than selecting play, stop, automatic or search modes. The player is shown in Fig. 2 and it will be demonstrated today. It is built for this reason only and has no commercial purpose.

The coding system

The main object of the encoding system is to obtain the required high quality properties in combination with a high information density on the disc.

As a digital encoding system is chosen Pulse Code Modulation (P.C.M.), offering the following advantages:

- It is an efficient encoding method requiring a low transmission bandwidth as compared e.g. with FM modulation.
- The noise in the transmission channel is not determined by the disc, but by the code chosen.
- The frequency response can be very flat and independent of the disc properties.
- Disc surface deteriorations, clearly audible on a conventional disc, can be made inaudible by applying an appropriate error correcting code.
- Besides music information, other data can be added in encoded form, such as text and programme information.

The text information like e.g. music titles, the name of the composer, conductor, etc. can be incorporated, and the potential exists for visual display of this information as well. Numerical data can be included during disc recording, which makes it possible to play the disc in programmed sequence.

To convert the analog signal into digital form the analog signal has to be sampled with a frequency which has to be at least two times the audio bandwidth, which is 20 kHz per channel, see Fig. 4. The sampling frequency chosen is 44.3 kHz and is derived from a 4.4 MHz crystal.

Player	
Number of channels	: P.C.M. 2 channels (more channels possible)
Frequency response	: 20 Hz - 20000 Hz
Dyn. range	: > 85 dB
S/N ratio	: > 85 dB
Harmonic distorsion	: less than 0.05 %
Wow/Flutter	: precision of Quartz - oscillator.
Quantisation	: 14 bits linear
Drop out compensation	: yes
Sampling rate	: 44.3 kHz

Fig. 4. Key characteristics of the prototype CD system.

The samples are uniformly quantized and converted into binary words. Each individual sample of sound information consists of 14 bits and so a 60 minute recording will total approximately 6 billion bits. The bits are laid out on the disc in the form of a helical track of microscopic pits and non pits. Digitally a pit represents 1 and the area between the pits nought. The 14 bits give a total of more than 16.000 levels and are required to achieve a signal to noise ratio of 85 dB. By the application of pre emphasis a signal to noise ratio of 92 dB is in fact obtained.

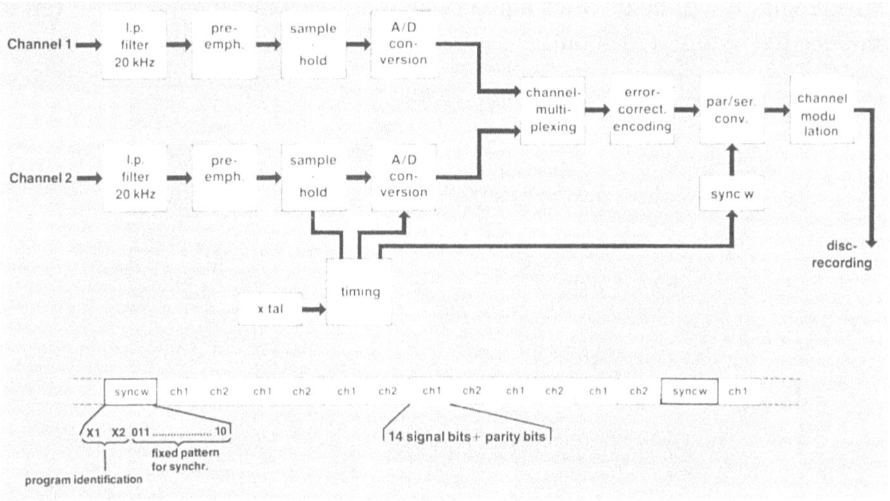

Fig. 5. Signal encoding system and frame format.

Fig. 5 shows the encoding system as applied to both channels. Following the functions of pre emphasis, sampling and conversion, time multiplexing of the two audio channels, in case of stereo, takes place. In the following stage error correcting parity bits are added to the 14 bit words to enable correction of bit errors. Word synchronization also occurs in this stage. The multiplexer has been so designed as to allow implementation of more than two channels in the future. Thereafter channel modulation occurs in which the bit stream is adapted to the properties of the read system and of the disc. The main requirements for the channel modulation are:

- D.C. free transmission, necessary for good tracking error signals.
- Good clock regeneration capability.
- No increase of transmission bandwidth.

The information (word) pattern is shown in the lower part of this figure. Each word per channel consists of 14 signal bits and the added parity bits. In the synchronization word (sync w) bits are reserved for text and programme information.

The optical read-out

Since the information has been deposited in the form of a helical track of pits and non pits in the disc, an efficient read out system had to be devised.

The information structure as it appears in the disc is shown in Fig. 6 at a magnification of 10.000 times. As the minimum length of the pits is less than 1 micron, the width a constant 0.6 micron and the depth a quarter of the wavelength, it will be obvious that a system of mechanical contact will fail to produce the required read out.

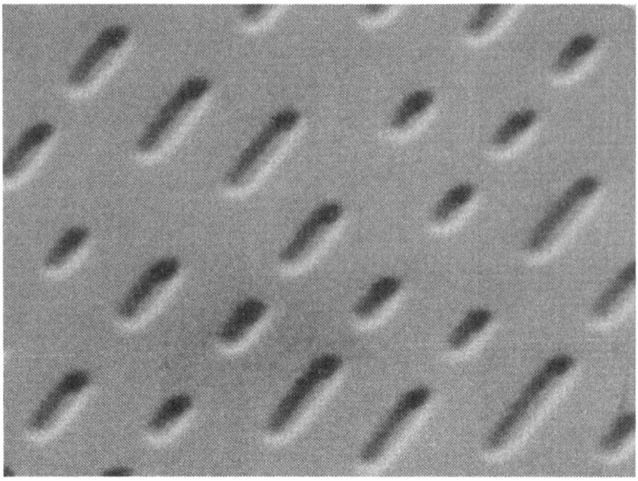

Fig. 6. Information structure as it appears in the disc.

Fig. 7 serves to illustrate this point and shows the comparison with the conventional gramophone record.

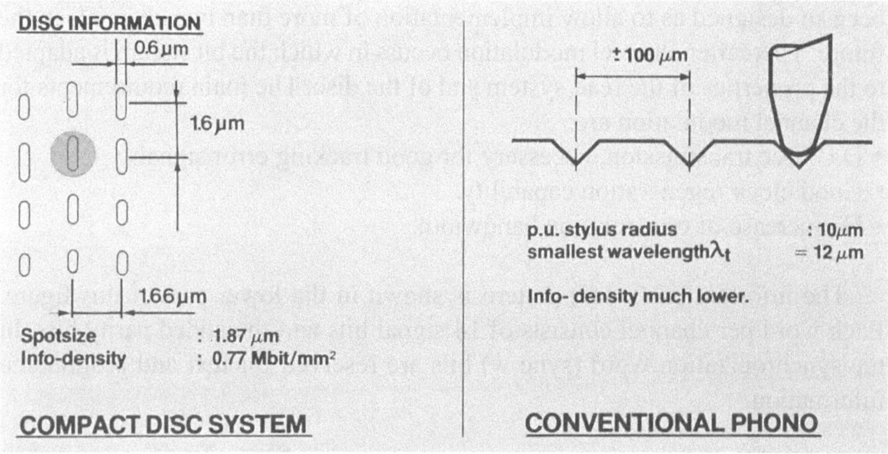

Fig. 7. Comparison of compact disc structure and structure of conventional gramophone.

The information layer is covered with a metallic reflective coating so that we can extract the information by means of reflected light. We achieve this by focussing the light from an Aluminum Gallium Arsenide (AlGaAs) laser onto the track. This diode laser is a light source of considerably less power than that

used for writing the master disc. The laser light, which is concentrated into a spot of 1.87 microns in diameter, follows the track thereby striking pits and non pits alternately. Due to this, light will be lost because it is diffracted over angles larger than the lens is capable of accepting. Thus the intensity of the reflected light is modulated by the physical structure of the disc and this is detected by a photodiode which, in turn, produces a modulated electrical signal.
The optical pick up unit is shown in Fig. 8.

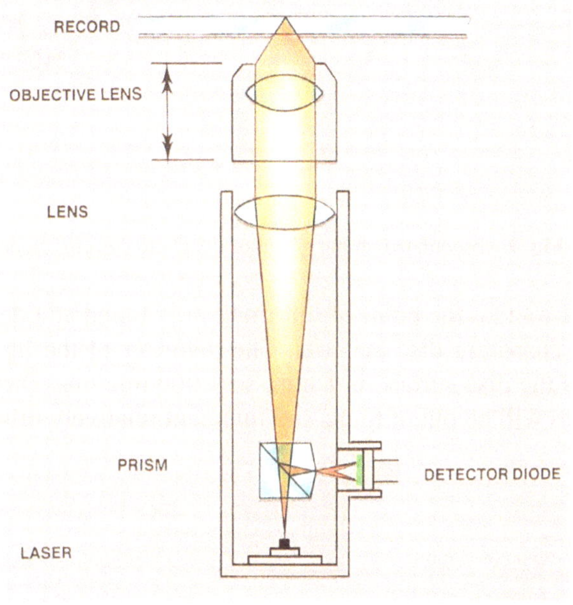

Fig. 8. Optical pick-up unit of the prototype CD system.

The divergent light beam emitted by the laser is converted into a parallel beam by means of a lens. The parallel beam is directed toward the objective lens. It is here that the beam is focussed onto the information track. The reflected, modulated light is directed at the detector diode by a prism, which serves as an output coupling mirror. A wedge is situated between this half mirror and the photo diode to split up the reflected beam into two parts, forming spots on different parts of the photo diode. The output currents of the diode parts contain the desired information signal as well as the error signals for radial tracking and focussing.
The optical pick up unit is only 45 mm in length, 12 mm in diameter and weighs 14 grams. It is mounted at the end of a moveable arm enabling it to follow the track in radial direction. The objective lens is mounted above the light-pen and with the help of a drive system of the principle of that of a loudspeaker it is possible to keep the spot focused on the information layer.

The modulated output signal of the photo detector diode in relation to the pits and non pits on the disc is shown in Fig. 9.

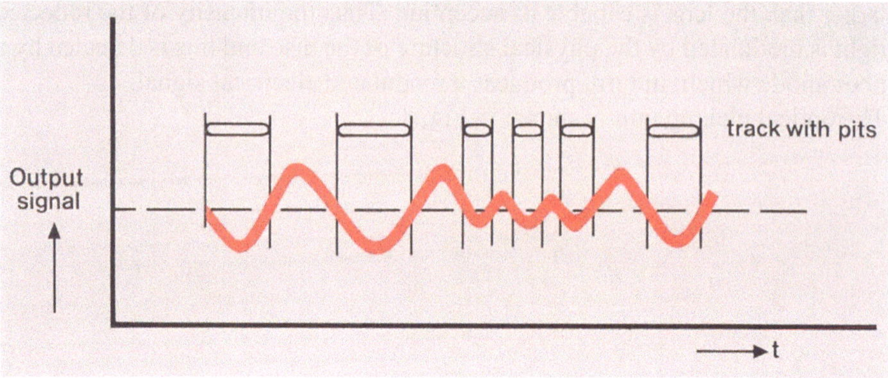

Fig. 9. Modulated output signal of the photo detector diode.

Fig. 10 shows that the point of information is found at a depth of 1.1 mm through the transparent disc material. The diameter of the light beam at the place it enters the disc surface is 1 mm, so 1.000 microns. Dust particles and small scratches will be out of focus and intercept relatively little of the beam.

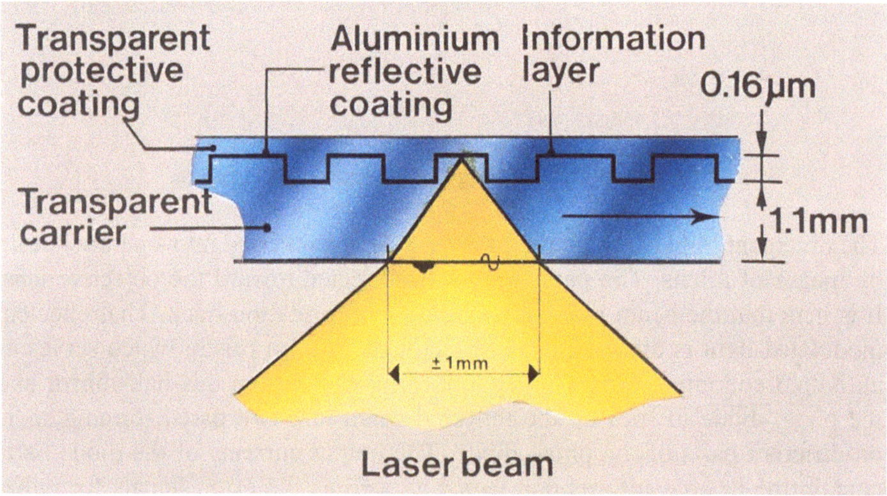

Fig. 10. Information read-out through a transparant coating.

The track following servo system

Track deviations from the circular and vertical unevenness of the rotating disc must be accounted for, since both the width of the track, 0.6 microns, and the depth of focus of the spot, 2 microns, are particularly critical. Because there is no mechanical contact with the track, these irregularities have to be controlled by servo systems, which receive their information from the optical pick up unit.

The focus error signal, as indicated in Fig. 11, results in a vertical movement of the objective lens. The track error signal derived from the disc, maintains the spot exactly on the track. The turntable speed varies with the detection radius to give a constant linear track velocity. In order to exactly reproduce the speed used during recording the motor servo, controls the turntable motor to make the detected digital coding signal equal to a standardized clock frequency. The track error signal and the arm position signal have a direct relation to each other as the random access facility enables the arm to be moved to a predetermined position. Therefore the tracking process is automatically cut out by a control logic system. This control logic system initiates also the correct function of user operated keys, such as start and stop.

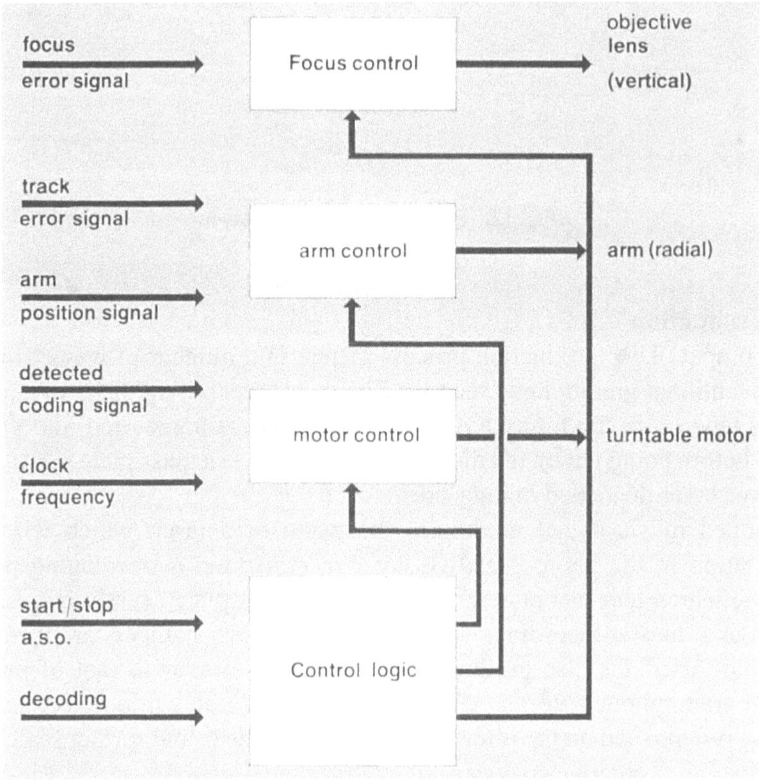

Fig. 11. Block diagram of the servo systems.

The block diagram, given in Fig. 12, shows the main functions of the player:

- The disc with the drive motor.
- The optical pick up unit giving the high frequency, focus and radial tracking signals.

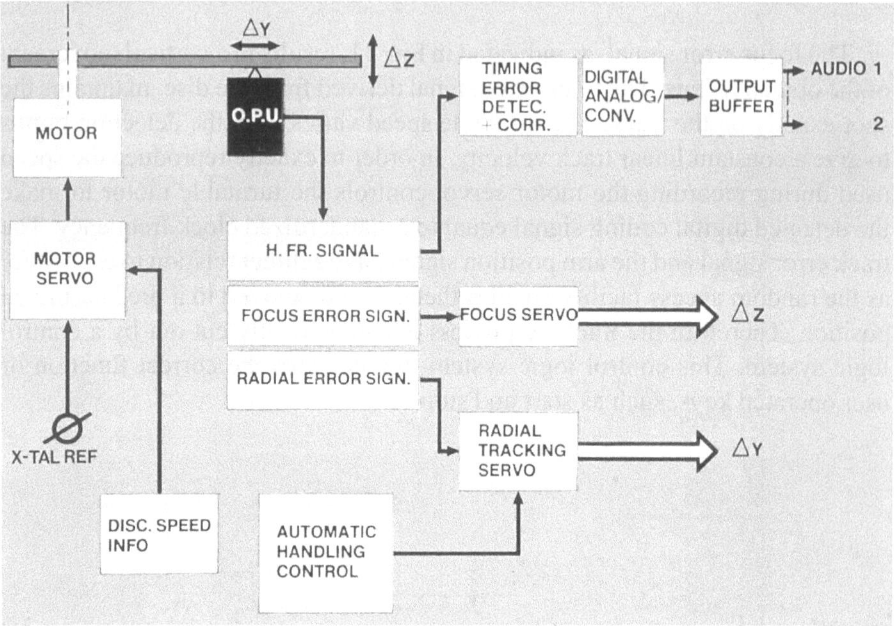

Fig. 12. Block diagram of the player.

Disc production

The Compact Disc production process differs in a number of ways from that of conventional gramophone records. The master recording, be it an analogue master tape or, in the future a digital master tape, is transferred into a coded signal before being put on the disc. The master disc is a glass plate with a photo sensitive layer deposited on one side.

The coded music signal modulates the beam of a laser, which writes the information in the photo sensitive layer in real time. A developing process follows, which leaves a pattern of pits in the glass plate exactly representing the original master recording. Via a galvanic process, stampers are then made which are used for disc production in a manner similar to that of pressing normal gramophone records. After pressing an extremely thin reflective metal coating is deposited on the information side of the disc, and further sealed with a transparent protective coating.

2.3 The Philips 'VLP' System

K. Compaan, P. Kramer

Drs K. Compaan is with the Philips Electro-Acoustics Division (ELA), Eindhoven;
Dr P. Kramer is with Philips Research Laboratories, Eindhoven.

Abstract

Television pictures are recorded on the Philips video long-playing ('VLP') record in a spiral track of pits in the surface. The pits have constant width and depth but the lengths and spacings are variable. The information is read out by a beam of light, which is reflected at the surface of the record. The reflected beam is modulated by deflection of the light through diffraction at the pits. To enable the 'VLP' playback unit to operate at the required accuracy, control systems have been developed for holding the speed of rotation of the record constant, focusing the read-out beam on the record surface and centring the beam on the spiral track without the assistance of mechanical guides. The player can be used to show the recorded pictures one at a time, and will also allow them to be shown in reverse motion, slow motion, or at faster speed.

Now that almost every home and many educational institutions have a television set it is natural to think of the possibility of using it, in combination with a playback unit, for reproducing programmes that have been permanently recorded in some way or another. This gives the user the freedom of being able to watch a programme he is interested in at a time convenient to himself - the same freedom he can enjoy with a shelf of books or a collection of gramophone records.

The 'VLP' system described here allows a colour-television programme of about 30 minutes duration to be reproduced from a recording on a 'gramophone record' 30 cm in diameter, the usual size for a long-playing record. The 'VLP' record can be produced simply and in quantity by the normal pressing techniques. The 'VLP' system is complementary to the video cassette recorder (VCR), which has been on the market for some time, but to some extent it offers an alternative to it. A programme can be recorded as desired with a cassette recorder, but it is more expensive to produce recorded tapes than it is to press 'VLP' records.

The development of the 'VLP' system is the result of the combined efforts of a team of specialists in very divergent fields. In this article we shall give a broad general survey of the system; the three short articles that follow will describe some of the components in more detail [1] [2] [3].

Reprinted with permission from Philips Tech. Rev. **33**, 178-180, 1973.

The diameter of the spot is of the same order of magnitude as the wavelength of the light used in the equipment, and it is therefore no longer possible to speak of a particular diameter. A diffraction pattern (an Airy disc) is formed at the focal plane of the lens; this pattern consists of a central maximum surrounded by successive dark and light rings. To produce a pattern in which the half-intensity diameter is 0.9 to 1.0 μm at the wavelength used, a lens with a numerical aperture of 0.4 is required.

The information is recorded on the record disc along a spiral track, which occupies the part of the disc between the 10 cm and 30 cm diameters. The speed at which the disc rotates has been made equal to the picture frequency, 25 s^{-1} for the European market and 30 s^{-1} for North America. As we shall see later, this offers some interesting possibilities. If the playing time is half an hour, these figures give a pitch of 2 μm for the track.

For following a track with such a small pitch an optical method is very suitable. In the 'VLP' player this scanning is done with a spot of light 1-2 μm in diameter, projected on to the track by a lens.

The information for the reproduction of a television picture is recorded as a succession of short grooves or pits of variable length and repetition frequency. The width of the pits is 0.8 μm, and the depth 0.16 μm (see Fig. 1). Since in pressing a gramophone record the surface roughness does not amount to more than 0.01 μm, it is clearly a practical possibility to make such a pattern in the surface of a pressed disc.

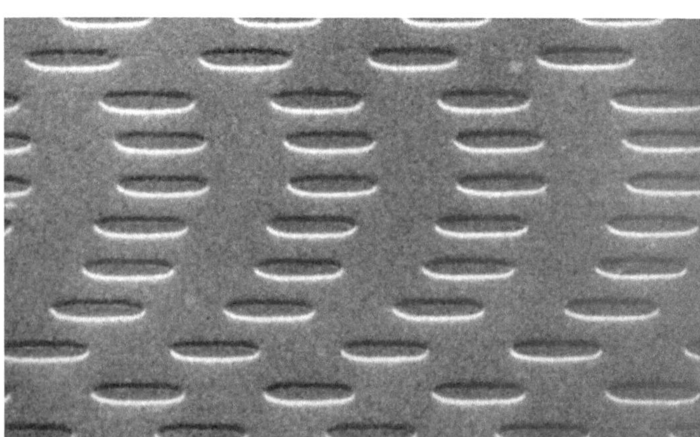

Fig. 1. Information layer of the 'VLP' disc.

If the spot of light falls on the surface of the disc between two of the pits, then most of the light will be reflected back into the objective lens. If on the other hand the spot falls on one of the pits, the light will be deflected by diffraction at the pit in such a way that most of it is not returned to the objective (Fig. 2). In this way the intensity of the light reflected through the aperture of the lens is modulated by the pattern of pits[1]. The intensity variations are converted into an electrical signal by a photodiode. The width and depth of the pits in the surface are arranged to give as large a modulation depth as possible.

To obtain a high signal-to-noise ratio in the detector signal, the reflected beam should have as high an intensity as possible. If the photocurrent is too low, the noise will no longer be mainly determined by the thermal noise in the detector, but by the shot noise in the photon current. We have therefore used an He-Ne laser as the light source. Also, to improve the reflectivity, the surface of the 'VLP' disc has been coated with a thin layer of evaporated metal.

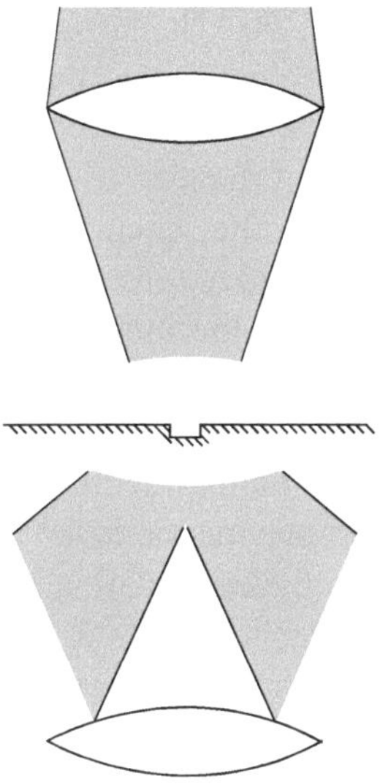

Fig. 2. Modulation of the light by a pit in the surface of a 'VLP' record. For clarity the system is drawn as if the record were transparent, with the beam incident from above and a second lens placed underneath the record to receive the light. The pit is also shown many times enlarged with respect to the rest of the figure. If the record surface is flat, all of the incident light is received by the lower lens. If there is a pit in the surface there will be diffraction, and some of the light will be deflected; when the pit is correctly dimensioned much of the incident light will be deflected away from the aperture of the lower lens. In practice the record surface is reflecting, and only one lens is required for concentrating the light on to the record and receiving the reflected light.

Some of the members of our team have developed a special technology that enables the He-Ne laser to be manufactured in quantity. This 1 mW laser has been built into the player in such a way that it can be of no possible danger to the user.

The information on the surface of the disc can be read out through a transparent protective layer. Any contamination or damage only affects the outer surface of this layer, and not the disc. The diameter of the beam at this outer surface is much larger than the spot, so that these imperfections have very little effect on the detector signal. This arrangement makes use of the very small depth of focus of an objective lens with a resolving power in the micron range.

To enable it to be encoded in the pattern of pits, the video signal undergoes a number of special processes[2]. The bandwidths of the brightness signal and the colour signal are both limited to some extent. The frequency of the colour-signal subcarrier, which is 4.43 MHz in the PAL system, is reduced to a value of 1 MHz, fixed with respect to the line frequency. This allows the original carrier frequency to be restored when the record is played, even if there are deviations caused by variations in the speed of revolution. The sound is treated as a frequency modulation of a 250 kHz carrier. The brightness signal, which modulates a 4.75 MHz carrier, determines the *repetition frequency* and the *average length* of the pits, while the preprocessed colour and sound signals give a *modulation of the length* of the pits.

Work has also been done on other encoding systems whose potentialities include the recording of a video signal with a wider bandwidth.

The master record from which the moulds are produced for pressing the 'VLP' records is cut by a laser in the specially prepared surface of a glass disc. This cutting is done at the same speed at which the records will be played. A scene can therefore be recorded on the record directly from the video camera or transferred without delay from a magnetic tape. The moulds are made in the usual way from the master by an electroplating process.

If a 'VLP' player is to give good results four special requirements have to be satisfied. In the first place, the speed of revolution of the record must be kept constant to an accuracy of 1 in 10^3, or the playback of the video signal will be unsatisfactory.

Secondly, the objective must remain focused on the surface of the record. Because of its large aperture the objective has only a very small depth of focus. Although the irregularities on the surface of the record are locally very small, the deviations over a wider area can be as much as 0.5 mm.

In the third place the beam of light must remain centred on the track, even though the track may be not truly circular (out-of-round) or eccentric. Deformation of the disc during pressing can lead to out-of-roundness; eccentricity of the spindle-hole in the record and play between it and the shaft of the playback unit can cause the track to rotate eccentrically. The player must be able to operate correctly even when the total deviation of the track from the ideal position is as much as 0.1 mm.

Finally, the complete optical system must move radially across the record at the rate at which the track advances ('tracking'), without the aid of a continuous groove or other mechanical guide in the disc or the player. To meet these requirements a number of control systems have been developed; these will be described in one of the following articles[3].

Fig. 3 shows a diagram of the 'VLP' player. The complete pick-up unit can move backwards and forwards on a carriage on rails underneath the record disc 1 to follow the track. The light from the laser 2 is focused at the record by the objective 3. The control systems mentioned above act on the objective and a pivoting mirror 4, thus keeping the beam focused and centred on the track. A prism 5 ensures that light reflected by the record falls on the detector 6.

The 'VLP' player can also be used to show the pictures in reverse motion, slow motion or at higher speed. This is possible because the record rotates synchronously with the picture frequency – 25 rps for the European version, 30 rps for the American one. Consequently at each rotation of the track the field-synchronizing pulses always fall within two fixed diametrically opposite sections of the record disc. (A television picture consists of two interlaced

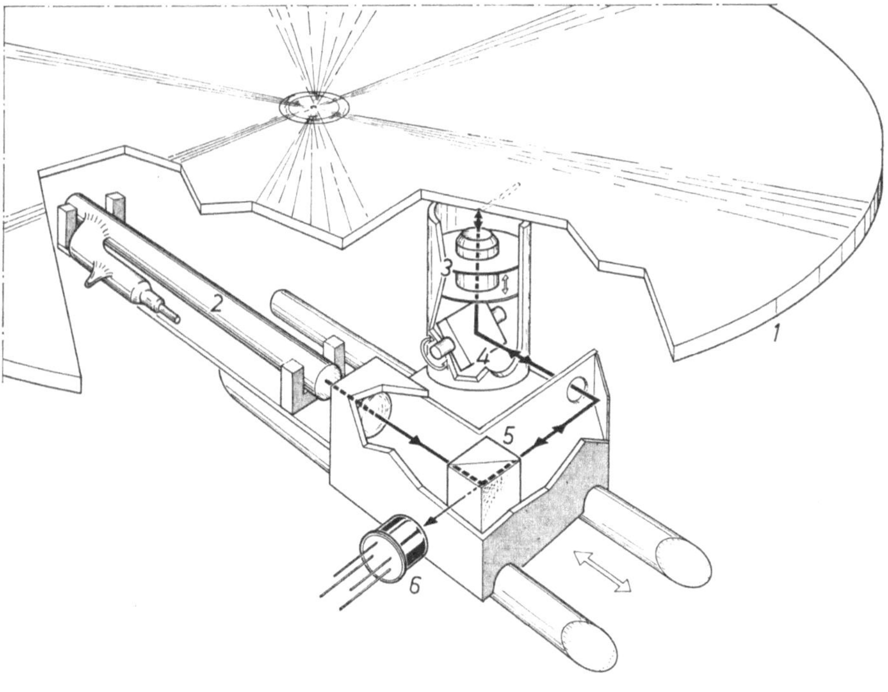

Fig. 3. Schematic diagram of the 'VLP' playback unit. The record 1 is scanned from below by light from the He-Ne laser 2. The objective 3 is held focused on the record by a system based on a loudspeaker mechanism. The pivoting mirror 4 ensures that the beam remains centred on the track; the mirror is operated by a rotating-coil arrangement. Incident and reflected light are separated by the prism 5. The detector 6 converts the reflected light into an electrical signal.

fields.) Wherever the spiral track crosses the two sectors it therefore contains the same information - the field-synchronizing signal. This means that inside the sector the beam can be allowed to change from one turn of the track to an adjacent one, without spoiling the picture. This is done by applying a control pulse at the correct moment to the control system for correct centring on the track. By continually repeating the same turn and thus the same picture in this way, a stationary picture will be obtained. By repeating each picture twice a picture in slower motion will be obtained, and by omitting every other picture the action of the scene will be reproduced at twice the speed. A picture in reverse motion is obtained by jumping back a turn at each half revolution.

Because of the accurate centring of the scanning beam on the track the cross-talk between successive turns is very small (<-30 dB), so that it is possible to record completely different pictures on successive turns. This will give a 'picture-book' of about 45 000 different pictures. Address coding allows any particular picture to be found rapidly.

The large number of pictures - which can be completely different if desired - that can be stored on the 'VLP' record, and the scope for manipulation of the recorded information, make the 'VLP' system one that clearly offers more than the simple dissemination of video information.

References

[1] G. Bouwhuis and P. Burgstede, Philips Tech. Rev. **33**, 186-189, 1973.

[2] W. van den Bussche, A. H. Hoogendijk and J. H. Wessels, Philips Tech. Rev. **33**, 181-185, 1973.

[3] P. J. M. Janssen and P. E. Day, Philips Tech. Rev. **33**, 190-193, 1973.

2.4 The error control system of Philips Compact Disc

L.B. Vries

Philips Research Laboratories, Eindhoven, The Netherlands

Abstract

The error control system of Philips Compact Disc consists of an error correction system or decoder followed by an error concealment unit, design of the error correction system was based on measured statistics of the disk errors. It was observed that the majority of disk errors are a mixture of predominant random errors and very scarce long bursts. A computer search was carried out to find the convolutional code of shortest constraint length that would meet a given performance specification at a desireable code rate.

Implementations made so far prove that a single chip realization of the Philips Compact Disc Decoder is very well feasible.

2.4.1 Introduction

The basic requirement to be imposed on an error control system for digital audio to be used in conjunction with an optical disc as storage medium is, that it should prevent that disc errors will lead to audible clicks during playback.

This goal can be accomplished by combining an error correcting and detecting system with an error concealment unit. The error correcting system (or decoder) requires that the information be redundantly encoded before written onto the disk, but that offers the possibility to correct for the vast majority of errors. The error concealment unit on its turn receives a warning from the decoder whenever it failed to decode reliably, on which command it replaces the received unreliable samples by estimated values obtained through linear interpolation between correct samples.

This general set-up can only work if the decoder does indeed succeed to lower the error-rate drastically enough and if it is capable of producing its warnings about uncorrectable data with even higher reliability. However an additional requirement is that uncorrectable data can be replaced by linearly interpolated samples, this means that uncorrectable samples should occur very well separated in time. A standard solution to solve this problem is to use an interleaving scheme, (see Fig. 1). In such a scheme the symbols of a codeword (or sequence) are interlaced with the symbols of (L_i-1) other codewords (or sequences) such that a burst of consecutive errors causes only small errors in

Courtesy Audio Engineering Society (www.aes.org). Reprinted with permission from:
64th AES convention, 1979, New York, paper G-8.

each of the codewords or sequences, this is usually called interleaving with degree L_i.

The nice advantage of applying interleaving for burst correction purposes is that it allows us to split the design problem of the coding system into two choices :

- what degree of interleaving should be selected;
- which kind of error correcting code should be chosen.

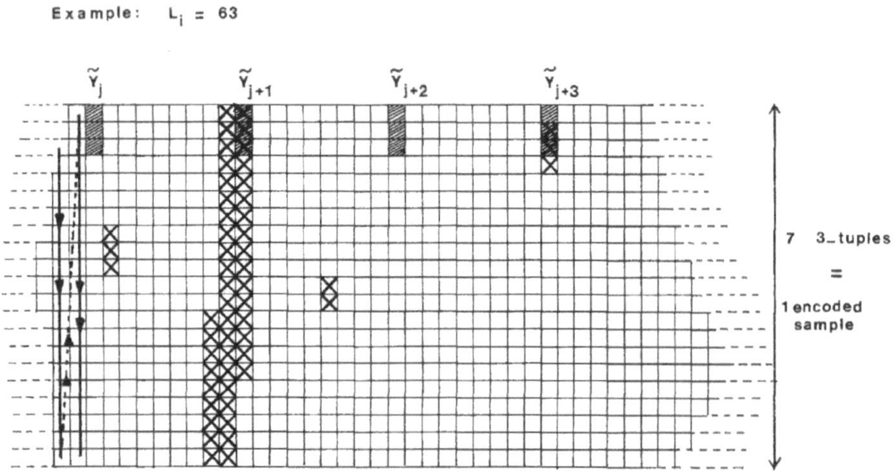

Fig. 1. Interleaving scheme.

It is intuitively appealing to determine the degree of interleaving on the basis of the maximum burst length, while the error correcting capability of the code should be based on the resulting average error rate of the channel, however in order to be able to cope with bursts and random errors, *multiple* error correction is a must. Before we can make a selection, some knowledge about the kind of errors we will encounter must be gathered.

Within Philips an extensive experience was built up on digital optical recording, mostly for applications on DRAW (direct read after write)[1]. The same kind of measurement techniques used in DRAW were applied for Compact Disc again. At present recording densities of 1.3 bits/µm along the track, measurements done by MG Carasso revealed that the majority of disk errors are randomly occurring errors of 1, 2 or 3 consecutive bit intervals long, they cause an error rate of 2.10^{-4}. Only a relatively small amount (fewer than 0.003) falls into the category 3-32 bit intervals, while an extremely low fraction of “calamities” varying from 33 to 200 bits occurs (only a few per disc).

An error correcting code was selected, which basic unit of information is a three bit character (hereafter called 3tuple), its error correcting capability will also be specified in terms of the number of character errors that it can correct for. Input to the encoder (Fig. 2) are 2-tuples of information bits. Because Compact Disc represents audio samples as 14 bit words, 7 of these 2-tuples are needed to transmit one audio sample.

A separate section is devoted to explain this error correction system which is of the convolutional type.

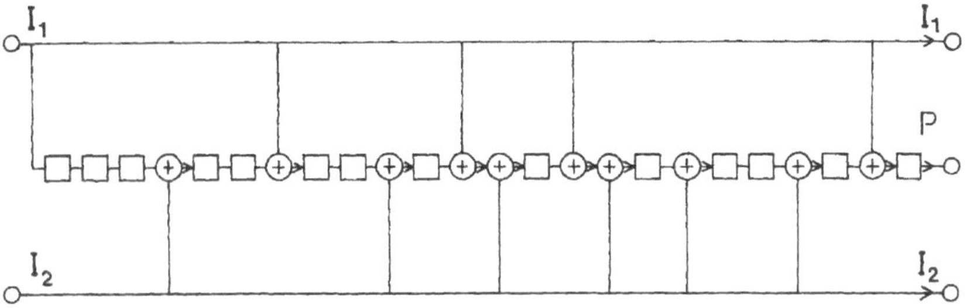

Fig. 2. Encoder.

2.4.2 Convolutional codes, a review of some elementary theory

We will start with the general definition of a convolutional code. An (n,k,v) linear convolutional code is a system which maps a sequence of k-tuples (of bits)

$$\ldots, x_{j-1}, x_j, x_{j+1}, \ldots$$

onto a sequence of n-tuples (of bits)

$$\ldots, y_{j-1}, y_j, y_{j+1}, \ldots \text{ where } n > k$$

according to the following recurrent expression

$$y_j = G_0 x_j + G_1 x_{j-1} + G_2 x_{j-2} + \ldots + G_v x_{j-v}$$

In this expression all matrices of type G_i, i=0, 1, 2, ...v are n × k-matrices with elements either 1 or 0. All calculations, like matrix-to-vector multiplication and vector addition, are carried out modulo 2. The number v representing the memory size of the encoder, is usually referred to as the *encoding constraint length*. Convolutional codes of this type are called *linear* because modulo 2 addition of the corresponding bits of two arbitrary sequences of the same code, generates a sequence which is itself a code sequence. A convolutional code is called *systematic* whenever encoding involves nothing else but the juxta position of a parity sequence of (n-k)-tuples to the original information sequence of k-tuples.

Example: Fig.2 shows the encoder of a systematic (n=3, k=2, v=14) convolutional code. We observe that at each instant a parity bit P is added to the information 2-tuple $(I_1, 1_2)$ to obtain the output 3-tuple (P, I_1, I_2). The matrices G_i of this code are as follows.

$$G_0 = \begin{bmatrix}00\\10\\01\end{bmatrix} \; G_1 = \begin{bmatrix}10\\10\\01\end{bmatrix} \; G_2 = \begin{bmatrix}01\\10\\01\end{bmatrix} \; G_3 = \begin{bmatrix}00\\10\\01\end{bmatrix} \; G_4 = \begin{bmatrix}01\\10\\01\end{bmatrix}$$

$$G_5 = \begin{bmatrix}11\\10\\01\end{bmatrix} \; G_6 = \begin{bmatrix}11\\10\\01\end{bmatrix} \; G_7 = \begin{bmatrix}01\\10\\01\end{bmatrix} \; G_8 = \begin{bmatrix}00\\10\\01\end{bmatrix} \; G_9 = \begin{bmatrix}10\\10\\01\end{bmatrix}$$

$$G_{10} = \begin{bmatrix}00\\10\\01\end{bmatrix} \; G_{11} = \begin{bmatrix}01\\10\\01\end{bmatrix} \; G_{12} = \begin{bmatrix}00\\10\\01\end{bmatrix} \; G_{13} = \begin{bmatrix}00\\10\\01\end{bmatrix} \; G_{14} = \begin{bmatrix}10\\10\\01\end{bmatrix}$$

Of course not all codes are practically useable, some codes are in that respect significantly better than others. A qualification good or bad, should depend on the error correcting capabilities offered by the code. Unfortunately for convolutional codes the formulation of this error correcting capability is somewhat involved, this is due to the fact that the *decoding* operation is a rccurrent proces. At some time instant, the decoder has to decide whether the received n-tuple $\tilde{y}_j$ contains an error or not. This decision then will be based on an observation made on a whole segment of consecutively received n-tuples,

$$\tilde{y}_j, \tilde{y}_{j+1}, \tilde{y}_{j+2}, \ldots, \tilde{y}_{j+m}, \text{ where } m \geq v.$$

If this decision is made for $\tilde{y}_j$, then upon the receipt of $\tilde{y}_{j+m+1}$, it can be done for $\tilde{y}_{j+1}$ and so on.

We now come to the definition of error correcting capability: A convolutional code is said to be t n-tuple error correcting with decoding constraint length m, if and only if, any error pattern in $\tilde{y}_j$ is recoverable from the sliding segment $\tilde{y}_j, \tilde{y}_{j+1}, \tilde{y}_{j+2}, \ldots, \tilde{y}_{j+m}$, given that at most t n-tuple errors have occurred in this segment.

The code used in our example for instance is a double error-correcting code with decoding constraint length m=14. The first codes to be discovered, were the single n-tuple error correcting codes, found by Berlekamp and Preparata independently [2], [3], these codes have k=n-1 and m=2n. Since for convolutional codes decoding decisions are based on a segment; given the parameters n, k and t, we would like to have the length of the segment called decoding constraint length as short as possible in order to get the best performance on our channel. In this sense the Berlekamp-Preparata codes are optimal t=1 codes.

2.4.3 Necessary and sufficient conditions for a convolutional code to have an error correcting capabillty of t n-tuples (Non interested readers may skip this section.)

The principal determiner of the error correcting capability is the so called free-distance of the code. What free distance means is explained in the following lines. To that purpose, let us consider two code-sequences y en z that were equal in the past ($y_i = z_i$ for $i < 0$) but differ from a certain moment on ($y_0 \neq z_0$). For instance: for the code of Fig. 2 these could be

									0	1	2	3	4	5	6	7	8	9	10	11	12	13	14	15	16	
	0	1	0	1	0	1	0	1	0	1	1	0	0	1	1	0	1	0	0	0	0	0	0	1	0	
y---	0	1	1	0	0	1	0	0	1	1	0	1	1	1	1	0	1	1	0	0	1	0	1	0	0	---
	1	0	0	1	1	1	0	1	1	0	0	0	0	1	0	1	0	0	0	0	1	1	0	1	1	
	0	1	0	1	0	1	0	1	0	0	1	0	0	1	1	1	1	0	0	0	1	0	1	0	1	
z---	0	1	1	0	0	1	0	0	0	1	1	1	1	1	1	1	1	0	0	0	0	0	1	0	0	---
	1	0	0	1	1	1	0	1	1	1	0	0	0	1	0	1	1	0	0	0	0	1	0	1	1	
									0	1	0	0	0	0	0	0	0	1	0	0	1	0	1	1	1	
(y+z)	----------------								1	0	1	0	0	0	0	0	0	0	0	0	1	0	0	0	0	---
									0	1	0	0	0	0	0	0	0	1	0	0	1	0	0	0	0	

We observe that if two code sequences differ from a certain moment on, they will differ in many more n-tuples, in the segment following time instant 0. One could say that the two code sequences, will *diverge* in the sense that the number of n-tuples in which they will differ initially grows. The guaranteed minimum number of differences that will eventually result taken over all pairs y and z then is the free-distance d_{free}. The necessary segment length that guarantees this number of differences to be observeable, then will be taken as decoding constraint length. It can easily be deduced that a code is error correcting if and only if

$$d_{free} \geq 2\,t{+}1.$$

To see this, assume a received sequence $\tilde{y}$, which differs for the first time from its transmitted sequence y at time 0, and which differs in at most (t-1) other positions from y in the segment following this instant. Now $\tilde{y}$ will be "closer" to y then to any other code sequence, because it differs in only t positions from y but will certainly differ in at least (t + 1) positions from any other code sequence. Which means that by going over all finite segment continuations of the code sequence specified up to $\tilde{y}_{-1}$, only one continuation will differ in at most t positions from $\tilde{y}$. Although this is not a practical decoding procedure, it delivers a constructive proof of the error-correcting capability.

Insight in, how two code sequences are bound to diverge can be obtained as follows. If y and z are code sequences then so must be (y + z). Now (y + z) is a sequence which is identially zero in the past, and whose first non zero term is $(y_0 + z_0)$. Such a sequence will be called initial sequence. Because $(y_i + z_i)$ is non zero if and only if $y_i \neq z_i$, the n-tuple distance of y and z grows in the same way as the number of nonzero n-tuples of (y + z) does. Therefore, the free distance can be obtained from inspection of all initial code sequences over a finite segment length. Because we do not have to compare pairs of code segments, this means an important reduction in computational effort to determine the free distance of a convolutional code. This property has been used in a computer search for rate 2/3 codes of short decoding constraint length.

2.4.4 Computer generated codes

To the best of our knowledge, n-tuple error correcting codes for t > 1 were not published as yet. Because multiple error correction was needed, the idea was worked out to find them via an exhaustive computer search. This way the rate 1/2 (n=2, k=1) case was exhaustively searched through, using 50 hours of minicomputer time. The result of this search can be summarized as follows: t=2 requires m=8, while t=3 requires m=15. It was observed that some t=3 codes could also have been obtained from certain t=2 codes by appending some matrices G_i, for $8 < i \leq 15$, to it. Thus for the rate 1/2 case it was demonstrated that for decoding constraint lengths up to 15 optimal t=3 codes could have been obtained from optimal t=2 codes, via the "extension" procedure. In order to save computational effort, the search for the systematic rate 2/3 t=2 codes was limited to an "extension" search. Starting point for this search was the original rate 2/3 BP-code.

<u>The decoder</u>

A standard feedback decoder for the code of Fig. 2 is depicted in Fig. 3, it consists of 2 data registers where the information bits of the received sequence are stored. Both registers are tapered of and modulo 2 added to form a reconstructed version of the parity. This *reconstructed* parity is modulo 2 added to the *received* parity to form the *syndrome*, now the j-th syndrome bit will depend on the received segment as follows:

$$\begin{aligned} s_j &= H_0 \tilde{y}_j + H_1 \tilde{y}_{j-1} + H_2 \tilde{y}_{j-2} + \dots && + H_m \tilde{y}_{j-m} \\ &= H_0 (y_j + e_j) + H_1 (y_{j-1} + e_{j-1}) + \dots && + H_m (y_{j-m} + e_{j-m}) \\ &= H_0 e_j + H_1 e_{j-1} + \dots && + H_m e_{j-m}. \end{aligned}$$

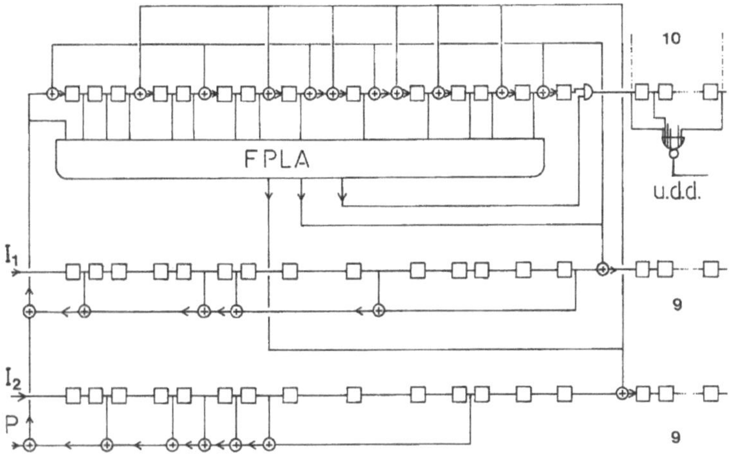

Fig. 3. Decoder.

Thus because the syndrome of a code sequence is identically zero, the syndrome of the received sequence will depend on the error sequence only. Due to the construction of the code double 3-tuple errors will be recoverable from the stored syndrome segment. Correction is only carried out for those info-bits, which are the "oldest" in their segment. If a correction is done, the syndrome is updated to remove the effect of the error that was just corrected for. In case of decoding errors either a correction is omitted or a false correction is executed, both cases will result into a wrong updating of the syndrome, which can be used for unreliable data detection. Because the u.d.d.-signal may show some detection delay, corrected data is delayed to ensure that erroneous decodings are covered by a u.d.d. warning. Interleaving is implemented by replacing every shift register in both encoder and decoder by a cascade of L_i shift registers. This transformation is also applied to the u.d.d. circuitry.

In the present experimental encoder and decoder the value of L_i is kept programmable, several multiples of 7 can be selected up to a maximum of 63. This way the calculated required value of the interleaving can be confirmed by practical experiments.

2.4.5 Decoder performance

If the interleaving is larger than the longest burst that occurs, thc errors occurring in one code sequence will be random like.

In this situation, each code sequence has to deal with a channel with an error rate P_{in}, where P_{in} is the average 3-tuple error rate of the channel. The decoder performance in those cases can then be expressed as a plot of the output error rate P_{out} of the decoder as a function of the input error rate P_{in} (see Fig. 4).

Such a performance curve can be characterized by its behaviour in the operation range and its behaviour in the breakdown range. In the operation range, were the error rate is low we have

$$P_{out} = 2450\, P_{in}^{3}.$$

This is because triple errors are the most likely events that will lead to erroneous decodings.

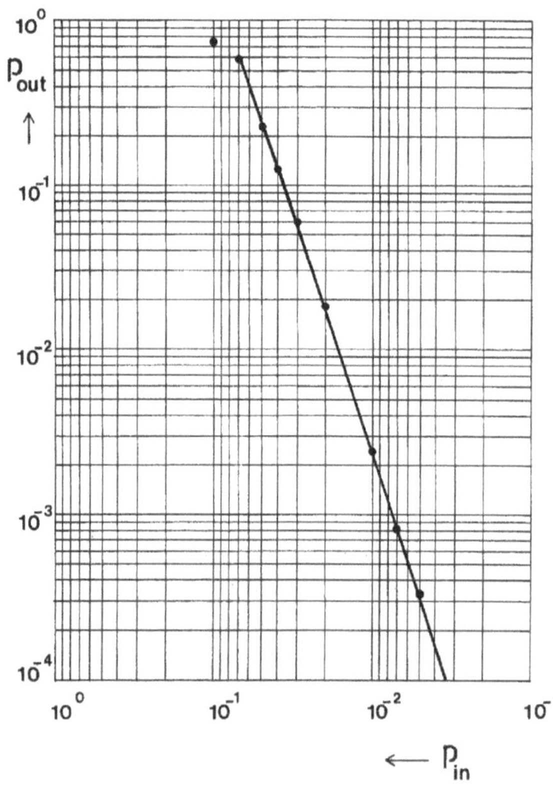

Fig. 4. Decoder performance.

However, because no probability can exceed the value 1, somewhere in the curve at the higher input error rates P_{out} has to deviate from this behaviour, specifically should saturate at a certain value. Of course a total breakdown has to occur at error rates exeeeding 2/15 = 0.133... .

Another noticeable point in this plot is, the point were the output error rate becomes larger than the input error rate, this occurs approximately at $P_{in} = 2.10^{-2}$, this error rate is about a hundred times worse than the error rate on the disk.

The following table demonstrates the robustness of the code: here P_s denotes an upperbound to the resulting average *sample interpolation rate.*

P_{in}	P_{out}	P_s
10^{-2}	$2.45\ 10^{-3}$	$1.715\ 10^{-2}$
10^{-3}	$2.45\ 10^{-6}$	$1.715\ 10^{-5}$
10^{-4}	$2.45\ 10^{-9}$	$1.715\ 10^{-8}$

Concluding remarks

In our discussion on error correction for Compact Disc, the coding problem was split into two parts:

- What degree of interleaving should be used in order to correct for the longest burst that can reasonably be expected at standardized disk quality?
- Which kind of multiple error correcting code should be selected in order to be capable to cope with mixtures of bursts and random errors?

Because the average error rate is low (order of 2.10^{-4}) it turns out that double 3-tuple error correction is sufficient. Interleaving then is based on the following consideration, let B_{disk} denote the longest burst length (in bit intervals) that can reasonably be expected, then the degree of interleaving L_i is selected high enough, such that correcting this burst "uses up" only half of the error correcting power of the code. This way, even if an equally long burst follows the first one, within the effective constraint length, tho combined event still remains correctable. For $L_i = 63$ this implies, taking a 12 bit sync word into account,

$$B_{disk} = 12 + 63 \text{ x } 3 = 201 \text{ bit intervals.}$$

As regards the choice of the code, emphasis was laid on minimizing the decoding constraint length. This is profitable, not only because it gives the best performance it also brings down the number of flipflops (or storage cells) that goes into the decoder. This number is of significant importance for the yield of future LSI chip realizations.

An interesting property of this error correcting code is, that although it was designed for random 3-tuple errors, its effective burst-to-guard space ratio B/G is only somewhat smaller then the optimum value of B/G that a pure burst corrector can theoretically attain. For the Compact Disc code we have

$$\left[\frac{B}{G}\right]_{CD} = \frac{2}{15},$$

while according to a theory of Forney[4], to achieve zero error-capacity on the classical bursty channel

$$\left[\frac{B}{G}\right] \leq \frac{1-k/n}{1+k/n} = \frac{1/3}{5/3} = \frac{3}{15}.$$

Thus the random error correcting capability is "paid for" by a reduction of $\frac{B}{G}$ by a factor of 2/3. This is certainly a sensible trade-off considering that the majority of errors is random.

References

[1] K. Bulthuis, M.G. Carasso, J. Heemskerk, P. Kivits, W. Kleuters and P. Zalm, "High Density Optical Recording and Retrieval of Data", IEEE-Spectrum, July 1979.

[2] E.R. Berlekamp, "Note on Recurrent Codes", IEEE-Trans. Inf. Th. 1964, pp. 257-258.

[3] F.P. Preparata, "Systematic Construction of Optimal Linear Recurrent Codes for Burst Error Correction", Calcolo (Pisa) 1964, pp. 147-153.

[4] G.D. Forney Jr., "Burst-Correcting Codes for the Classic Bursty Channel", IEEE-Trans. Comm., **15**, 772-781 (1971).

2.5 A Monolithic 14-bit D/A converter

R.J. van de Plassche, D. Goedhart

Abstract
A monolithic 14-bit D/A converter using "dynamic element matching" to obtain a high accuracy and good long-term stability is described. Over a temperature range from - 50° to 70°C the nonlinearity is less that one-half least significant bit <½ LSB). Dynamic tests show a distortion at a level of about - 90 dB with respect to the maximum sinewave output. Nearly no glitches are found, so the converter can be operated without a deglitcher circuit. The chip, with a size of 3.1 x 3.2 mm, contains all elements needed, except the output amplifier and digital input latches.

2.5.1 A Monolithic 14-Bit D/A Converter

Monolithic D/A converters are the subject of growing interest due to the rapidly expanding market for digital signal-processing systems. The introduction of digital signal processing in sound recording and reproduction systems imposes stringent requirements on the dynamic behavior of the converters. Many of these systems require a 14- to 16-bit resolution to obtain a high signal-to-noise ratio and a good linearity.

In integrated D/A converters an R-2R ladder network with terminating transistors is widely used to generate binary weighted currents. These currents are switched by the bit switches and the conversion from digital information into an analog signal is performed. In Fig. 1 an example of such a converter is shown. There are two main design problems. The first problem, to which most attention has been paid, is the weighting accuracy problem of the bit currents. The second one, which determines the dynamic performance, is the switching of the accurately weighted currents without glitches. Returning to the accuracy problem, the table in Fig . 1 shows that D/A converters up to 10 bits can be integrated without too many problems. Twelve-bit D/A converters are available on the market[1], but laser trimming of thin-film resistors or Zener zapping techniques are required to achieve the accuracy. How successfully these techniques can be applied to 14- or 16-bit converters is still questionable, and some people have doubts about the long-term stability. Furthermore, in large-volume production, trimming costs cannot be ignored. In this paper a monolithic 14-bit D/A converter is described which uses a different scheme to achieve a high weighting accuracy and good long-term stability. This approach, called "dynamic element matching"[2], needs no trimming and combines a passive division with a time-division concept. Moreover, it is insensitive to element aging.

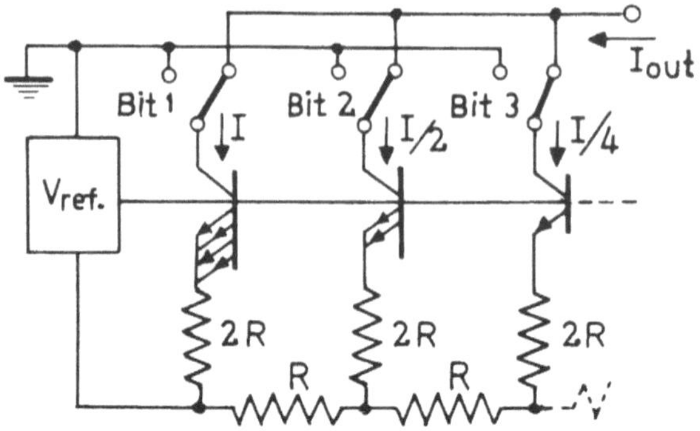

(a)

Fab. process	Matching tolerance			
	σ(%)		mean(%)	
	10μ	40μ	10μ	40μ
Diffusion	0.44	0.23	− 0.1	0.07
Thin film	0.24	0.11	− 0.1	−0.06
Ion implant	0.34	0.12	0.05	0.05

RESISTOR LINEWIDTH 10μ and 40μ

(b)

Fig.1. (a) Standard R-2R ladder-network D/A converter. **(b)** Matching tolerances of different resistor types .

2.5.2 Basic Divider Scheme

A simplified diagram of the divider is shown in Fig. 2(a). It consists of a passive current divider and a set of switches driven by a clock generator *f*. The total current 2*I* is divided by the passive current divider into two nearly equal parts : $I_1 = I + \Delta I$, $I_2 = I - \Delta I$. The currents I_1 and I_2 are now interchanged during equal time intervals with respect to output terminals 3 and 4. At these terminals currents then flow whose average values are exactly equal and have a dc value *I*. Fig. 2(b) shows the currents as a function of time. A small ripple current $2\Delta I$ of frequency f is present on the output currents too. This ripple gives a measure of the matching performance of the passive divider. With a simple low-pass filter this ripple can be suppressed and an exact l-to-2 current ratio is obtained. If the time intervals differ by a value Δt, there is an error in the division ratio equal to:

$$\frac{\Delta I_{3.4}}{I_{3.4}} = \frac{\Delta t}{t} \cdot \frac{\Delta I}{I}$$

With $(\Delta I/I) \cong 1$ percent and $(\Delta t/t) \cong 0.1$ percent an accuracy of $\cong 10^{-5}$ can be obtained. In a practical circuit a minimum supply voltage of 2 V is needed for good operation of the system. By cascading divider stages an accurate binary weighted current network is formed at the cost of an increase in supply voltage. In a 14-bit current network this leads to an impractically large supply voltage. Therefore, an improved divider scheme must be used to give more weighted currents in one interchanging operation.

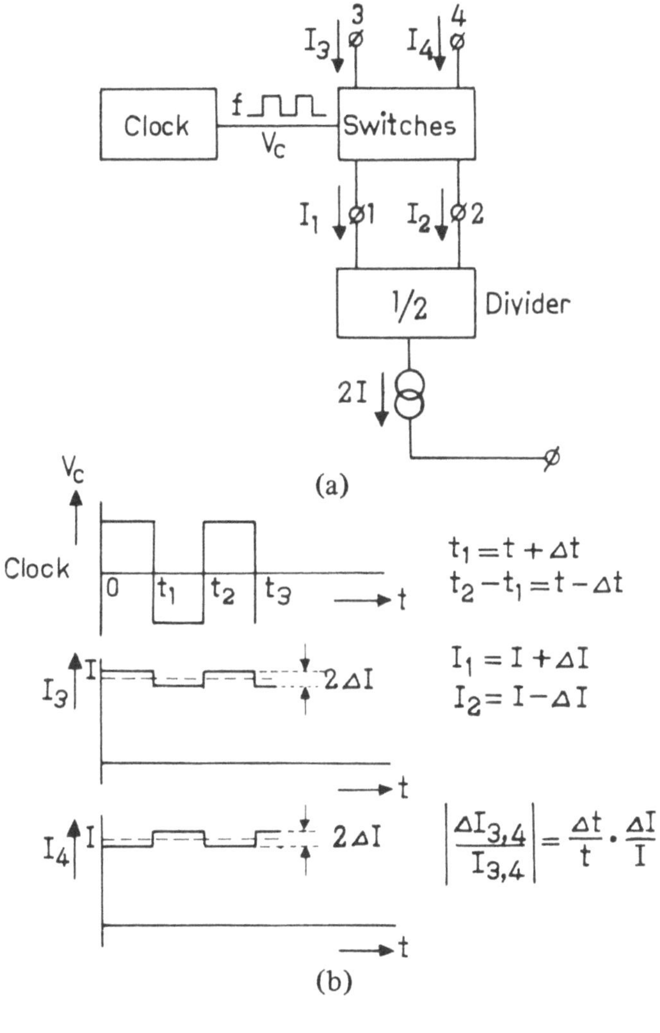

Fig. 2. (a) Basic current divider. **(b)** Currents as a function of time.

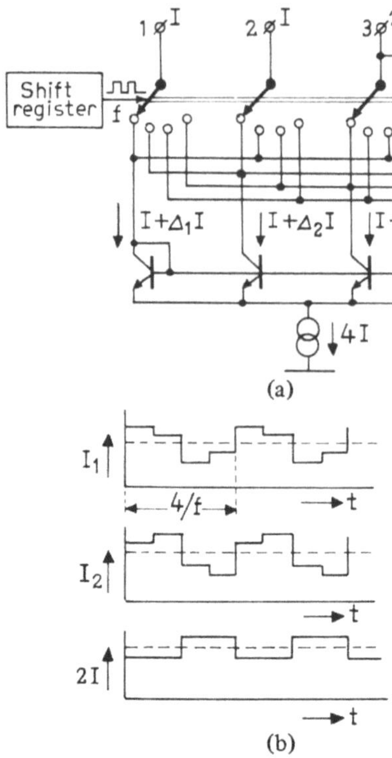

Fig. 3. (a) Improved current divider. **(b)** Currents as a function of time.

2.5.3 Improved Divider Scheme

In the improved divider circuit the passive current divider is extended to divide a current $4I$ into four nearly equal parts:

$I_1 = I + \Delta_1 I$, $I_2 = I + \Delta_2 I$, $I_3 = I + \Delta_3 I$ and $I_4 = I + \Delta_4 I$ [see Fig. 3(a)]. Note that $\Delta_1+\Delta_2+\Delta_3+\Delta_4=0$. These currents are now fed into a switching network that interchanges all currents during equal time intervals. These time intervals are generated by a 4-bit shift register. At the output of the switching network, the currents are combined to give values of $2I$, I, and I. The output currents as a function of time are shown in Fig. 3(b). The figure shows that the currents with a value I have a ripple with the same frequency as the clock generator f, while the current with a value $2I$ has a ripple with a frequency $f/2$. Timing errors have the same influence on accuracy as in the system shown in Fig. 2(a).

Fig. 4 shows the circuit diagram of a practical divider. Transistors T_1, T_2, T_3, and T_4, with the resistors R, divide the current $4I$ into four nearly equal currents I. These currents are fed to the interchanging network consisting of Darlington switches to minimize base current loss. In the layout of the circuit, two currents are directly summed by combining the collector islands, which results in an output current 2*I*.

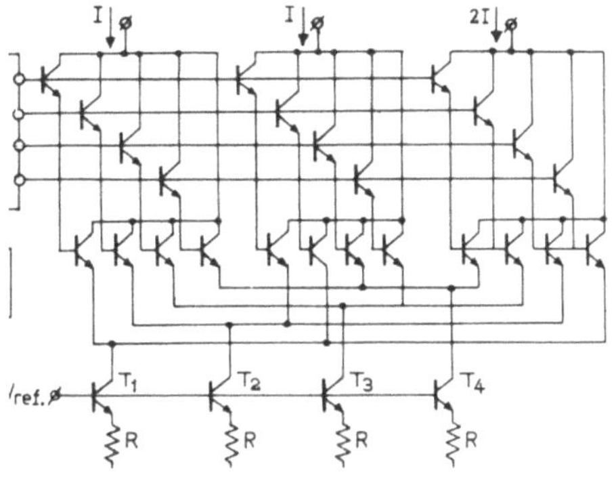

Fig. 4. Practical 2-bit/switching-level current divider.

A four-stage shift register provides the signals for the interchanging of the currents. The only design criterion for a high division accuracy is a high current gain for the switching transistors.

2.5.4 Binary Weighted Current Network

By cascading current-division stages, a binary weighted current network is formed (see Fig. 5).

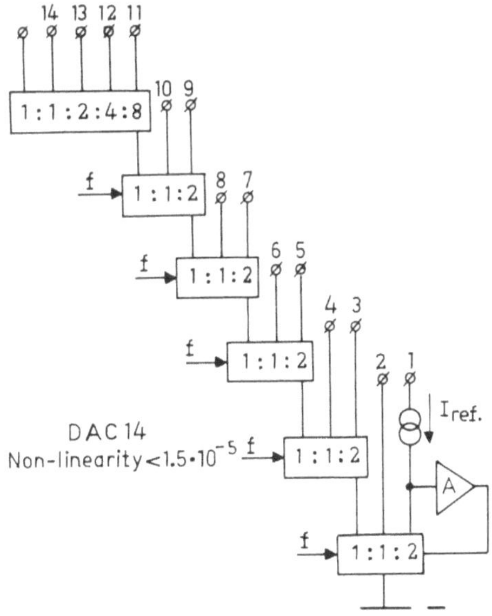

Fig. 5. Binary weighted current network.

In the first stage a combination with the reference current source I_{ref} and a current amplifier is used as an accurate current mirror . The reference current itself is used as the most significant bit current (MSB), which has the advantage that filtering is not required. There is a tradeoff between circuit yield and minimum supply voltage. To obtain 14-bit accuracy, a choice between the number of switched and nonswitched current dividers must be made. A high circuit yield is found with five switched stages followed by a 4-bit passive divider using emitter scaling.

2.5.5 Filtering and Switching

How the output currents of a switched divider stage are filtered and switched to the output line is shown in detail in Fig. 6.

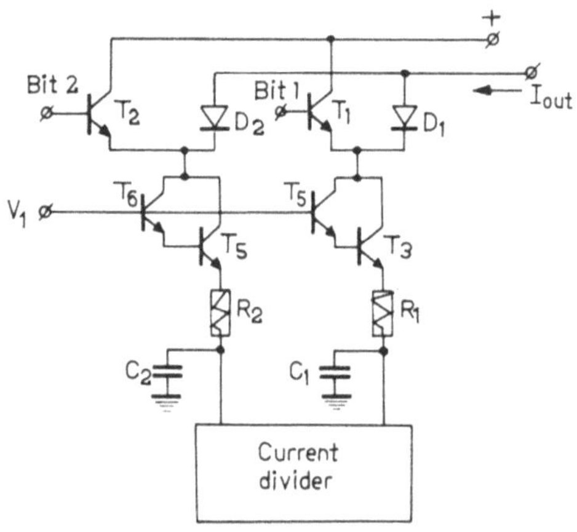

Fig. 6. Detail of the filtering and switching circuit part.

A first-order filtering operation is used (C_1R_1, C_2R_2) for which external capacitors are added to the chip (C_1 ,C_2). Additional Darlington cascode stages (T_3, T_4 and T_5, T_6) isolate the filtering operation from the switching of the binary weighted bit currents. The individual filtering of the bit currents minimizes the noise of the converter output current. Bit switching is performed with a diode transistor configuration (T_1, D_1, and T_2, D_2), yielding rather fast and accurate switching with no loss of base currents.

2.5.6 Practical D/A Converter

The circuit diagram of the complete D/A converter is shown in Fig. 7. The 14-bit binary weighted current network , the reference current source, cascode stages with filtering elements, and the bit switches are easily recognized. The shift register for the interchanging consists of a gated master-slave flipflop driven by an emitter-coupled multivibrator (bottom left side). Provisions are available for obtaining individual filtering of the ripple currents of the most significant bits. When this filtering is used, the conversion speed is determined only by the speed of the bit switches.

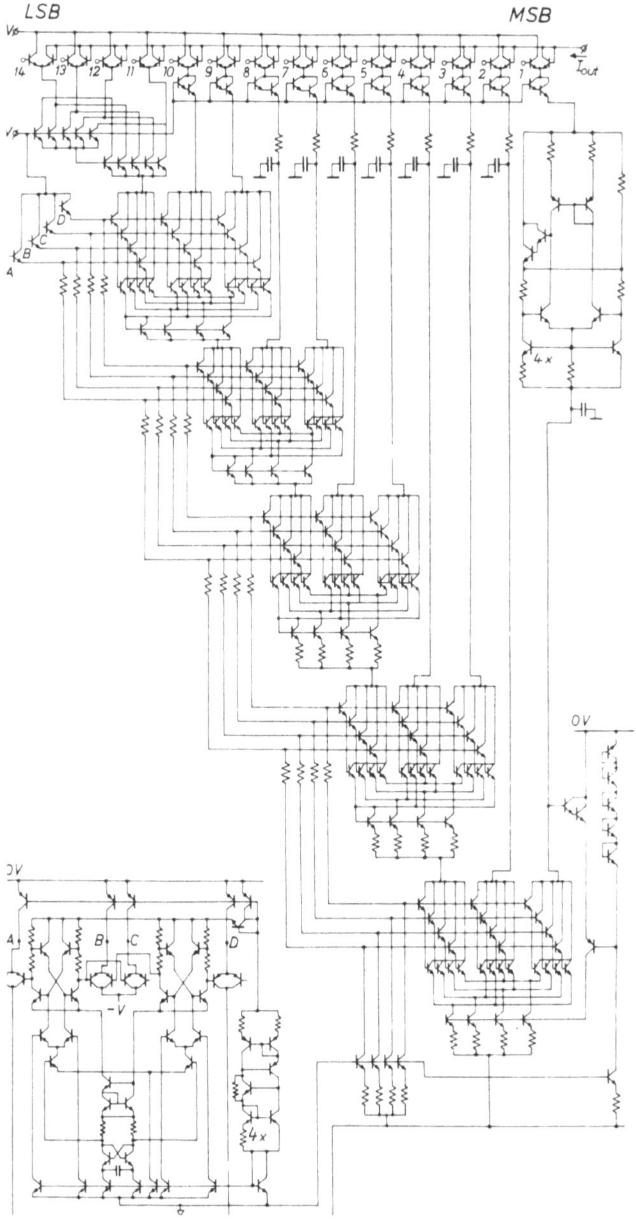

Fig. 7. Complete circuit diagram of a 14-bit D/A converter.

2.5.7 Measurements

An important parameter of a D/A converter is the linearity. If the linearity is better than one-half a least significant bit (½ LSB), the converter is automatically monotonic. Fig. 8 shows the results of a linearity measurement as a function of temperature. Over a temperature range from - 50° to 70°C the nonlinearity is less than 3.10^{-5} = ½ LSB. With the test scheme in Fig. 9 some dynamic tests were carried out as follows.

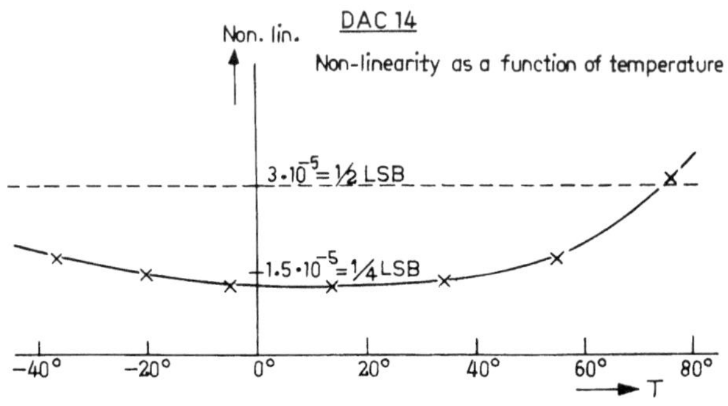

Fig. 8. Nonlinearity of the 14-bit D/A converter as a function of temperature.

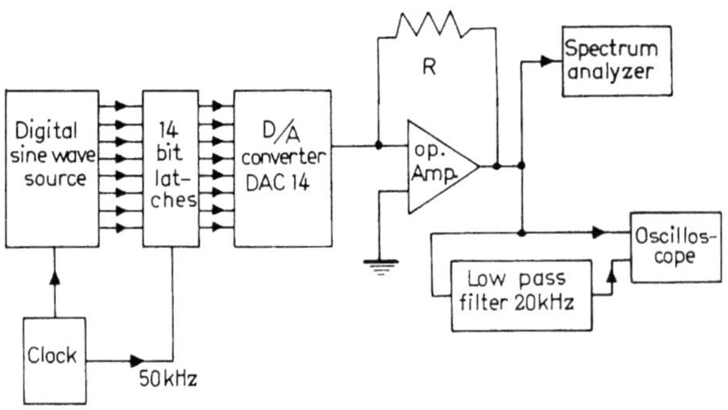

Fig. 9. Measurement scheme to determine distortion and output pulse response.

Out of a digital sine-wave source 14-bit words at a clock rate of 50 kHz are latched. The outputs of the latches directly drive the switches of the D/A converter. The output current of the converter is converted into a voltage by means of a very high-speed operational amplifier with feedback resistor R. The output signal of the operational amplifier is analyzed with a spectrum analyzer and an oscilloscope. Spectrum analyzer results are shown in Fig. 10(a)-(c). Sine-wave frequencies in these cases are about 600 Hz, 9 kHz, and 18 kHz, respectively. The results show that the distortion is at a level of about -90 dB with respect to the maximum sine-wave output. This -90 dB level corresponds to the limit of the spectrum analyzer, too.

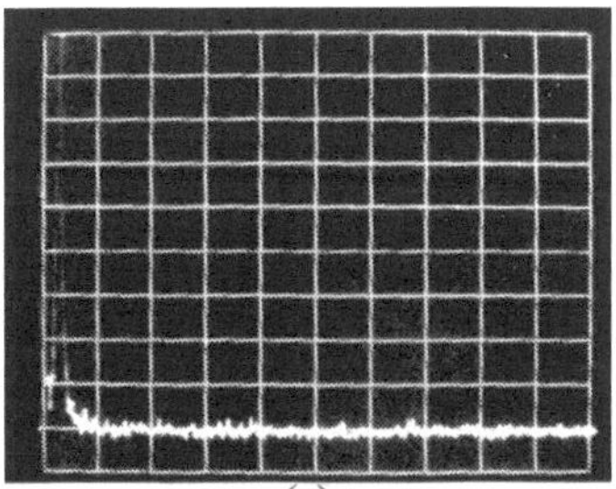
(a)

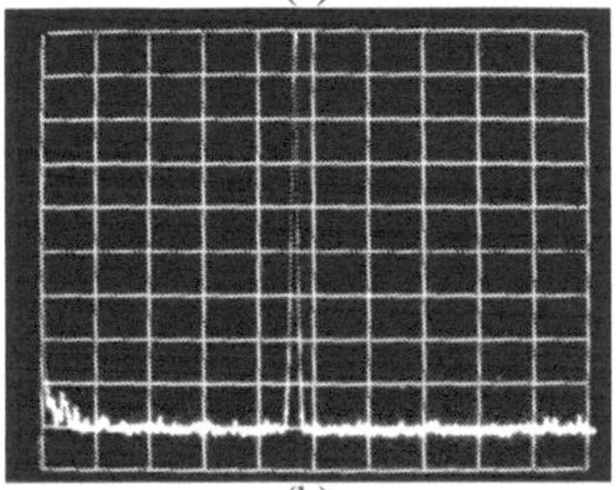
(b)

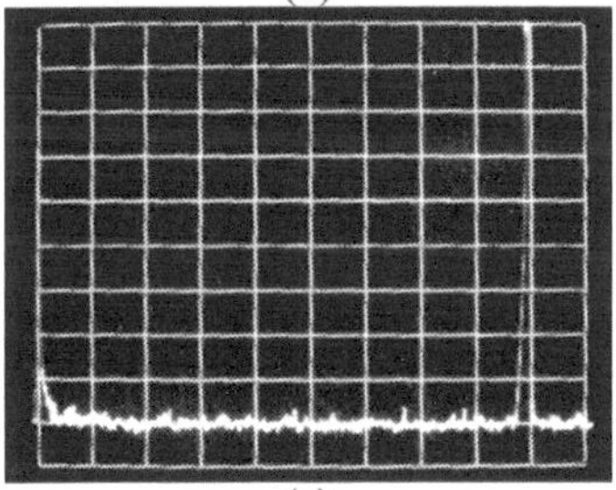
(c)

Fig. 10. (a) Distortion of an output sine wave of about 1 kHz. Horizontal 2 kHz/cm. Bandwidth 30 Hz. Vertical 10 dB/cm. **(b)** Same for an output 9 kHz. **(c)** Same for an output of about 18 kHz.

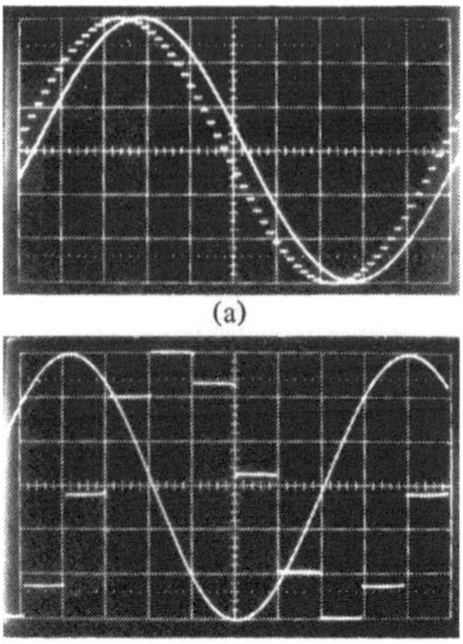

(a)

(b)

Fig. 11. (a) Filtered and nonfiltered output signals for a 1 kHz output frequency. **(b)** Same for an output frequency of 6.3 kHz.

D/A Converter data:	
Resolution	14 bits
Linearity	±¼ LSB at T = 25°
	±½ LSB - 50°<T<70°
Output current	2mA
Conversion speed	10μsec to ½ LSB
Temp. coeff. of output current	5 ppm/°C
Chip size	3.1x3.2 mm
Optimum interchanging freq.	2.5 kHz
Power supply	+5V and -15V

Table I. D/A Converter Specifications

The results of the oscilloscope display are shown in Figs. 11(a) and (b) for sine-wave frequencies of 1 kHz and 6.3 kHz, respectively. A synchronization mechanism between sine-wave and clock frequency is needed to obtain a stable display . This reduces the number of output frequencies that can be displayed. The delay between the stepped and the filtered sine wave is introduced by the low-pass filter. The photographs show no glitches and a good step response .

2.5.8 D/A CONVERTER DATA

Some converter data are shown in Table I. Note that the given settling time corresponds to a D/A converter with filtering applied to the bits. A photomicrograph of the chip is shown in Fig. 12.

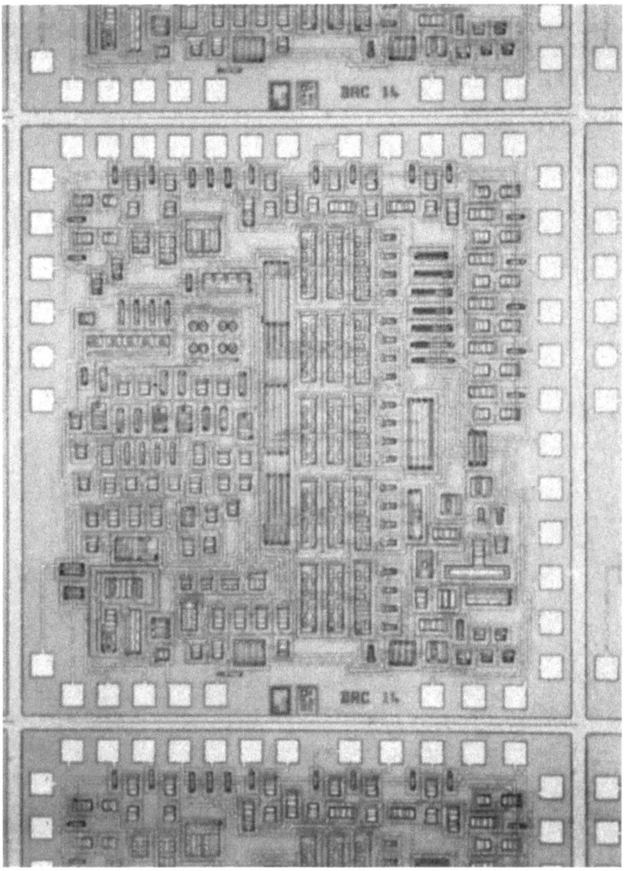

Fig. 12. Photomicrograph of the D/A converter chip.

CONCLUSION

The dynamic element matching method provides a simple, accurate, and reliable design procedure for high-accuracy monolithic D/A converters. The method requires no costly trimming procedures and is insensitive to process variations and aging of components. The good long-term stability and the low noise of the filtered bit currents are major advantages of the system. The good dynamic performance of the converter described makes it very suitable for sound-reproduction and recording systems.

Acknowledgment
The authors wish to thank A. Schmitz for the processing of the circuits.

References

[1] D. T. Comer, A monolithic 12-bit DAC, *IEEE Trans. Circuits Syst.*, vol. CAS-25, 504-509 (1978).

[2] R.J. van de Plassche, "Dynamic element matching for high accuracy monolithic D/A converters", *IEEE J. Solid-State Circuits*, vol. SC-11, 795-800 (1976).

Chapter 3

THE CD SYSTEM AS STANDARDIZED BY PHILIPS AND SONY

3.1 Introduction to publications of the Compact Disc digital audio system

J.B.H. Peek and J.P. Sinjou

Preface

It should be stressed that this introduction does not intend to mention and recognize all people within Philips and Sony who contributed to establish the CD digital audio system standard in 1980. This standard is based on the collaborative work of many persons, both from Philips and from Sony, and it would be impossible to properly acknowledge all these individuals in the space of only a few pages. More information on the persons involved can be found in the doctoral thesis (in German) by Jürgen Lang ("Das Compact Disc Digital Audio System", 1996, RWTH, Aachen, ISBN 3-00-001052-1). This introduction only aims at describing some important decisions that were made between the successful demonstration of the CD prototype on March 8, 1979, and the establishment of the Philips-Sony CD standard in June 1980.

The Philips-Sony partnership

Already at an early stage in the development of the CD prototype, the Philips Board of Management emphasized that the directors of the Philips Audio product division should aim at realizing a world standard for the CD. With this in mind the directors of Audio decided that, to achieve this goal, a first step would be to find a strong industrial partner that would be interested to cooperate with Philips in attaining a common CD system standard. Therefore, Philips went public with its digital audio disc innovation in a press conference on March 8, 1979. And as a follow-up, the directors of Audio decided to approach several Japanese companies and ask them if they would be interested to receive a delegation of Audio so that the Philips CD prototype could be shown and demonstrated. A positive response of these companies was received and from March 14 till March 23, 1979, the following companies and organizations were visited in succession: JVC, Sony, Pioneer, Hitachi, MEI (Matsushita) and the DAD (Digital Audio Disc Committee). The DAD had been installed by the Japanese Ministry of Industrial Trade and Industry with the task to evaluate various digital audio disc systems and to recommend a world standard. On the last day of the visit, J. van Tilburg, the general director of Audio, received a phone call from A. Morita, the president of Sony. Morita said that, after consulting the management of Sony, he had decided to cooperate with Philips. The vice-president of Sony, N. Ohga, would come to Eindhoven to discuss the contract.

The Philips-Sony collaboration

With Sony, Philips had an ideal partner. Sony not only had an excellent position in products related to digital recording of audio on magnetic tape, but they also had developed a prototype optical digital audio player and disc. The diameter of Sony's disc, however, was 30 cm, much larger than the 11.5 cm diameter of the Philips CD disc. To determine a common standard for the CD system, Philips and Sony agreed to a sequence of meetings to be held alternately in Eindhoven and Tokyo. During these meetings, the technical experts from Philips and Sony had to settle issues like the playing time of a disc, its diameter, the audio sampling frequency, the signal quantization (bits/sample), and the signal format to be used. Determining the signal format implied that Philips and Sony had to agree on the purpose and the interpretation of the successive bits in each block of data on the disc, including the modulation code and the error correcting code to be used.

The first meeting, of in total six meetings, was held in Eindhoven on August 27 and 28, 1979. The last meeting was in Tokyo on 17 and 18 June, 1980. The way in which the final error correcting code gradually emerged, illustrates the interchange of ideas between Philips and Sony engineers.

At their first meeting, Philips and Sony each proposed a different error correcting code. Philips proposed the rate 2/3 convolutional code that was used in the prototype CD system. This code was developed by L. Vries and is described in his paper [L.B. Vries, "The Error Control System of Philips Compact Disc", AES Preprint 1548, New York, November 1979], reprinted here in Sect. 2.4. Sony proposed a rate 2/3 code too, a b-adjacent code with

16-bit symbols (Sony opted for 16-bit quantization) and minimum distance 3, in combination with a simple parity-check code for error detection.

On February 5, 1980, after several in-depth and open discussions with Philips experts, Sony proposed a revised code that they called a cross b-adjacent code. This code, again with 16-bit symbols, was a combination of a couple of single-error or double-erasure correcting codes with a convolutional delay interleave in between [T. Doi, "Error Correction for Digital Audio Recordings", AES Premiere Conference, New York 1982, June 3-6, p.170]. L. Vries and L. Driessen, both engineers from Philips, subsequently analyzed this revised code. Their mathematical analysis of the performance of the revised Sony code appeared in an internal Philips Research Technical Note [L. Driessen, L. Vries, "The Performance of Sony's Cross-B-Adjacent Code on a Memoryless Channel", Technical Note Nr. 54, 1980]. This Technical Note was submitted as a discussion paper for the next Philips-Sony meeting. As a consequence of their analysis it became clear to Vries and Driessen that Sony's revised code had a better performance than the convolutional code as originally proposed by Philips.

While analyzing Sony's code, Driessen and Vries saw possibilities to enhance the correction and detection capabilities, both for errors and erasures, without changing the rate of the code. Instead of 16-bit symbols, Driessen (educated in algebraic coding theory) suggested to use 8-bit symbols corresponding to the Galois Field GF(2^8), which implied that a codeword (still having the same number of information and parity bits) doubled in length (counted in symbols) and that the minimum distance increased from 3 to 5. This modification offered a better protection against random errors and short burst errors, while keeping the same protection against long burst errors. A further attractive feature of the proposed improved code was that it allowed several decoding strategies, thereby increasing the freedom for each manufacturer to choose a distinguishing decoding strategy. Although implementing the improved code on a chip at the time was much more complicated than implementing Sony's revised code, the proven advantages were so convincing that Sony accepted the suggested improvements without any changes. Later, after having reached agreement on the lengths of the two interleaved codes and the interleave scheme itself, the improved rate 3/4 code was called CIRC (Cross Interleaved Reed Solomon Code) and it became the Philips-Sony error correcting coding standard for CD in June 1980.

As mentioned before, Philips and Sony also had to agree on a modulation code, which is needed to adapt the incoming bit stream to the characteristic of the CD storage channel. At their first meeting on August 27, 1979, Philips and

Sony each proposed a different modulation code. Philips presented the M3 code also used in their CD prototype, whereas Sony proposed a code called 3PM. Both codes were DC-free and run-length-limited. A DC-free code produces an encoded bit stream with very little spectral content at low frequencies, as required to prevent disturbance of the servo systems. A run-length-limited code produces runs of ones and zeros that are constrained to have a prescribed minimum and maximum length. The choice of the minimum run-length permits the power spectrum of the encoded data sequence to be adapted to the low-pass transfer characteristic of the CD-channel, thereby facilitating bit detection. A proper choice also helps to reduce the impact of various disc artifacts. The maximum run-length ensures that the encoded data stream contains enough timing information to permit reliable clock recovery.

First comparative experiments showed that the M3 code performed better if the disc was scratched or contaminated, while the 3PM code could achieve a higher data density in a clean, well-aligned environment. With this result in mind, the engineers from both sides proposed new modulation codes, supported by practical test results. Experimental data and test discs were exchanged. At some point in the discussions, the successive codes were called ASAP1, ASAP2, ASAP3, indicating the urgency of the project (ASAP=As Soon As Possible). The iterations were stopped as soon as further iterations did not bring significant further improvements, and the resulting code was later dubbed EFM (Eight to Fourteen Modulation). The EFM code has a minimum run-length of 3 bit intervals, a maximum run-length of 11 bit intervals, a code rate of 8/17, and state-independent low-frequency content. Both parties felt that the intensive period of cross testing and of mutual improvements had resulted in "the best of both worlds": a code which combined a high rate (or, equivalently, a high information density on the disc) with a high robustness against disc and player errors.

In June 1980, Philips and Sony decided to apply for two patents, one on CIRC and the other one on EFM. These patents were later granted by the U.S. patent office and are registered as:

1) K. Odaka, Y. Sako, I. Ikuo, T. Doi (all from Sony), L. Vries (Philips), "Error correctable data transmission method", U.S. Patent 4,413,340.
2) K. Immink, J. Nijboer (both from Philips), H. Ogawa, K. Odaka (both from Sony), "Method of coding binary data", U.S .Patent 4,501,000.

Together with the patent of P. Kramer ("Reflective optical record Carrier", U.S. Patent 5068846), mentioned also in Sect. 2.1, these patents are essential to the CD system.

Fig. 1. On the last day, August 18, 1980, at the end of the six meetings, a photograph was taken in Tokyo. It shows a happy smiling team. From left to right:
2nd row: Heemskerk, Harada, Miyaoka, Vries, Nijboer, Tsurushima, Doi, Ogawa, Naruse, Odaka.
Front row: Sinjou, Bögels, Nakajima, Mizushima.

In the so-called 'Red Book' the Philips-Sony standard is described in detail. This book mentions important parameters, such as the playing time of approximately 60 minutes, the 44.1 kHz sampling frequency, the 16-bit signal quantization, and the 12 cm diameter of a disc. The standard of the 'Audio recording-Compact disc digital audio system' is available at the International Electrotechnical Commission (IEC) in Geneva as document 60908 (second edition 1999).

The diameter of 11.5 cm of the CD prototype disc changed to 12 cm in the standard because of a personal wish of N. Ohga. The reason was that with a diameter of 12 cm a particular performance of Beethoven's ninth symphony with a length of 74 minutes could be recorded on a disc.

After the CD standard was established in 1980, many papers were published, not only by Philips and Sony authors separately but also by Philips and Sony authors jointly. The amazing success of the CD after 1982, when the CD player and disc came on the market, also resulted in many books and papers that explained various aspects of the CD system.

Collected papers in this chapter

On the pages following this introduction a number of papers on the CD, by or with Philips authors, are reprinted. In 1982, a special issue of the 'Philips Technical Review' (Vol. 40, No. 6, 1982) was completely dedicated to the CD system. The papers contained in this issue are all reprinted in this chapter. The special issue started with an introduction to the integral CD system with the title "The Compact Disc Digital Audio System". The next three papers explain various subsystems in a CD player. The first of these, "Compact Disc: system aspects and modulation", describes EFM. The second paper, "Error correction and concealment in the Compact Disc system", explains the error correction subsystem CIRC and also the method to conceal those errors that CIRC could not correct but only detect. In the third paper, "Digital-to-analog conversion in playing a Compact Disc", it is shown how the performance of a 14-bit D/A converter, in combination with digital signal processing, can be made equivalent to a 16-bit D/A converter.

A large part of the success of the CD system can be attributed to the attractive small shining disc of which by now about 220 billion have been sold. The disc-mastering process, a key step before mass production of CD discs, is outlined in the paper "Compact Disc (CD) Mastering - An Industrial Process" that is reprinted in Sect. 3.6.

From a system point of view, the successive digital signal processing operations in a CD player are designed on the basis of communications concepts. These concepts encompass demodulation, error correction and detection, interpolation to conceal uncorrected but detected errors, and bandwidth expansion to ease D/A conversion. These ideas are explained in a paper "Communications Aspects of the Compact Disc Digital Audio System" that is reprinted in Sect. 3.7.

3.2 The Compact Disc Digital Audio system

M.G. Carasso, J.B.H. Peek, J.P. Sinjou

Drs M. G. Carasso and Dr Ir J. B. H. Peek are with Philips Research Laboratories, Eindhoven; J. P. Sinjou is with the Philips Audio Division, Eindhoven.

Abstract
Digital processing of the audio signal and optical scanning in the Compact Disc system yield significant advantages: insensitivity to surface damage of the disc, compactness of disc and player, excellent signal-to-noise ratio and channel separation (both 90 dB) and a flat response over a wide range of frequencies (up to 20 000 Hz). The Compact Disc, with a diameter of only 120 mm, gives a continuous playing time of an hour or more. The analog audio signal is converted into a digital signal suitable for transcription on the disc. After the digital signal has been read from the disc by an optical 'pick-up' the original audio signal is recreated in the player.

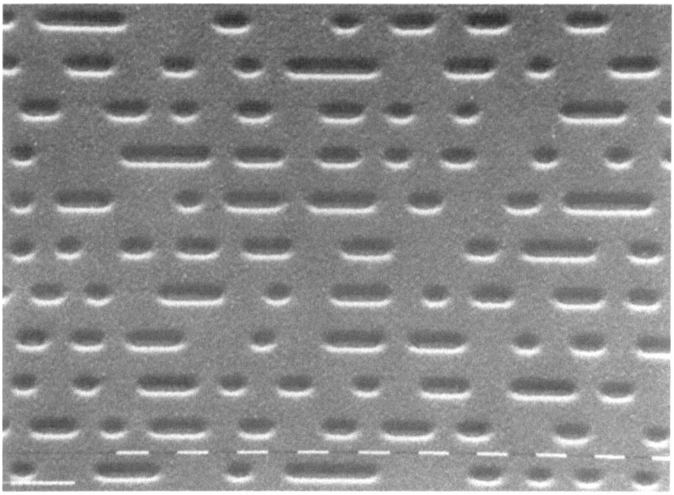

The information on the Compact Disc is recorded in digital form as a spiral track consisting of a succession of pits. The pitch of the track is 1.6 μm, the width 0.6 μm and the depth of the pits 0.12 μm. The length of a pit or the land between two pits has a minimum value of 0.9 and a maximum value of 3.3 μm. The scale at the bottom indicates intervals of 1 μm.

Reprinted with permission from Philips Tech. Rev. **40**, 151-155, 1982.

3.2.1 Introduction

During the many years of its development the gramophone has reached a certain maturity. The availability of long-play records of high quality has made it possible to achieve very much better sound reproduction in our homes than could be obtained with the machine that first reproduced the sound of the human voice in 1877. A serious drawback of these records is that they have to be very carefully handled if their quality is to be preserved. The mechanical tracking of the grooves in the record causes wear, and damage due to operating errors cannot always be avoided. Because of the analog recording and reproduction of the sound signal the signal-to-noise ratio may sometimes be poor (< 60 dB), and the separation between the stereo channels (< 30 dB) leaves something to be desired.

For these and other problems the Compact Disc system offers a solution. The digital processing of the signal has resulted in signal-to-noise ratios and a channel separation that are both better than 90 dB. Since the signal information on the disc is protected by a 1.2 mm transparent layer, dust and surface damage do not lie in the focal plane of the laser beam that scans the disc, and therefore have relatively little effect. Optical scanning as compared with mechanical tracking means that the disc is not susceptible to damage and wear. The digital signal processing makes it possible to correct the great majority of any errors that may nevertheless occur. This can be done because error-correction bits are added to the information present on the disc. If correction is not possible because there are too many defects, the errors can still be detected and 'masked' by means of a special procedure. When a Compact Disc is played there is virtually no chance of hearing the 'tick' so familiar from conventional records.

With its high information density and a playing time of an hour, the outside diameter of the disc is only 120 mm. Because the disc is so compact, the dimensions of the player can also be small. The way in which the digital information is derived from the analog music signal gives a frequency characteristic that is flat from 20 to 20 000 Hz. With this system the well-known wow and flutter of conventional players are a thing of the past.

Another special feature is that 'control and display' information is recorded, as 'C&D' bits. This includes first of all 'information for the listener', such as playing time, composer and title of the piece of music. The number of a piece of music on the disc is included as well. The C&D bits also contain information that indicates whether the audio signal has been recorded with pre-emphasis and should be reproduced with de-emphasis[1]. In the Compact Disc system a pre-emphasis characteristic has been adopted as standard with time constants of 15 and 50 μs. In some of the versions of the player the 'information for the listener' can be presented on a display and the different sections of the music on the disc can be played in the order selected by the user.

In the first article of a series of four on the Compact Disc system we shall deal with the complete system, without going into detail. We shall consider the disc, the processing of the audio signal, reading out the signal from the disc and the reconstitution of the audio signal. The articles that follow will examine the system aspects and modulation, error correction and the digital-to-analog conversion.

3.2.2 The disc

In the Laser Vision system[2], which records video information, the signal is recorded on the disc in the form of a spiral track that consists of a succession of pits. The intervals between the pits are known as 'lands'. The information is present in the track in analog form. Each transition from land to pit and vice versa marks a zero crossing of the modulated video signal. On the Compact Disc the signal is recorded in a similar manner, but the information is present in the track in digital form. Each pit and each land represents a series of bits called channel bits. After each land/pit or pit/land transition there is a '1', and all the channel bits in between are '0'; see Fig. 1.

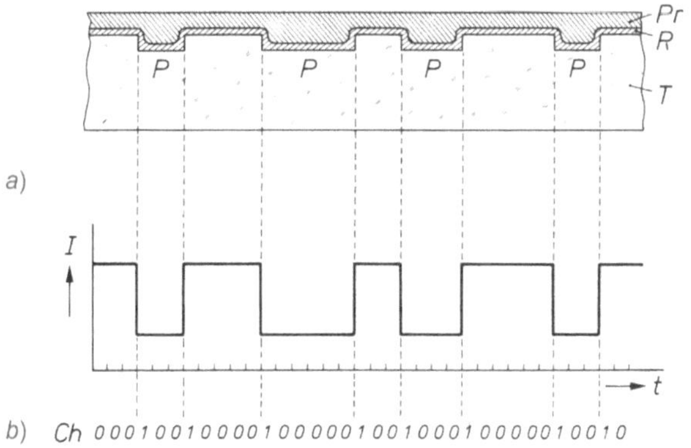

Fig. 1. *a)* Cross-section through a Compact Disc in the direction of the spiral track. T transparent substrate material, R reflecting layer, Pr protective layer. P the pits that form the track. *b*) I the intensity of the signal read by the optical pick-up (see Fig. 2), plotted as a function of time. The signal, shown in the form of rectangular pulses, is in reality rounded and has sloping sides[3]. The digital signal derived from this waveform is indicated as a series of channel bits *Ch*.

The density of the information on the Compact Disc is very high: the smallest unit of audio information (the audio bit) covers an area of 1 μm² on the disc, and the diameter of the scanning light-spot is only 1 μm. The pitch of the track is 1.6 μm, the width 0.6 μm and the depth 0.12 μm. The minimum length of a pit or the land between two pits is 0.9 μm, the maximum length is 3.3 μm. The side of the transparent carrier material T in which the pits P are impressed - the upper side during playback if the spindle is vertical - is covered with a reflecting layer R and a protective layer Pr. The track is optically scanned from below the disc at a constant velocity of 1.25 m/s. The speed of rotation of the disc therefore varies, from about 8 rev/s to about 3.5 rev/s.

3.2.3 Processing of the audio signal

For converting the analog signal from the microphone into a digital signal, pulse-code modulation (PCM) is used. In this system the signal is periodically sampled and each sample is translated into a binary number. From Nyquist's sampling theorem the frequency of sampling should be at least twice as high as the highest frequency to be accounted for in the analog signal. The number of bits per sample determines the signal-to-noise ratio in the subsequent reproduction.

In the Compact Disc system the analog signal is sampled at a rate of 44.1 kHz, which is sufficient for reproduction of the maximum frequency of 20 000 Hz. The signal is quantized by the method of uniform quantization; the sampled amplitude is divided into equal parts. The number of bits per sample (these are called audio bits) is 32, i.e. 16 for the left and 16 for the right audio channel. This corresponds to a signal-to-noise ratio of more than 90 dB. The net bit rate is thus $44.1 \times 10^3 \times 32 = 1.41 \times 10^6$ audio bits/s. The audio bits are grouped into 'frames', each containing six of the original samples.

Successive blocks of audio bits have blocks of parity bits added to them in accordance with a coding system called CIRC (Cross-Interleaved Reed-Solomon Code)[4]. This makes it possible to correct errors during the reproduction of the signal. The ratio of the number of bits before and after this operation is 3:4. Each frame then has C&D (Control and Display) bits, as mentioned earlier, added to it; one of the functions of the C&D bits is providing the 'information for the listener'. After the operation the bits are called data bits.

Next the bit stream is modulated, that is to say the data bits are translated into channel bits, which are suitable for storage on the disc; see fig. 1*b*. The EFM code (Eight-to-Fourteen Modulation) is used for this: in EFM code blocks of eight bits are translated into blocks of fourteen bits[5]. The blocks of fourteen bits are linked by three 'merging bits'. The ratio of the number of bits before and after modulation is thus 8:17.

For the synchronization of the bit stream an identical synchronization pattern consisting of 27 channel bits is added to each frame. The total bit rate after all these manipulations is 4.32×10^6 channel bits/s. Table I gives a survey of the successive operations with the associated bit rates, with their names.

From the magnitude of the channel bit rate and the scanning speed of 1.25 m/s it follows that the length of a channel bit on the disc is approximately 0.3 μm.

Name	Bit rate in 10^6 bits/s	Operations
Audio signal		PCM (44.1 kHz)
Audio bit stream	1.41	CIRC (+ parity bits)
		Addition of C&D bits
Data bit stream	1.94	EFM
		Addition of merging bits
		Addition of synchronization patterns
Channel bit stream	4.32	

Table I. Names of the successive signals, the associated bit rates and operations during the processing of the audio signal.

The signal produced in this way is used by the disc manufacturer to switch on and off the laser beam that illuminates the light-sensitive layer on a rotating glass disc (called the 'master'). A pattern of pits is produced on this disc by means of a photographic developing process. After the surface has been coated with a thin silver layer, an electroplating process is applied to produce a nickel impression, called the 'metal father'. From this 'father disc' impressions called 'mother discs' are produced in a similar manner. The impressions of the mother discs, called 'sons' or 'stampers', are used as tools with which the pits P are impressed into the thermoplastic transparent carrier material T of the disc; see Fig. 1.

3.2.4 Read-out from the disc

As we have seen, the disc is optically scanned in the player. This is done by the AlGaAs semiconductor laser described in an earlier article in this journal[6]. Fig. 2 shows the optical part of the 'pick-up'. The light from the laser *La* (wavelength 800 nm) is focused through the lenses L_2 and L_1 on to the reflecting layer of the disc. The diameter of the light spot S is about 1 μm. When the spot falls on an interval between two pits, the light is almost totally reflected and reaches the four photodiodes D_1-D_4 via the half-silvered mirror M. When the spot lands on a pit - the depth of a pit is about ¼ of the wavelength in the transparent substrate material - interference causes less light to be reflected and an appreciably smaller amount reaches the photodiodes. When the output signals from the four photodiodes are added together the result is a fairly rough approximation[3] to the rectangular pulse pattern present on the disc in the form of pits and intervals.

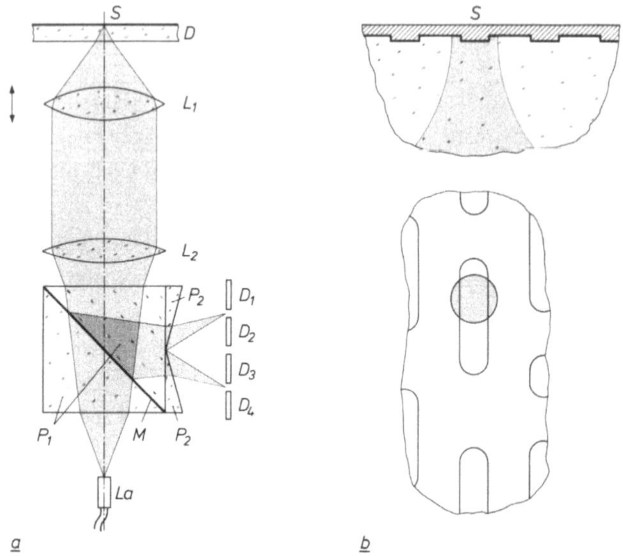

Fig. 2 *a)* Diagram of the optical pick-up. D radial section through the disc. S laser spot, the image on the disc of the light-emitting part of the semiconductor laser *La*. L_1 objective lens, adjustable for focussing. L_2 lens for making the divergent laser beam parallel. *M* half-silvered mirror formed by a film evaporated on the dividing surface of the prism combination P_1. P_2 beam-splitter prisms. D_1 to D_4 photodiodes whose output currents can be combined in various ways to provide the output signal from the pick-up and also the tracking-error signal and the focusing-error signal. (In practice the prisms P_2 and the photodiodes D_1 to D_4 are rotated by 90° and the reflection at the mirror *M* does not take place in a radial plane but in a tangential plane.) *b)* A magnified view of the light spot S and its immediate surroundings, with a plan view. It can clearly be seen that the diameter of the spot (about 1 μm) is larger than the width of the pit (0.6 μm).

The optical pick-up shown in Fig. 2 is very small (about 45 × 12 mm) and is mounted in a pivoting arm that enables the pick-up to describe a radial arc across the disc, so that it can scan the complete spiral track. Around the pivotal point of the arm is mounted a 'linear' motor that consists of a combination of a coil and a permanent magnet. When the coil is energized the pick-up can be directed to any required part of the track, the locational information being provided by the C&D bits added to each frame on the disc. The pick-up is thus able to find independently any particular passage of music indicated by the listener. When it has been found, the pick-up must then follow the track accurately - to within ± 0.1 μm - without being affected by the next or previous track. Since the track on the disc may have some slight eccentricity, and since also the suspension of the turntable is not perfect, the track may have a maximum side-to-side swing of 300 μm. A tracking servosystem is therefore necessary to ensure that the deviation between pick-up and track is smaller than the permitted value of ± 0.1 μm and in addition to absorb the consequences of small vibrations of the player.

The tracking-error signal is delivered by the four photodiodes D_1 to D_4. When the spot *S*, seen in the radial direction, is situated in the centre of the track, a symmetrical beam is reflected. If the spot lies slightly to one side of the track, however, interference effects cause asymmetry in the reflected beam. This asymmetry is detected by the prisms P_2, which split the beam into two components. Beyond the prisms one component has a higher mean intensity than the other. The signal obtained by coupling the photodiodes as (D_1+D_2) - (D_3+D_4) can therefore be used as a tracking-error signal.

As a result of ageing or soiling of the optical system, the reflected beam may acquire a slowly increasing, more or less constant asymmetry. Owing to a d.c. component in the tracking-error signal, the spot will then always be slightly off-centre of the track. To compensate for this effect a second tracking-error signal is generated. The coil that controls the pick-up arm is therefore supplied with an alternating voltage at 600 Hz, with an amplitude that corresponds to a radial displacement of the spot by ± 0.05 μm. The output sum signal from the four photodiodes - which is at a maximum when the spot is in the centre of the track - is thus modulated by an alternating voltage of 600 Hz. The amplitude of this 600 Hz signal increases as the spot moves off-centre. In addition the sign of the 600 Hz error signal changes if the spot moves to the other side of the track. This second tracking-error signal is therefore used to correct the error signal mentioned earlier with a direct voltage. The output sum signal from the photodiodes, which is processed in the player to become the audio signal, is thus returned to its maximum value.

The depth of focus of the optical pick-up at the position of *S* (see Fig. 2) is about 4 μm. The axial deviation of the disc, owing to various mechanical effects, can have a maximum of 1 mm. It is evident that a servosystem is also necessary to give correct focusing of the pick-up on the reflecting layer. The objective lens L_1 can therefore be displaced in the direction of its optical axis by a combination of a coil and a permanent magnet, in the same way as in a loudspeaker. The focusing-error signal is also provided by the row of photodiodes D_1 to D_4. If the spot is sharply focused on the disc, two sharp images are precisely located between D_1 and D_2 and between D_3 and D_4. If the spot is not sharply focused on the disc, the two images on the photodiodes are not sharp either, and have also moved closer together or further apart. The signal obtained by connecting the photodiodes as $(D_1+D_4)-(D_2+D_3)$ can therefore be used for controlling the focusing servosystem. The deviation in focusing then remains limited to ± 1 μm.

3.2.5 Reconstitution of the audio signal

The signal read from the disc by the optical pick-up has to be reconstituted to form the analog audio signal.

Fig. 3 shows the block diagram of the signal processing in the player. In *DEMOD* the demodulation follows the same rules that were applied to the EFM modulation, but now in the opposite sense. The information is then temporarily stored in a buffer memory and then reaches the error-detection and correction circuit *ERCO*. The parity bits can be used here to correct errors, or just to detect errors if correction is found to be impossible[4]. These errors may originate from defects in the manufacturing process, damage during use, or fingermarks or dust on the disc. Since the information with the CIRC code is 'interleaved' in time, errors that occur at the input of *ERCO* in one frame are spread over a large number of frames during decoding in *ERCO*. This increases the probability that the maximum number of correctable errors per frame will not be exceeded. A flaw such as a scratch can often produce a train of errors, called an error burst. The error-correction code used in *ERCO* can correct a burst of up to 4000 data bits, largely because the errors are spread out in this way.

If more errors than the permitted maximum occur, they can only be detected. In the CIM block (Concealment: Interpolation and Muting) the errors detected are then masked. If the value of a sample indicates an error, a new value is determined by linear interpolation between the preceding value and the next one. If two or more successive sample values indicate an error, they are made equal to zero (muting). At the same time a gradual transition is created to the values preceding and succeeding it by causing a number of values before the error and after it to decrease to zero in a particular pattern.

In the digital-to-analog converters *DAC*[7] the 16 bit samples first pass through interpolation filters F and are then translated and recombined to recreate the original analog audio signal A from the two audio channels L and R. Since samples must be recombined at exactly the same rate as they are taken from the analog audio signal, the DACs and also *CIM* and *ERCO* are synchronized by a clock generator C controlled by a quartz crystal.

Fig. 3 also illustrates the control of the disc speed n_D. The bit stream leaves the buffer memory at a rate synchronized by the clock generator. The bit stream enters the buffer memory, however, at a rate that depends on the speed of revolution of the disc. The extent to which n_D and the sampling rate are matched determines the 'filling degree' of the buffer memory. The control is so arranged as to ensure that the buffer memory is at all times filled to 50% of its capacity. The analog signal from the player is thus completely free from wow and flutter, yet with only moderate requirements for the speed control of the disc.

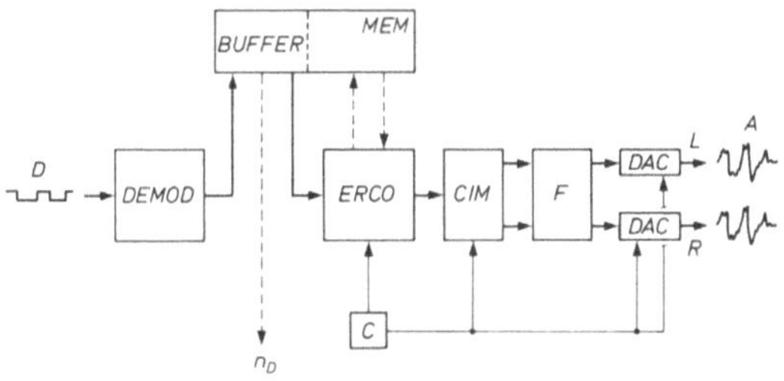

Fig. 3. Block diagram of the signal processing in the player. *D* input signal read by the optical pick-up; see Fig. 2. A the two output analog audio signals from the left (*L*) and the right (*R*) audio channels. *DEMOD* demodulation circuit. *ERCO* error-correction circuit. *BUFFER* buffer memory, forming part of the main memory *MEM* associated with *ERCO. CIM* (Concealment: Interpolation and Muting) circuit in which errors that are only detected since they cannot be corrected are masked or 'concealed'. *F* filters for interpolation. *DAC* digital-to-analog conversion circuits. Each of the blocks mentioned here are fabricated in VLSI technology. *C* clock generator controlled by a quartz crystal. The degree to which the buffer memory capacity is filled serves as a criterion in controlling the speed of the disc.

References

[1] See F. W. de Vrijer, Modulation, Philips tech. Rev. **36**, 305-362 (1976), in particular pages 323 and 324.

[2] See Philips tech. Rev. **33**, (Sect. 3.3).

[3] See Fig. 3 of the article by J. P. J. Heemskerk and K. A. Schouhamer Immink, Sect. 3.3

[4] See H. Hoeve, J. Timmermans and L. B. Vries, Error correction and concealment in the Compact Disc system, Sect. 3.4

[5] See J. P. J. Heemskerk and K. A. Schouhamer Immink, Compact Disc: system aspects and modulation, Sect. 3.3.

[6] J. C. J. Finck, H. J. M. van der Laak and J. T. Schrama, Philips tech. Rev. **39**, 37 (1980).

[7] See D. Goedhart, R. J. van de Plassche and E. F. Stikvoort, Digital-to-analog conversion in playing a Compact Disc, Sect. 3.5.

3.3 Compact Disc: system aspects and modulation

J.P.J. Heemskerk, K.A. Schouhamer Immink

Dr J. P. J. Heemskerk is with the Philips Audio Division, Eindhoven;
Ir K. A. Schouhamer Immink is with Philips Research Laboratories, Eindhoven.

Abstract

The Compact Disc system can be considered as a transmission system that brings sound from the studio into the living room. The sound encoded into data bits and modulated into channel bits is sent along the 'transmission channel' consisting of write laser — master disc — user disc — optical pick-up. The maximum information density on the disc is determined by the diameter d of the laser light spot on the disc and the 'number of data bits per light spot'. The effect of making d smaller is to greatly reduce the manufacturing tolerances for the player and the disc. The compromise adopted is $d \approx 1$ µm, giving very small tolerances for objective and disc tilt, disc thickness and defocusing. The basic idea of the modulation is that, while maintaining the minimum length for 'pit' and 'land' (the 'minimum run length') required for satisfactory transmission, the information density can be increased by increasing the number of possible positions per unit length for pit edges (the bit density). Because of clock regeneration there is also a maximum run length, and the low-frequency content of the transmission channel must be kept as low as possible. With the EFM modulation system used each 'symbol' of eight data bits is converted into 14 channel bits with a minimum run length of 3 and a maximum run length of 11 bits, plus three merging bits, chosen such that, when the symbols are merged together, the run-length conditions continue to be satisfied and the low-frequency content is kept to the minimum.

In this article we shall deal in more detail with the various factors that had to be weighed one against the other in the design of the Compact Disc system. In particular we shall discuss the EFM modulation system ('Eight-to-Fourteen Modulation'), which helps to produce the desired high information density on the disc.

Fig. 1 represents the complete Compact Disc system as a 'transmission system' that brings the sound of an orchestra into the living room. The orchestral sound is converted at the recording end into a bit stream B_i, which is recorded on the master disc. The master disc is used as the 'pattern' for making the discs for the user. The player in the living room derives the bit stream B_o - which in the ideal case should be identical to B_i - from the disc and reconverts it to the orchestral sound. The system between *COD* and *DECOD* is the actual *transmission channel*; B_i and B_o consist of 'channel bits'.

Reprinted with permission from Philips Tech. Rev. **40**, 157-164, 1982.

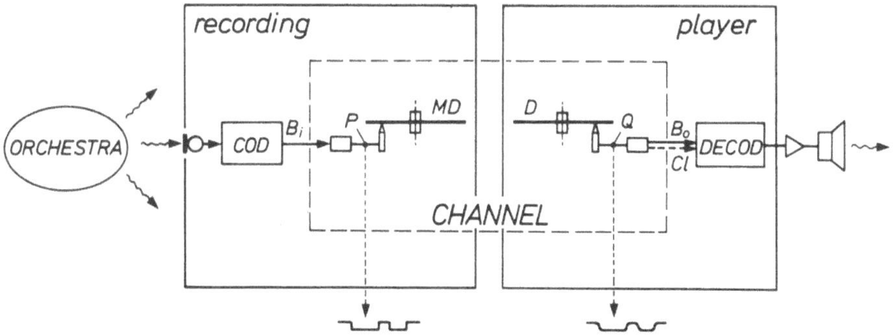

Fig. 1. The Compact Disc system, considered as a transmission system that brings sound from the studio into the living room. The transmission channel between the encoding system (*COD*) at the recording end and the decoding system (*DECOD*) in the player, 'transmits' the bit stream B_i to *DECOD* via the write laser, the master disc (*MD*), the disc manufacture, the disc (*D*) in the player and the optical pick-up; in the ideal case B_o is the same as B_i. The bits of B_o, as well as the clock signal (*Cl*) for further digital operations, have to be detected from the output signal of the pick-up unit at *Q*.

Fig. 2 shows the encoding system in more detail. The audio signal is first converted into a stream B_1 of 'audio bits' by means of pulse-code modulation. A number of bits for 'control and display' (C&D) and the parity bits for error correction are then added to the bit stream[1][2]. This results in the 'data bit stream' B_2. The modulator converts this into channel bits (B_3). The bit stream B_i is obtained by adding a synchronization signal. The number of data bits n that can be stored on the disc in given by:

$$n = \eta A/d^2,$$

where A is the useful area of the disc surface, d is the diameter of the laser light spot on the disc and η is the 'number of data bits per spot' (the number of data bits that can be resolved per length d of track). A/d^2 is the number of spots that can be accomodated side by side on the disc. The information density n/A is thus given by:

$$n/A = \eta/d^2.$$

The spot diameter d is one of the most important parameters of the channel. The modulation can give a higher value of η. We shall now briefly discuss some of the aspects of the channel that determine the specification for the modulation system.

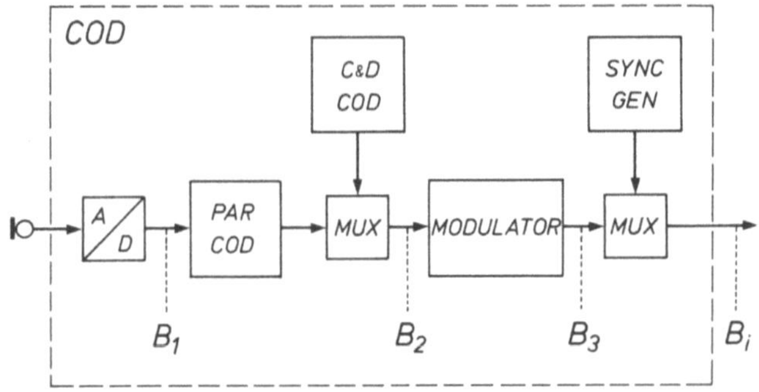

Fig. 2. The encoding system (*COD* in Fig. 1). The system is highly simplified here; in practice for example there are two audio channels for stereo recording at the input, which together supply the bit stream B_1 by means of PCM, and the various digital operations are controlled by a 'clock', which is not shown. The bit stream B_1 is supplemented by parity and C&D (control and display) bits (B_2), modulated (B_3), and provided with synchronization signals (B_i). *MUX*: multiplexers. Fig. 9 gives the various bit streams in more detail.

The channel

The bit stream B_i in Fig. 1 is converted into a signal at *P* that switches the light beam from the write laser on and off. The channel should be of high enough quality to allow the bit stream B_i to be reconstituted from the read signal at *Q*.

To achieve this quality all the stages in the transmission path must meet exacting requirements, from the recording on the master disc, through the disc manufacture, to the actual playing of the disc. The quality of the channel is determined by the player and the disc: these are mass-produced and the tolerances cannot be made unacceptably small.

We shall consider one example here to illustrate the way in which such tolerances affect the design: the choice of the 'spot diameter' *d*. We define *d* as the half-value diameter for the light intensity; we have

$$d = 0.6\, \lambda / NA,$$

where λ is the wavelength of the laser light and *NA* is the numerical aperture of the objective. To achieve a high information density (1) *d* must be as small as possible. The laser chosen for this system is the small CQL10[3], which is inexpensive and only requires a low voltage; the wavelength is thus fixed; $\lambda \approx$ 800 nm. This means that we must make the numerical aperture as large as possible. With increasing *NA*, however, the manufacturing tolerances of the player and the disc rapidly become smaller. For example, the tolerance in the local 'skew' of the disc (the 'disc tilt') relative to the objective-lens axis is proportional to NA^{-3}. The tolerance for the disc thickness is proportional

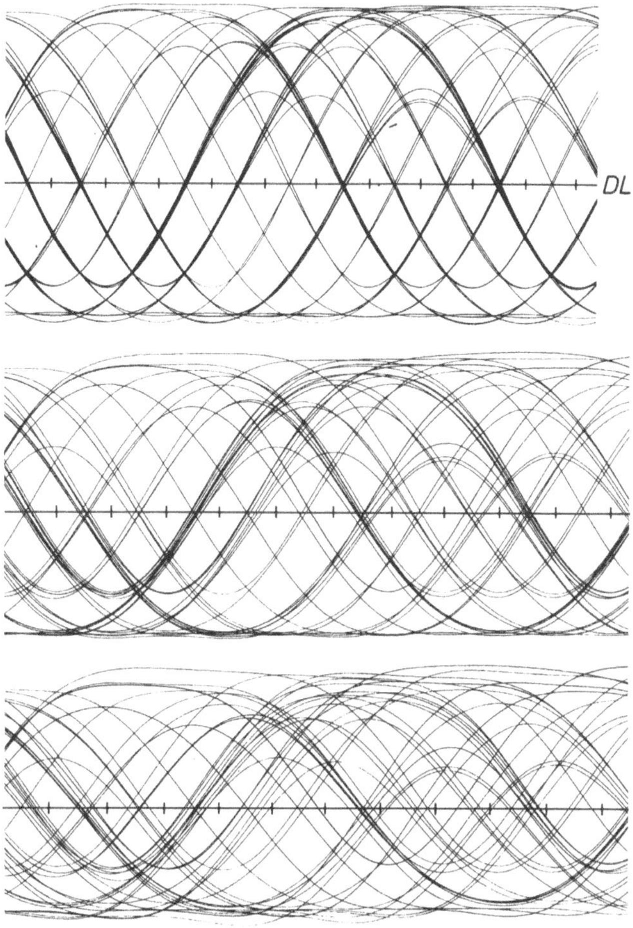

a

b

c

Fig. 3 *a)* Eye pattern. The figures give the read signal (at *Q* in Fig. 1) on an oscilloscope synchronized with the bit clock. At the decision times (marked by dashes) it must be possible to determine whether the signal is above or below the decision level (*DL*). The curves have been calculated for *a)* an ideal optical system, *b)* a defocusing of 2 μm, *c)* a defocusing of 2 μm and a disc tilt of 1.2°. The curves give a good picture of experimental results.

to NA^{-4}, and the depth of focus, which determines the focusing tolerance, is proportional to NA^{-2}. After considering all these factors in relation to one another, we arrived at a value of 0.45 for *NA*. We thus find a value of 1 μm for the spot diameter *d*.

The quality of the channel is evaluated by means of an 'eye pattern', which is obtained by connecting the point *Q* in Fig. 1 to an oscilloscope synchronized with the clock for the bit stream B_0; see Fig. 3*a*. The signals originating from different pits and lands are super-imposed on the screen; they are strongly rounded, mainly because the spot diameter is not zero and the pit walls are not vertical. If the transmission quality is adequate, however, it is always possible to determine whether the signal is positive or negative at the 'clock times' (the dashes in Fig. 3*a*), and hence to reconstitute the bit stream. The lozenge pattern around a dash in this case is called the 'eye'. Owing to channel imperfections the eye can become obscured; owing to 'phase jitter' of the signal relative to the clock an eye becomes narrower, and noise reduces its height. The signals in Fig. 3a were calculated for a perfect optical system. Fig. 3*b* shows the effect of defocusing by 2 μm and Fig. 3*c* shows the effect of a radial tilt of 1.2° in addition to the defocusing. In Fig. 3*b* a correct decision is still possible, but not in Fig. 3*c*.

This example also gives some idea of the exacting requirements that the equipment has to meet. A more general picture can be obtained from Table I, which gives the manufacturing tolerances of a number of important parameters, both for the player and for the disc. The list is far from complete, of course.

With properly manufactured players and discs the channel quality can still be impaired by dirt and scratches forming on the discs during use. By its nature the system is fairly insensitive to these[1], and any errors they may introduce can nearly always be corrected or masked[2]. In the following we shall see that the modulation system also helps to reduce the sensitivity to imperfections.

Player	Objective-lens tilt ± 0.2 °
	Tracking ±0.1 μm
	Focusing ± 0.5 μm
	R.M.S. wavefront noise of read laser beam 0.05 λ (40 nm)
Disc	Thickness 1.2 ± 0.1 mm
	Flatness ± 0.6° (at the rim corresponding to a sag of 0.5 mm)
	Pit-edge positioning ± 50 nm
	Pit depth 120 ± 10 nm

Table I: Manufacturing tolerances.

Bit-stream modulation

The playing time of a disc is equal to the track length divided by the track velocity *v*. For a given disc size the playing time therefore increases if we decrease the track velocity in the system (the track velocity of the master disc and of the user disc). However, if we do this the channel becomes 'worse': the eye height decreases and the system becomes more sensitive to perturbations. There is therefore a lower limit to the track velocity if a minimum value has been established for the eye height because of the expected level of noise and perturbation. We shall now show that we can decrease this lower limit by an appropriate bitstream modulation.

We first consider the situation without modulation. The incoming data bit stream is an arbitrary sequence of ones and zeros. We consider a group of 8 data bits in which the change of bit value is fastest (Fig. 4*a*). Uncoded recording (1: pit; 0: land, or vice versa) then gives the pattern of Fig. 4*b*. This results in the rounded-off signal of Fig. 4*c* at *Q* in Fig. 1; Fig. 4*d* gives the eye pattern. The signal in Fig. 4c represents the highest frequency (f_{m1}) for this mode of transmission, and we have $f_{m1} = ½ f_d$, where f_d is the data bit rate. The half eye height a_1 is equal to the amplitude A_1 of the highest-frequency signal.

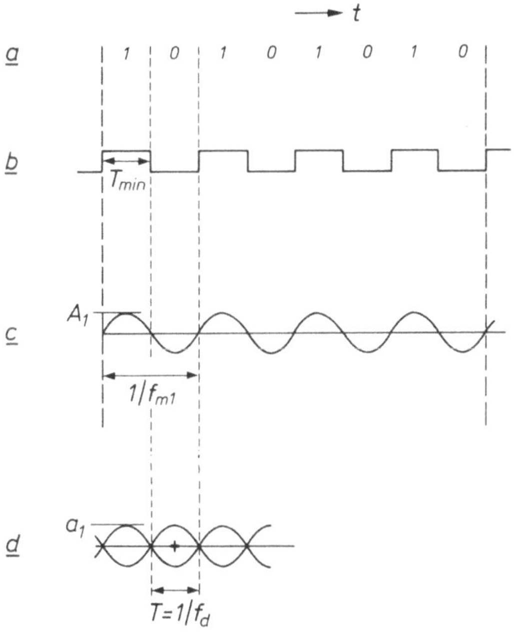

Fig. 4. Direct recording of the data bit stream on the disc. *a*) Data bit stream of the highest frequency that can occur. *b*) Direct translation of the bit stream into a pattern of pits. *c*) The corresponding output signal (at *Q* in Fig. 1); its amplitude A_1 is found with the aid of Fig. 5. *d*) The eye pattern that follows from (*c*). T_{min} minimum pit or land length; f_{m1} highest frequency; *T* data bit length; f_d data bit rate. We have $T_{min} = T$; $f_{m1} = ½ f_d$.

The relation between the eye height and the track velocity now follows indirectly from the 'amplitude-frequency characteristic' of the channel; see Fig. 5. In this diagram *A* is the amplitude of the sinusoidal signal at *Q* in Fig. 1 when a sinusoidal unit signal of frequency *f* is presented at *P*. With the aid of Fourier analysis and synthesis the output signal can be calculated from *A*(*f*) for any input signal. The line in the diagram represents a channel with a perfect optical system. In the first part of this section we shall take this for granted. The true situation will always be less favourable. The 'cut-off frequency' is determined by the spot diameter and the track velocity v; in the ideal case $f_c = (2NA/\lambda)v$.

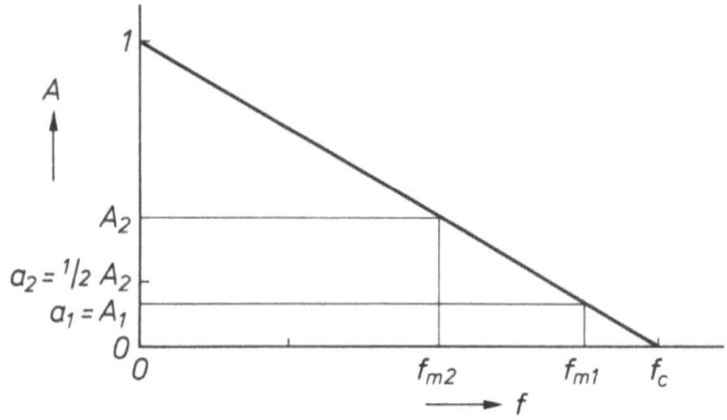

Fig. 5. Amplitude-frequency characteristic of the channel. The diagram gives the amplitude A of the sinusoidal signal at Q (Fig. 1) when a sinusoidal unit signal is presented at P as a function of the frequency f. The transfer is 'cut off' at the frequency f_c, which is given by $f_c = (2NA/\lambda)v$. The line shown applies to an ideal optical system; in reality A is always somewhat lower; the cut-off frequency is then effectively lower. The 'maximum frequencies' f_{m1}, f_{m2}, the amplitudes A_1, A_2 and the 'half eye heights' a_1, a_2 relate to the 'direct' and 'modulated' writing of the data bits on the disc; see Figs 4 and 6.

For a given track velocity we now obtain the half eye height a_1 in Fig. 4 directly from Fig. 5: it is equal to the amplitude A_1 at the frequency f_{m1}. If v, and hence f_c, is varied, the line in Fig. 5 rotates about the point 1 on the A-axis. For a given minimum value of a_1, the figure indicates how far f_c can be decreased; this establishes the lower limit for v. In particular, if the minimum value for a_1 is very small, f_c can be decreased to a value slightly above f_{m1} ($= \frac{1}{2} f_d$).

Fig. 6 gives the situation **with** modulation: an imaginary 8→16 modulation, which is very close to EFM, however. Each group of 8 incoming data bits (Fig. 6*a*) is converted into 16 channel bits (Fig. 6*a*′). This is done by using a 'dictionary' that assigns unambiguously but otherwise arbitrarily to each word of 8 bits a word of 16 bits, but in such a way that the resultant channel bit stream only produces pits and lands that are at least three channel bits long (Fig. 6*b*). On the time scale the minimum pit and land lengths ('the minimum run length' T_{min}) have become 1½ times as long as in Fig. 4, but a simple calculation shows that about as much information can nevertheless be transmitted as in Fig. 4 (256 combinations for 8 data bits), because there is a greater choice of pit-edge positions per unit length (see Fig. 6*b* and *b*′); the 'channel bit length' T_c has decreased by a half.

With the modulation we have managed to reduce the highest frequency (f_{m2}) in the signal (see Fig. 6*c*, left; $f_{m2} = \frac{1}{3} f_d = \frac{2}{3} f_{m1}$). Therefore f_c and v can be reduced by a factor of 1½ for the case in which a very small eye height is tolerable (see Fig. 5); this represents an increase of 50% in playing time.

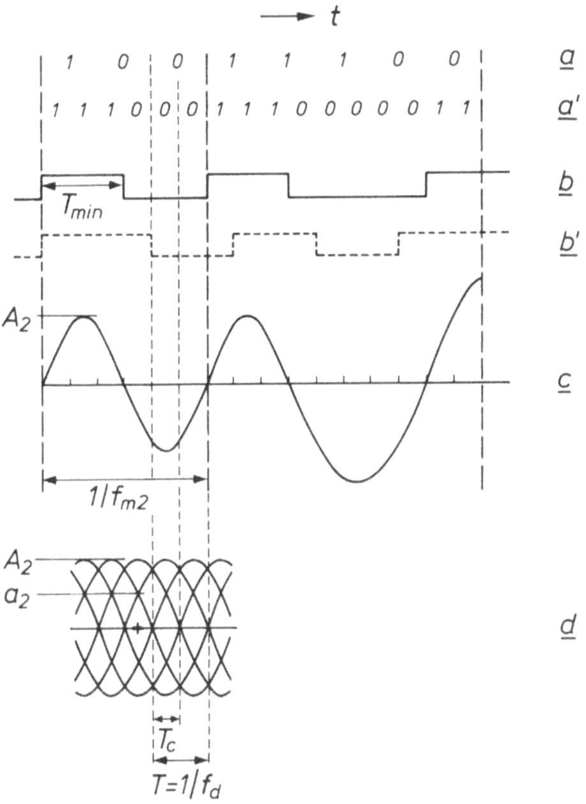

Fig. 6. Eight-to-sixteen modulation. Each group of 8 data bits (a) is translated with the aid of a dictionary into 16 channel bits (*a'*), in such a way that the run length is equal to at least three channel bits. *b*) Pattern of pits produced from the bit stream (*a'*). *b'*) pattern of pits obtained with a different input signal. *c*) The read signal corresponding to (*b*); its amplitude is again determined from Fig. 5.*d*) The resultant eye pattern. The half eye height ($a2$) here is only half the amplitude (A_2) of the approximately sinusoidal signal of maximum frequency (f_{m2}).

The modulation also has its disadvantages. In the first place the half eye height (a_2) in this case is only half of the amplitude (A_2) of the signal at the highest frequency (see Fig. 6*d*). This has consequences if the minimum eye height is not very small. For example, the modulation becomes completely unusable if the half eye height in Fig. 5 has to remain larger than ½ ($a_2 > ½$ implies $A_2 > 1$); uncoded recording is then still possible ($A_1 = a_1$). In the second place, the tolerance for time errors and for the positioning of pit edges, together with the eye width (T_c), has decreased by a half. In designing a system, the various factors have to be carefully weighed against one another.

To show qualitatively how a choice can be made, we have plotted the half eye height in Fig.7 as a function of the 'linear information density' σ (the number of incoming data bits per unit length of the track; $\sigma = f_d/v$) for three systems: '8→8 modulation' (i.e. uncoded recording), 8→16 modulation, and a system that also has about the same information capacity (256 combinations for 8 data bits) in which, however, the minimum run length has been increased still further, again at the expense of eye width of course ('8→24 modulation', $T_{min} = 2T$, $T_c = ⅓T$). The figure is a direct consequence of the reasoning above, with the assumption that the cut-off frequency is 20% lower than the ideal value $(2NA/\lambda)v$, as a first rough adjustment to what we find in practice for the function $A(f)$.

In qualitative terms, the 8→16 system has been chosen because the nature of the noise and perturbations is such that the eye can be smaller than at *A* in Fig. 7, but becomes too small at *C*. An improvement is therefore possible with 8→16 modulation, but not with 8→24 modulation.

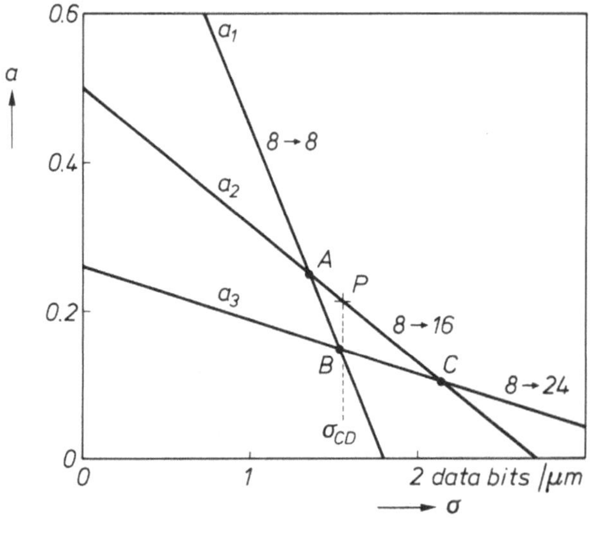

Fig. 7. Half eye height a as a function of the linear information density σ, for 8→8, 8→16 and 8→24 modulation. These systems are characterized by the following values for the channel bit length T_c and the minimum run length T_{min}:
8→8: $T_c = T$, $T_{min} = T$ (Fig. 4),
8→16: $T_c = ½\,T$, $T_{min} = T$ (Fig. 6),
8→24: $T_c = ⅓\,T$, $T_{min} = 2\,T$,
where T is the data bit length. The straight lines give the relations that follow from Fig. 5:
$a_1 = c_1(1 - f_{m1}/f_c) \rightarrow a_1 = 1 - \sigma/1.8$,
$a_2 = c_2(1 - f_{m2}/f_c) \rightarrow a_2 = 0.5(1 - \sigma/2.7)$,
$a_3 = c_3(1 - f_{m3}/f_c) \rightarrow a_3 = 0.26(1 - \sigma/3.6)$,
where σ is the numerical value of the linear information density, expressed in data bits per μm. The c's are the ratios of the half eye height to the amplitude, and the f_m's the maximum frequencies for the three systems ($c_1 = 1$, $c_2 = \sin 30° = 0.5$, $c_3 = \sin 15° = 0.26$, $f_{m1} = ½\,f_d$, $f_{m2} = ⅓\,f_d$, $f_{m3} = ¼f_d$; f_d is the data bit rate). The second set of equations follows from the first set by substituting $0.8 \times (2NA/\lambda)\upsilon$ for f_c, with $NA = 0.45$, $\lambda = 0.8$ μm, $\upsilon = f_d/\sigma$. The factor 0.8 is introduced as a rough first-order correction to the 'ideal' amplitude characteristic.

For our Compact Disc system we have $\sigma = 1.55$ data bits/µm ($f_d = 1.94$ Mb/s, $v = 1.25$ m/s[1]); the operating point would therefore be at *P* in Fig. 7. The model used is however rather crude and in better models *A*, *B* and *C* lie more to the left, so that *P* approaches *C*. But 8 → 16 modulation is still preferable to 8 → 24 modulation, even close to *C*, since the eye width is 1½ times as large as for 8 → 24 modulation.

EFM is a refinement of 8→16 modulation. It has been chosen on the basis of more detailed models and many experiments. At the eye height used, it gives a gain of 25% in information density, compared with uncoded recording.

Further requirements for the modulation system

In developing the modulation system further we still had two more requirements to take into account.

In the first place it must be possible to regenerate the *bit clock* in the player from the read-out signal (the signal at *Q* in Fig. 1). To permit this the number of pit edges per second must be sufficiently large, and in particular the 'maximum run length' T_{max} must be as small as possible.

The second requirement relates to the 'low-frequency content' of the read signal. This has to be as small as possible. There are two reasons for this. In the first place, the servosystems for track following and focusing[1] are controlled by low-frequency signals, so that low-frequency components of the information signal could interfere with the servo-systems. The second reason is illustrated in Fig. *8*, in which the read signal is shown for a clean disc (*a*) and for a disc that has been soiled, e.g. by fingermarks (*b*). This causes the amplitude and average level of the signal to fall. The fall in level causes a completely wrong read-out if the signal falls below the decision level. Errors of this type are avoided by eliminating the low-frequency components with a filter (*c*), but the use of such a filter is only permissible provided the information signal itself contains no low-frequency components. In the Compact Disc system the frequency range from 20 kHz to 1.5 MHz is used for information transmission; the servosystems operate on signals in the range 0-20 kHz.

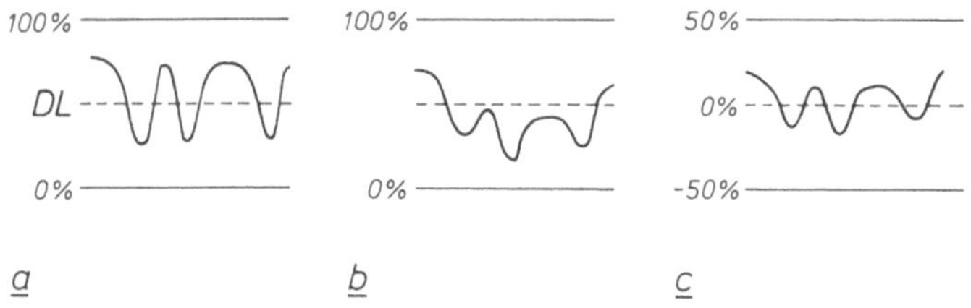

Fig. 8. The read-out signal for six pit edges on the disc, a) for a clean disc, b) for a soiled disc, c) for a soiled disc after the low frequencies have been filtered out. DL decision level. Because of the soiling, both the amplitude and the signal level decrease; the decision errors that this would cause are eliminated by the filter.

The EFM modulation system

Fig. 9 gives a schematic general picture of the bit streams in the encoding system. The information is divided into 'frames'. One frame contains 6 sampling periods, each of 32 audio bits (16 bits for each of the two audio channels). These are divided into symbols of 8 bits. The bit stream B_1 thus contains 24 symbols per frame. In B_2 eight parity symbols have been added and one C&D symbol, resulting in 33 'data symbols'.

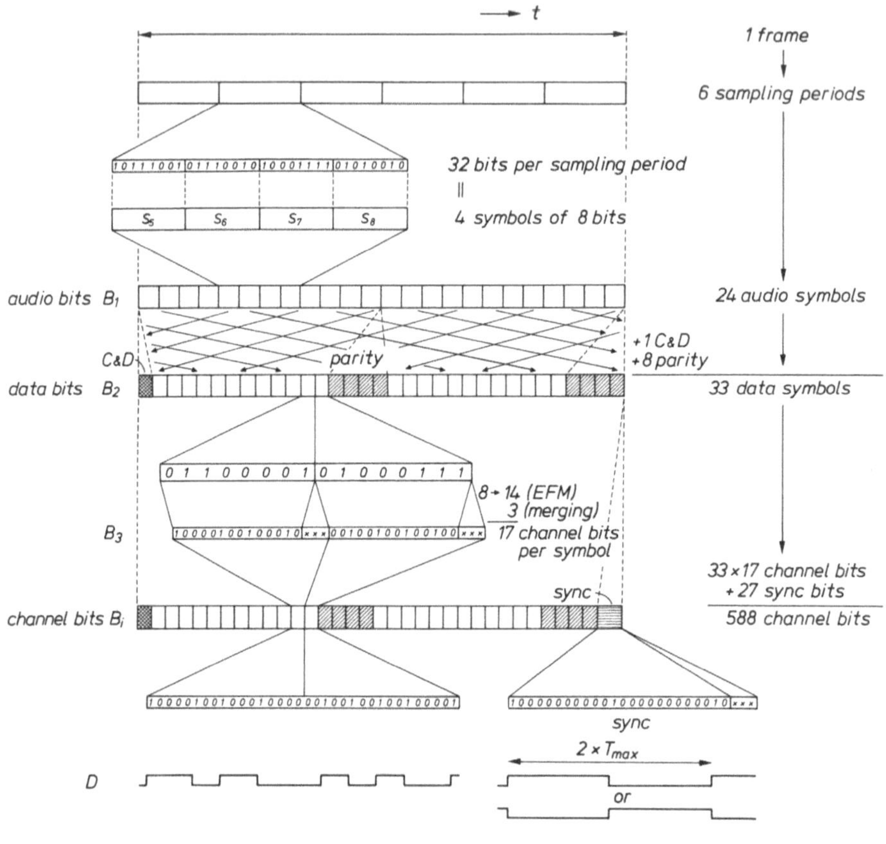

Fig. 9. Bit streams in the encoding system (Fig. 2). The information is divided into frames; the figure gives one frame of the successive bit streams. There are six sampling periods for one frame, each sampling period giving 32 bits (16 for each of the two audio channels). These 32 bits are divided to make four symbols in the 'audio bit stream' B_1. In the 'data bit stream' B_2 eight parity and one *C&D* symbols have been added to the 24 audio symbols. To scatter possible errors, the symbols of different frames in B_1 are interleaved, so that the audio signals in one frame of B_2 originate from different frames in B_1. The modulation translates the eight data bits of a symbol of B_2 into fourteen channel bits, to which three 'merging bits' are added (B_3). The frames are marked with a synchronization signal of the form illustrated (bottom right); the final result is the 'channel bit stream' (B_i) used for writing on the master disc, in such a way that each '1' indicates a pit edge (D).

The modulator translates each symbol into a new symbol of 14 bits. Added to these are three 'merging bits', for reasons that will appear shortly. After the addition of a synchronization symbol of 27 bits to the frame, the bit stream B_i is obtained. B_i therefore contains $33 \times 17 + 27 = 588$ channel bits per frame. Finally, B_i is converted into a control signal for the write laser. It should be noted that in B_i '1' or '0' does not mean 'pit' or 'land', as we assumed for simplicity in Fig. 6, but a '1' indicates a pit edge. The information is thus completely recorded by the positions of the pit edges; it therefore makes no difference to the decoding system if 'pit' and 'land' are interchanged on the disc.

Opting for the translation of series of 8 bits following the division into symbols in the parity coding has the effect of avoiding error propagation. This is because in the error-correction system an entire symbol is always either 'wrong' or 'not wrong'. One channel-bit error that occurs in the transmission spoils an entire symbol, but — because of the correspondence between modulation symbols and data symbols — never more than one symbol. If a different modulation system is used, in which the data bits are not translated in groups of 8, but in groups of 6 or 10, say, then the bit stream B_2 is in fact first divided up into 6 or 10 bit 'modulation symbols'. Although one channel-bit error then spoils only one modulation symbol, it usually spoils two of the original 8 bit symbols.

In EFM the data bits are translated 8 at a time into 14 channel bits, with a T_{min} of 3 and a T_{max} of 11 channel bits (this means at least 2 and at the most 10 successive zeros in B_i). This choice came about more or less as follows. We have already seen that the choice of about 1½ data bits for T_{min}, with about 16 channel bits on 8 data bits, is about the optimum for the Compact Disc system[4]. A simple calculation shows that at least 14 channel bits are necessary for the reproduction of all the 256 possible symbols of 8 data bits under the conditions $T_{min} = 3$, $T_{max} = 11$ channel bits. The choice of T_{max} was dictated by the fact that a larger choice does not make things very much easier, whereas a smaller choice does create far more difficulties.

With 14 channel bits it is possible to make up 267 symbols that satisfy the run-length conditions. Since we only require 256, we omitted 10 that would have introduced difficulties with the 'merging' of symbols under these conditions, and one other chosen at random. The dictionary was compiled with the aid of computer optimization in such a way that the translation in the player can be carried out with the simplest possible circuit, i.e. a circuit that contains the minimum of logic gates.

The merging bits are primarily intended to ensure that the run-length conditions continue to be satisfied when the symbols are 'merged'. If the run length is in danger of becoming too short we choose '0's for the merging bits; if it is too long we choose a '1' for one of them. If we do this we still retain a large measure of freedom in the choice of the merging bits, and we use this freedom to minimize the low-frequency content of the signal. In itself, two merging bits would be sufficient for continuing to satisfy the run-length conditions. A third is necessary, however, to give sufficient freedom for effective suppression of the low-frequency content, even though it means a loss of 6% of the information density on the disc. The merging bits contain no audio information, and they are removed from the bit stream in the demodulator.

Fig. 10 illustrates, finally, how the merging bits are determined. Our measure of the low-frequency content is the 'digital sum value' (DSV); this is the difference between the totals of pit and land lengths accumulated from the beginning of the disc. At the top are shown two data symbols of B_2 and their translation from the dictionary into channel symbols (B_3). From the T_{min} rule the first of the merging bits in this case must be a zero; this position is marked '*X*'. In the two following positions the choice is free; these are marked '*M*'. The three possible choices *XMM*=000, 010 and 001 would give rise to the patterns of pits as illustrated, and to the indicated waveform of the DSV, on the assumption that the DSV was equal to 0 at the beginning. The system now opts for the merging combination that makes the DSV at the end of the second symbol as small as possible, i.e. 000 in this case. If the initial value had been −3, the merging combination 001 would have been chosen.

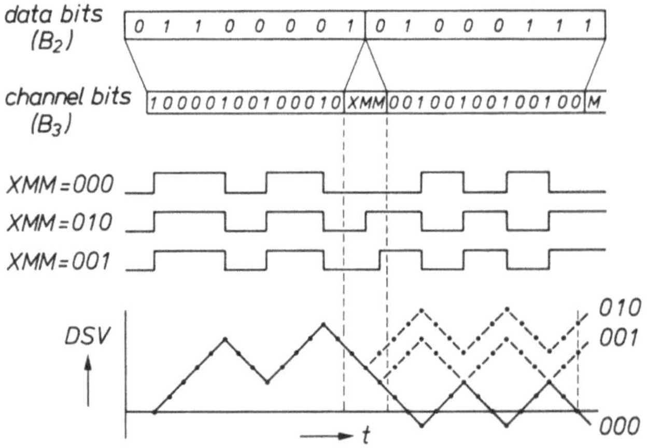

Fig. 10. Strategy for minimizing the digital sum value (DSV). After translation of the data bits into channel bits, the symbols are merged together by means of three extra bits in such a way that the run-length conditions continue to be satisfied and the DSV remains as small as possible. The first run-length rule (at least two zeros one after the other) requires a zero at the first position in the case illustrated here, while the choice remains free for the second and third positions. In this case there are thus three merging alternatives: 000, 010 and 001. These alternatives give the patterns of pits shown in the diagram and the illustrated DSV waveform. The system chooses the alternative that gives the lowest value of DSV at the end of the next symbol. The system looks 'one symbol ahead'; strategies for looking further ahead are also possible in principle.

When this strategy is applied, the noise in the servo-band frequencies (< 20 kHz) is suppressed by about 10 dB. In principle better results can be obtained, within the agreed standard for the Compact Disc system, by looking more than one symbol ahead, since minimization of the DSV in the short term does not always contribute to longer-term minimization. This is not yet done in the present equipment.

References

[1] M. G. Carasso, J. B. H. Peek and J. P. Sinjou, The Compact Disc Digital Audio system, (Sect. 3.2).

[2] H. Hoeve, J. Timmermans and L. B. Vries, Error correction and concealment in the Compact Disc system (Sect. 3.4)..

[3] J. C. J. Finck, H. J. M. van der Laak and J. T. Schrama, A semiconductor laser for information read-out, Philips tech. Rev. **39**, 37-47 (1980).

[4] A more detailed discussion is given in K. A. Immink, Modulation systems for digital audio discs with optical readout, Proc. IEEE Int. Conf. on Acoustics, speech and signal processing, Atlanta 1981, pp. 587-589.

3.4 Error correction and concealment in the Compact Disc system

H. Hoeve, J. Timmermans, L.J. Vries

Ir H. Hoeve and J. Timmermans are with the Philips Audio Division, Eindhoven; Ir L. B. Vries is with Philips Research Laboratories, Eindhoven.

Abstract
After an example showing how errors in a digital signal can be corrected, the article deals with the theory of block codes. The treatment of random errors and error bursts is discussed. Error correction in the Compact Disc system uses a Cross-Interleaved Reed-Solomon Code (CIRC), which is a combination of a (32,28) and a (28,24) code. One of the two decoders in the CIRC decoding circuit corrects single errors, the other corrects double errors. The residual errors are interpolated linearly to a length of up to 12 000 bits, and longer errors are muted. The interpolation and the signal muting take place in a separate chip, whose configuration is briefly discussed.

3.4.1 Introduction

When analog signals such as audio signals are transmitted and recorded via an intervening system such as a gramophone record it is difficult to properly correct signal errors that have occurred in the path between the audio source and the receiving end. With suitably coded digital signals, however, a practical means of error correction does exist. We shall demonstrate this with the following example[1].

Suppose that a message of 12 binary units (bits) has to be transmitted (a stream of digital information can always be divided into groups of a particular size for transmission). The 12 bits x_{ij} are arranged as follows in a matrix, in which all x_{ij} can only have the value 0 or 1:

$$\begin{matrix} x_{11} & x_{12} & x_{13} & x_{14} \\ x_{21} & x_{22} & x_{23} & x_{24} \\ x_{31} & x_{32} & x_{33} & x_{34} \end{matrix}$$

To discover at the receiving end whether the message read there contains an error, and, if so, what the error is, one extra bit (called a 'parity bit') is added to

Reprinted with permission from Philips Tech. Rev. **40**, 166-172, 1982.

each row and column: x_{15}, x_{25}, x_{35} and x_{41}, x_{42}, x_{43}, x_{44} respectively. These parity bits provide a check on the correctness of the message received. The values assigned to them are such that x_{i5} (i = 1, 2, 3) makes the number of ones in

row i even, for example, while x_{4j} (j = 1, 2, 3, 4) makes the number of ones in column j even. Next, a further parity bit (x_{45}) is added that has a value such that the number of ones in the block is made even. This results in the following matrix of four rows and five columns:

$$\begin{matrix} x_{11} & x_{12} & x_{13} & x_{14} & x_{15} \\ x_{21} & x_{22} & x_{23} & x_{24} & x_{25} \\ x_{31} & x_{32} & x_{33} & x_{34} & x_{35} \\ x_{41} & x_{42} & x_{43} & x_{44} & x_{45} \end{matrix}$$

It is easy to verify that the number of ones in the last row is also even, and so is the number of ones in the last column. If now a bit, say x_{23}, is incorrectly read at the receiving end, then the number of ones in the second row and the number of ones in the third column will no longer be even, and once this has been ascertained, a 0 at position x_{23} can be changed into a 1, or vice versa, thus correcting the error.

So as to be able in this way to correct one error in 12 information bits, it is necessary to send a total of 20 bits instead of 12: the 'code word' of n=20 bits consists of k=12 information bits and $n-k$=8 parity bits. The (n,k) code used here, a (20,12) code, makes it possible to correct single errors and also, as can easily be verified, to detect various multiple bit errors.

The 'rate' of an error-correcting code is taken to be the ratio of the number of information bits to the total number of bits per code word: k/n. The (20,12) code does not have a high rate, because it requires a relatively large number of parity bits. For the Compact Disc this would entail a considerable reduction in the playing time.

The theory of error-correcting codes[2] gives design methods that entail a minimal addition of parity bits when certain correction criteria are satisfied. An important concept in this theory is the 'distance' and in particular the 'minimum distance' d_m between two code words of n bits. Distance here is taken to be the number of places in which the bits of the two code words differ from each other. In the above example the minimum distance d_m is equal to 4: if one single bit of the k information bits changes, then the two parity bits of the associated row and column change at the same time, as does the one at the bottom right-hand corner, x_{45}, so that the entire code word has changed at four places. Theory tells us that to correct all the combinations of t errors occurring within one word, the minimum distance must be at least $2t+1$. To correct single errors, therefore, the minimum distance need be no greater than three. Examples of this are the single-error-correcting Hamming codes[4].

The statement that a code word x, which is received as a different word z because of terrors, can be restored to its original form if the minimum distance is $2t + 1$, can be seen from Fig. 1. A decoder provided with a list in which all the code words are stored can compare z with each of these code words and thus recover the correct code word unambiguously.

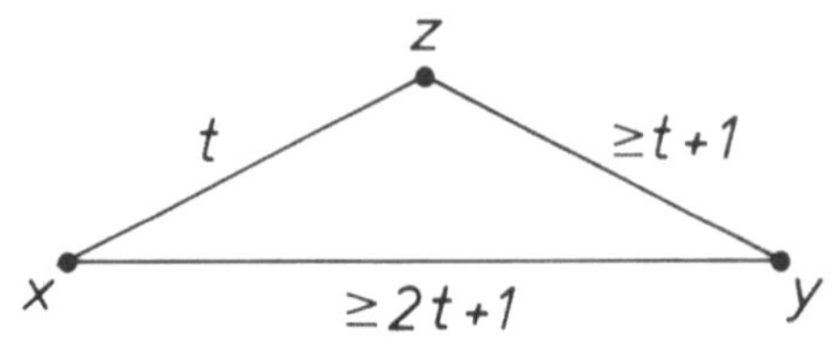

Fig. 1. The original transmitted code word x is received as z owing to t bit errors. Any code word y differing from x lies at a distance $\geq 2t + 1$ from x. To cause z to change into y it is necessary to change at least $t + 1$ bits. It follows that x is the only code word that has a distance t from z.

3.4.2 On the theory of block codes

In the foregoing we have shown with a simple example that it is possible to correct errors. Error-correcting systems do have their limitations, of course. To make this clear we shall consider how error-correcting codes should be designed to guarantee a specific measure of correction, with as few extra bits as possible added to the digital information to be transmitted. It will help if we first say something about the theory of block codes.

So that known and efficient error-correcting codes can be applied, groups of bits are formed by adding together a fixed number s of consecutive bits; these groups are called symbols. With these symbols we now set to work in the same way as with the bits in the foregoing: the information symbols are grouped together to form blocks with a length of k symbols. For error-correction we now add parity symbols to expand each block of k information symbols into a code word of n symbols. The $n - k$ parity symbols to be added are calculated from the k information symbols, and this is done in such a way as to make the error correction as effective as possible. Thus, of the very large number of possibly different words of n symbols only a small fraction, i.e. $2^{(k-n)s}$, become code words (see Fig. 2).

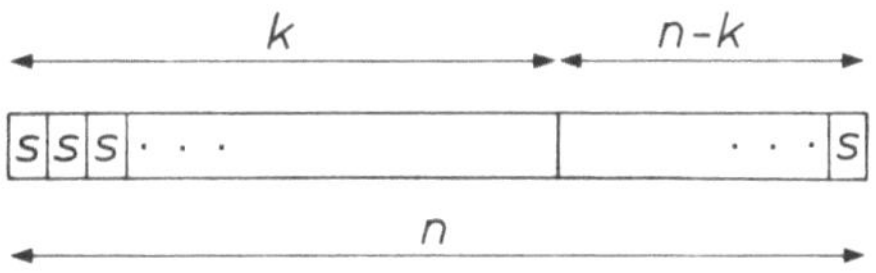

Fig. 2. A code word of length n consists of an information block of k symbols and a parity block of n-k symbols; each symbol comprises s bits. The number of possible words of n symbols is 2^{ns}. The parity bits are fixed for each combination of the ks information bits in accordance with established encoding rules . The number of code words is thus 2^{ks}. It follows that the fraction $2^{(k-n)s}$ of the number of possible words consists of code words.

For a given encoding system both n and k are fixed.

As already mentioned in the article on modulation in the Compact Disc system[3], the start of each word is marked by a synchronization symbol. (A word marked by a synchronization symbol is called a 'frame'.) The error-correcting system therefore knows when a new word begins, and the only errors it has to deal with are errors that occur in the transmission of data.

There are two kinds of errors: those that are distributed at random among the individual bits, the random errors, and errors that occur in groups that may cover a whole symbol or a number of adjacent symbols; these are called 'bursts' of errors. They can occur on a disc as a result of dirt or scratches, which interfere with the read-out of a number of adjacent pits and lands.

The best code for correcting random errors is the one that, for given values of n and k, is able to correct the largest number of independent errors within one code word. In the detection and correction of errors the symbols have to undergo a wide variety of operations. Large k-values (as with the Compact Disc) require extremely complex computing hardware. Practice has shown that the only acceptable solution to this problem is to choose a convenient code. And the only usable codes that enter into consideration, so far as we know at present, are the 'linear codes'.

A code is linear if it obeys the following rule:

If $x = (x_1, \ldots, x_n)$ and $y = (y_1, \ldots, y_n)$ are code words, then their sum $x + y = (x_1 + y_1, \ldots, x_n + y_n)$ is also a code word.

In this sum the symbol $x_i + y_i$ is produced - irrespective of the number of bits s per symbol - by a modulo-2 bit addition. The special feature of the linear code is thus that each sum of code words yields another code word, i.e. a word of n symbols, which also belongs to the small fraction of symbol combinations permitted in the code.

It is this linearity feature that makes it possible to cut down considerably on the extent of the decoding equipment. The *Reed-Solomon codes*[2] are examples of such a linear code. They are also extremely efficient, since for every $s > 1$

and $n \leq 2^S - 1$ there exists a Reed-Solomon code with

$$d_m = n - k + 1.$$

Together with the general condition $d_m \geq 2t + 1$ mentioned earlier, which the minimum distance must satisfy for the correction of t errors, this yields $n - k \geq 2t$. Put in another way: to correct t symbol errors it is sufficient to add $2t$ parity symbols. (By 'distance' between two words we mean here the number of positions in which there are different symbols in the two words; it does not matter how many corresponding bits differ from each other within the corresponding symbols.)

In practice a less cumbersome algorithm will generally be used for error correction than the comparison with the aid of a list of all the code words, as described at the end of Sect. 3.4.1. We shall not consider the details of the algorithm here. We shall, however, try to give some idea of the manner in which error bursts are tackled with block codes. To do this we must introduce the concept of 'erasure'.

The position (i) of a particular symbol (x_i) in a transmitted code word (x) is called an erasure position if a decoder-independent device signals that the value of x_i is not reliable. This value is then erased, and in the decoding procedure the correct value has to be calculated. The decoding is now simpler and quicker because the positions at which errors can occur are known. (We assume for the moment that no errors occur outside the erasure positions.) The advantage of correcting by means of the erasures is expressed quantitatively by the following proposition:

If a code has a minimum distance d_m, then $d_m - 1$ erasures can be reconstituted.

Since the number of errors that can be corrected without erasure information is $\frac{1}{2}(d_m - 1)$ at most, the advantage of correcting by means of erasures is clear. In the Compact Disc system the value of the analog signal to be reproduced is converted at every sampling instant into a binary number of 16 bits per audio channel. For error correction the digital information to be transmitted is divided into groups of eight bits, so that in each sampling operation four information symbols (consisting of audio bits) are generated. In fact, eight parity symbols are added to each block of 24 audio symbols[4]. The calculation of the parity symbols will not be dealt with here.

3.4.3 Cross-Interleaved Reed-Solomon Code

The error-correcting code used in the Compact Disc system employs not one but two Reed-Solomon codes (C_1,C_2), which are interleaved 'crosswise' (Cross-Interleaved Reed-Solomon Code, CIRC). For code C_1 we have: $n_1 = 32$, $k_1 = 28$, $s = 8$, and for C_2: $n_2 = 28$, $k_2 = 24$, $s = 8$. The rate of the CIRC we use is $(k_1/n_1)(k_2/n_2) = 3/4$.

For both C_1 and C_2 we have $2t=n-k=4$, so that for each the minimum distance d_m is equal to $2t + 1=5$. This makes it possible to directly correct a maximum of two (= t) errors in one code word or to make a maximum of four (= $d_m - 1$) erasure corrections. A combination of both correction methods can also be used.

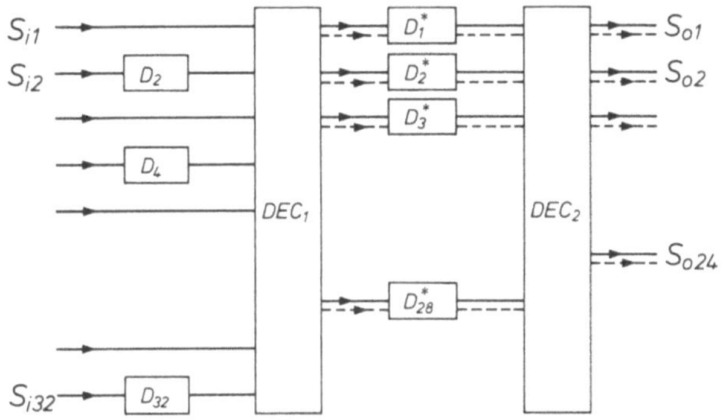

Fig. 3. Schematic representation of the decoding circuit for CIRC. The 32 symbols (S_{i1}, ..., S_{i32}) of a frame (24 audio symbols and 8 parity symbols) are applied in parallel to the 32 inputs. The delay lines D_{2i} (i = 1, ..., 16) have a delay equal to the duration of one symbol, so that the information of the 'even' symbols of a frame is cross-interleaved with that of the 'odd' symbols of the next frame. The decoder DEC_1 is designed in accordance with the encoding rules for a Reed-Solomon code with $n_1 = 32$, $k_1 = 28$, $s = 8$. It corrects one error, and if multiple errors occur passes them on unchanged, attaching to all 28 symbols an erasure flag, sent via the dashed lines. Owing to the different lengths of the delay lines D_j^* (j = 1, ..., 28), errors that occur in one word at the output of DEC_1 are 'spread' over a number of words at the input of DEC_2. This has the effect of reducing the number of errors per DEC_2 word. The decoder DEC_2 is designed in accordance with the encoding rules for a Reed-Solomon code with $n_2 = 28$, $k_2 = 24$, $s = 8$. It can correct a maximum of four errors by means of the erasure-positions method. If there are more than four errors per word, 24 symbol values are passed on unchanged, and the associated positions are given an erasure flag via the dashed lines. S_{o1}, ..., S_{o24} outgoing symbols.

Decoding circuit

The error-correction circuit[5] is shown schematically in Fig. 3; Fig. 4 is a photograph of the actual IC. The circuit consists of two decoders, *DEC*, and a number of delay lines, *D* and *D**. The input signal is a sequence of frames[6].

The 32 symbols of a frame are applied in parallel to the 32 inputs. In passing through the delay lines D_2,D_4, ..., D_{32}, each of length equal to the duration of one symbol, the even symbols of a frame with the odd symbols of the next frame form the words that are fed to the decoder DEC_1. (The symbols of the frames are 'cross-interleaved'. In fact they are 'deinterleaved', because the 'interleaving'[4] has already taken place, before the information was recorded on the disc.) If there are no errors in the transmission path, the decoder DEC_1 will receive code words that correspond to the encoding rules for C_1, and it will pass on 28 symbols unchanged. DEC_1 is designed for correcting one error. If it receives a word with a double or triple error, that event is detected with certainty; all the symbols of the received word are passed on unchanged, and all 28 positions are provided with an erasure flag. The same happens in principle for events from 4 to 32 errors, but here there is a small probability ($\approx 2^{-19}$) that this detection will fail. We shall return to this probability later.

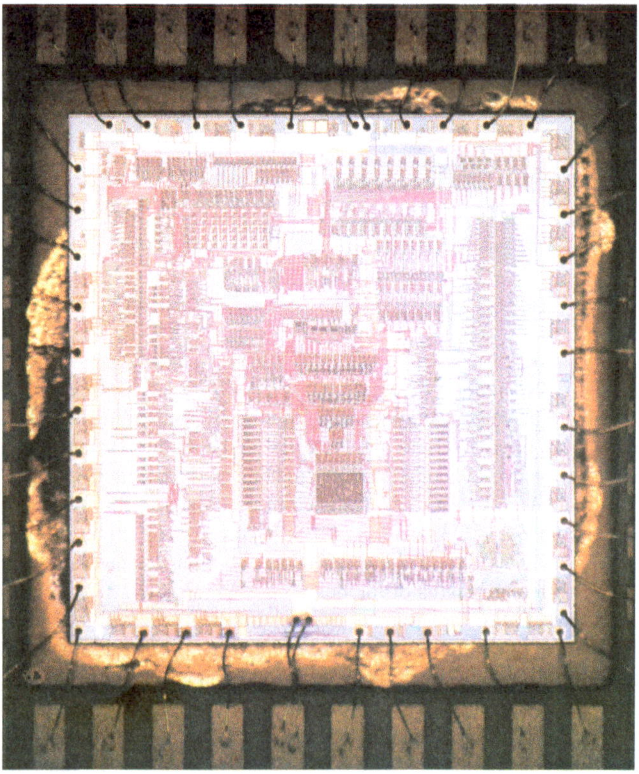

Fig. 4. The integrated circuit for error detection and correction is fabricated in n-channel MOS silicon-gate technology. It has an area of 45 mm^2 and contains about 12 000 gates.

The symbols arrive via the delay lines $D_1^*,\ldots, D_{28}^*$, which differ from each other in length, at the input of DEC_2 in different words. If there are no errors present, DEC_2 will receive words that correspond to the encoding rules for C_2, and it will pass on 24 audio symbols unchanged. DEC_2 can correct up to four errors, by means of erasure decoding. (In the current Compact Disc system full use is not made of this facility: DEC_2 is arranged in such a way that only two errors are corrected.) If DEC_2 receives a word containing five or more errors with given erasure positions, it will pass on 24 symbols unchanged, but provided with an erasure flag at the appropriate positions; this flag has in fact already been assigned by DEC_1. A value for the erroneous samples can still be calculated with the aid of a linear interpolation.

As already mentioned, DEC_1 has been designed to allow the correction of single errors, and the detection of double and triple errors. The probability that DEC_1 will not detect quadruple or higher multiple errors is only about 2^{-19}. It may seem strange that the possibility of correcting two random errors is not utilized: in fact it would considerably increase the chance of DEC_1 failing to detect quadruple or higher multiple errors.

The probability P of quadruple or higher multiple errors passing DEC_1 without being detected can be approximated by the expression

$$P = \frac{1 + n_1(2^s - 1)}{2^{s(n_1 - k_1)}} \approx 2^{-19}.$$

The numerator contains the number of error patterns with one error or none. (The factor $(2^s - 1)$ is the number of possibilities for one bit error per symbol; such a symbol can occur at n_1 positions. The value 1 is added because zero errors can be achieved in exactly one way.) This complete expression is to be related to the number of possibilities for filling in the parity: $2^{s(n_1 - k_1)}$). For proof of this equation the reader is referred to the literature[7].

When a disc is used for the recording and read-out of digital signals there are few random errors; most errors then occur as bursts. This is because the dimensions of a pit are small in relation to the most common mechanical imperfections such as dirt and scratches. It is therefore very important that multiple errors of this type cannot pass DEC_1 without being indicated with a high degree of certainty.

Since the bursts are 'spread out' over several words at the input of DEC_2, the number of errors per word hardly ever exceeds the limit value $d_m - 1 = 4$. In this way most error bursts are fully corrected.

3.4.4. Specifications of CIRC

In assessing the quality of our CIRC decoder for Compact Disc applications its ability to correct both error bursts and random errors is of great importance.

The quality characteristics for the correction of bursts are the maximum fully correctable burst length and the maximum interpolation length. The first is determined by the design of the CIRC decoder and in our case amounts to about 4000 data bits, corresponding to a track length on the disc of 2.5 mm. The maximum interpolation length is the maximum burst length at which all erroneous symbols that leave the decoder uncorrected can still be corrected by linear interpolation between adjacent sample values. This 'length' is about 12000 data bits; see the next section.

Random errors can also introduce multiple errors within one code word now and again; we shall return to this presently. The greater the relative number of errors ('bit error rate', BER) at the receiving end, the greater is the probability of uncorrectable errors. A measure for the performance of this system is the number of sample values that have to be reconstituted by interpolation for a given BER value per unit time. This number of sample values per unit time is called the sample interpolation rate. The lower this rate is at a given BER value, the better the quality of the system for random-error correction.

Aspect	Specification
Maximum *completely* correctable burst length	≈ 4000 data bits (i.e. ≈ 2.5 mm track length on the disc)
Maximum interpolatable burst length in the *worst* case	≈ 12 300 data bits (i.e. ≈ 7.7 mm track length)
Sample interpolation rate	One sample every 10 hours at BER = 10^{-4}; 1000 samples per minute at BER = 10^{-3}
Undetected error samples (clicks)	Less than one every 750 hours at BER = 10^{-3}; negligible at BER $\leqslant 10^{-4}$
Code rate	3/4
Structure of decoder	One special LSI chip plus one random-access memory (RAM) for 2048 words of 8 bits
Usefulness for future developments	Decoding circuit can also be used for a four-channel version (quadraphonic reproduction)

Table I. Specifications of CIRC.

An objective assessment of the quality of the error-correcting system also requires an indication of the number of errors that pass through unsignalled and are therefore not corrected by the system. These unsignalled and uncorrected errors may produce a clearly audible 'click' in the reproduction.

The main features of the CIRC system are summarized in Table I. Details of the calculation relating to the quality can be found in the literature[7].

3.4.5 Concealment of residual errors

The purpose of error concealment is to make the errors that have been detected but not corrected by the CIRC decoder virtually inaudible. Depending on the magnitude of the error to be concealed, this is done by interpolation or by muting the audio signal[8].

Two consecutive 8 bit symbols delivered by the decoder together form a 16 bit sample value. Since a sample value in the case of a detected error carries an erasure flag, the concealment mechanism 'knows' whether a particular value is reliable or not. A reliable sample value undergoes no further processing, but an unreliable one is replaced by a new value obtained by a linear interpolation between the (reliable) immediate neighbours. Sharp 'clicks' are thus avoided; all that happens is a short-lived slight increase in the distortion of the audio signal. With alternate correct and wrong sample values, the bandwidth of the audio signal is halved during the interpolation (10 kHz).

If the decoder delivers a sequence of wrong sample values, a linear interpolation does not help. In that case the concealment mechanism deduces from the configuration of the erasure flags that the signal has to be muted. This is done by rapidly turning the gain down and up again electronically, a procedure that starts 32 sampling intervals before the next erroneous sample values arrive. To achieve this the reliable values are first sent through a delay line with a length of 32 sampling intervals, while the unreliable values are processed immediately. The gain is kept at zero for the duration of the error and then turned up again in 32 sampling intervals. The gain variation follows a cosine curve (from 0 to 180° and from 180 to 360°) to avoid the occurrence of higher-frequency components. This also means that there are no clicks when the audio signal is muted, as in switching the player on and off, during an interval in playing or during the search procedure.

Maximum burst-interpolation length

Two associated 16 bit sample values, one from the left and one from the right audio channel, together form a sample set. If these sets were fed to the concealment circuit in the correct sequence, it would not be possible to interpolate more than one set from their reliable neighbouring sets. This would mean that in the case of an error longer than the maximum correctable burst length signal muting would very soon have to be applied.

By interleaving the sample sets it becomes possible to interpolate new sets for a given length of consecutive erroneous sets. This is done by alternating groups of 'even' sample sets with groups of 'odd' sets. Such a group, odd or even, can be interpolated from its neighbouring group or groups. The maximum burst-interpolation length is thus equal to the length of such a group. In our system we have grouped the twelve 16 bit sample values of a frame in the way shown in Fig. 5. The odd and even groups are separated by the parity values *Q*. Since these are not necessary for the reconstitution of the original signal and may therefore permissibly be unreliable, they increase the interpolation length. The maximum length with this grouping is certainly seven or even eight sample values, for some error patterns.

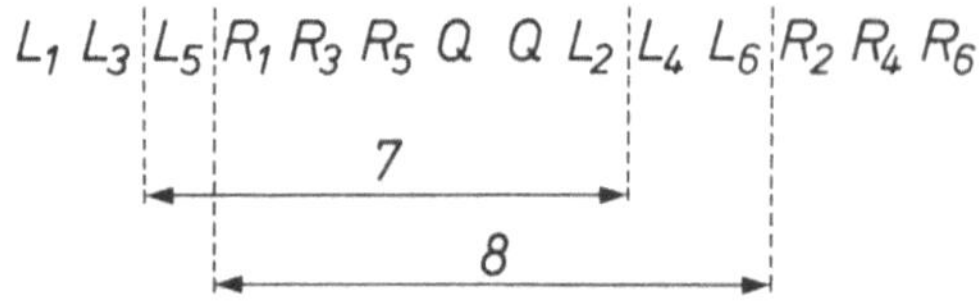

Fig. 5. Grouping of the sample values within a frame; L_i values for the left channel, R_i values for the right channel. For each sequence of seven unreliable values, new values can be calculated with certainty from reliable neighbours (e.g. if L_5 to L_2 are unreliable, the new values are interpolated from R_6 of the preceding frame and from the reliable values of the above frame). Given a favourable situation, new values can in fact be derived for eight consecutive values (e.g. values for R_1 up to L_6 from R_6 of the preceding frame, the reliable values of the above frame and L_1 of the succeeding frame).

The delay lines corresponding to D_i^* (see Fig. 3) in the encoder[4] have placed eight frames between two successive sample values, after interleaving. The maximum burst length that can always be interpolated is therefore 56 frames. This presupposes, of course, that we are working with sample values consisting of two immediately consecutive symbols; the distance between all successive symbols is four frames, however. This is also the work of the delay lines D_i^*.

The delay lines corresponding to D_i (again Fig. 3) in the encoder[4] now ensure, however, that this distance is alternatively three and five frames, after interleaving. The distance of five frames is responsible for a decrease in the maximum interpolation length from 56 to 51 frames. We have tacitly assumed here that the burst also comes within a block of eight frames. If we discount this assumption, there is still a reduction of a length of 1 frame - 2 symbols. The maximum burst length that can be interpolated with certainty has now become 50 frames + 2 symbols.

So far we have taken no account of random errors that can be interpolated; this is the subject of the next and final section. At this point we shall simply mention the effect of the interpolation of such errors on the maximum interpolation length.

To achieve good results in the treatment of random errors, the symbols are finally sent through a further set of delay lines Δ_i with a length of two frames. These delay lines, which serve purely and simply for 'restoring' uncorrected random errors, cause in their turn a reduction of the interpolation length by two frames. The final maximum burst length that is guaranteed capable of interpolation is thus 48 frames + 2 symbols, which corresponds to 12 304 bits.

Interpolation of random errors

If the symbols S_{oi} (Fig. 3) after the decoder DEC_2 were already in the correct sequence, a pattern of errors might arise that would rule out any possibility of interpolation, even though there were no long error bursts. This would happen if DEC_1 failed to detect an error but DEC_2 had detected it, resulting in the rejection of the entire frame at the output of DEC_2. As described in Sect. 3.4.3, however, the chance of DEC_1 failing is very small.

Since we prefer not to have to mute the audio signal, the concealment network contains a set of delay lines Δ_i, with a length of two frames, which ensure that the symbols of a single or double completely rejected frame from DEC_2 can still be interpolated from the reliable adjacent frames (see *fig. 6*). The probability that three completely rejected frames will occur within the interpolatable length determined by Δ_i is negligible.

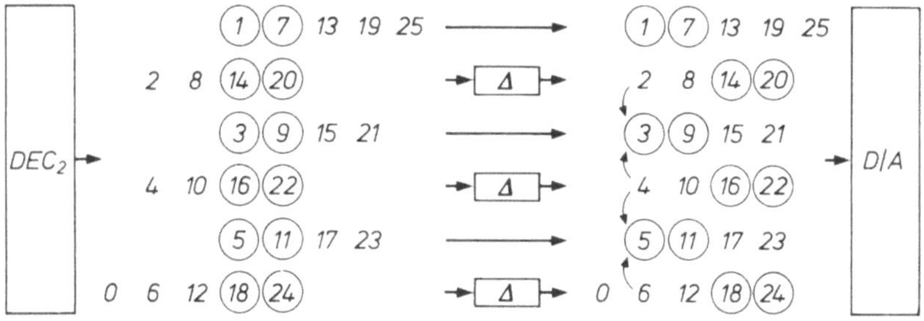

Fig. 6. The effect of the delay lines Δ_i with a length equal to the duration of two frames on the signal from the decoder DEC_2. Each number represents a sample set, and a circle around a number is an erasure flag. A frame, consisting of 24 symbols or 6 sample sets, is represented by a complete column. The succession of frames on the left in the figure (sample sets that are irrelevant in the present context have been omitted) comes direct from DEC_2 and comprises a pattern of random errors, causing the total rejection of two consecutive frames (1, 14, 3, … 11, 24). It can be seen, however, that the chosen grouping enables a new value from reliable neighbours to be interpolated for each unreliable sample set, e.g. a value for 5 from 4 and 6. After passing through the delay lines Δ_i with a length equal to the duration of two frames, the sample sets are applied in the correct sequence to the D/A converter. If a frame in the succession of frames on the right in the figure were to be completely rejected, no interpolation would be possible.

After the symbols have passed through the delay lines Δ_i, they are in the correct sequence. Most of the errors have been corrected and the signal is ready for the digital-to-analog conversion[9].

References

[1] This example is taken from S. Lin, An introduction to error-correcting codes, Prentice-Hall, Englewood Cliffs 1970.

[2] See for example F. J. MacWilliams and N. J. A. Sloane, The theory of error-correcting codes, North-Holland, Amsterdam 1978.

[3] See J. P. J. Heemskerk and K. A. Schouhamer Immink, Compact Disc: system aspects and modulation Philips Tech. Rev. **40**, 157-164 (1982), (Sect. 3.3).

[4] The calculation and addition of the parity symbols take place in the encoding circuit *PAR COD* in fig. 2 of the article of note [3]. Delay lines are used for interleaving the audio and parity symbols.

[5] This circuit corresponds to the *ERCO* chip in fig. 3 of the article by M. G. Carasso, J. B. H. Peek and J. P. Sinjou, Philips Tech. Rev. **40**, 151-155 (1982), (Sect. 3.2).

[6] In fig. 9 of [3] a frame of this kind is represented by the bit stream B_2, from which the C&D symbol has already been removed.

[7] L. M. H. E. Driessen and L. B. Vries, Performance calculations of the Compact Disc error correcting code on a memoryless channel, in: 4th Int. Conf. on Video and data recording, Southampton 1982 (IERE Conf. Proc. No. 54), pp. 385–395.

[8] Error concealment takes place in the *CIM* chip in Fig. 3 of the article of note [5].

[9] See D. Goedhart, R. J. van de Plassche and E. F. Stikvoort, Digital-to-analog conversion in playing a Compact Disc, Philips Tech. Rev. **40**, 174-179 (1982), (Sect. 3.5).

3.5 Digital-to-analog conversion in playing a Compact Disc

D. Goedhart, R.J. van de Plassche, E.F. Stikvoort

Ing. D. Goedhart is with the Philips Audio Division, Eindhoven; Ir R. J. van de Plassche and Ir E. F. Stikvoort are with Philips Research Laboratories, Eindhoven.

Abstract
The 16 bit words from the error-correcting circuit are converted into an analog signal by a 16 bit conversion system. This system consists of a digital transversal filter, in which the signal is oversampled 4 times (sampling rate 176.4 kHz) and then filtered in such a way that signals at frequencies above 20 kHz are attenuated by 50 dB after digital-to-analog conversion. The filter is followed by a noise shaper, which rounds off to 14 bits with negative feedback of the rounding-off error of the preceding sample. Next there is a 14 bit digital-to-analog converter, which is followed by a low-pass third-order Bessel filter. The signal-to-noise ratio of the complete system is about 97 dB. Even though the lowpass filter has a sharp cut-off the system is phase linear. The entire system, except for a few operational amplifiers, is contained in three integrated circuits; one for the digital filter (for both of the stereo channels) and two for the two digital-to-analog converters.

3.5.1 Introduction

The last stage in the series of operations on the signal in the Compact Disc system is the return from the digital code to the analog signal, which has the same shape as the acoustic vibration that was picked up by the microphone.

After decoding and error correction the digital signal has the form of a series of 16 bit words. Each word represents the instantaneous numerical value of the measured sound pressure in binary form, and is therefore a sample of the acoustic signal. There are 44 100 of these words per second.

The digital-to-analog converter in the Compact Disc player generates an electric current of the appropriate magnitude for each word and keeps it constant until the next word arrives. The electric current thus describes a 'staircase' curve that approximates to the shape of the analog signal (Fig. 1a). In terms of frequency, the steps in the staircase represent high frequencies, which extend beyond the band of the analog audio signal (20 Hz - 20 kHz). These high frequencies have to be suppressed by a lowpass filter; in the Compact

Reprinted with permission from Philips Tech. Rev. **40**, 174-179, 1982.

Disc player their level should be reduced to at least 50 dB below that of the maximum audio signal.

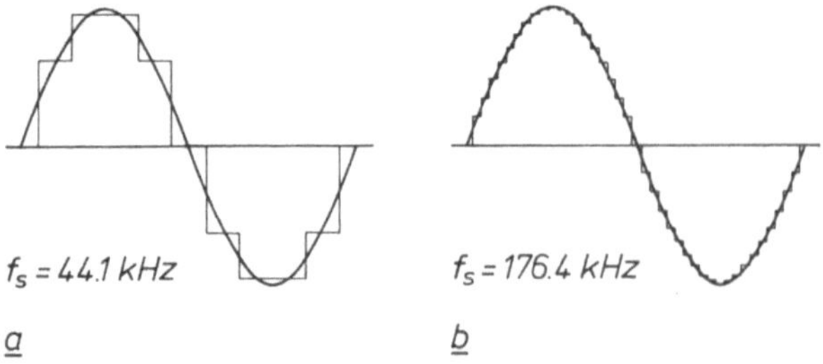

Fig. 1. A sinusoidal signal at 4.41 kHz sampled with a sampling rate f_s of 44.1 kHz (*a*) and with a frequency four times higher (*b*). In (*b*) the 'staircase' curve approximates more closely to the analog waveform, and the high frequencies present in the staircase signal are more easily filtered out.

If this high attenuation of the frequencies above the audio band is to be achieved solely with an analog lowpass filter, the filter must meet a very tight specification. It was decided to avoid this problem in the Philips Compact Disc player by introducing a filter operation, earlier in the digital stages. This was done by 'oversampling' by a factor of four: a digital filter, operating at four times the sampling rate (4 × 44.1 kHz = 176.4 kHz) delivers signal values at this increased frequency, thus refining the staircase curve (Fig. 1b) and making it easier to filter out the high frequencies. As a result it is possible to make do with a relatively simple lowpass filter of the third order after the digital-to-analog conversion.

The conversion of the 16 bit words into an analog signal is performed in the Philips Compact Disc player by a 14 bit digital-to-analog converter available as an integrated circuit and capable of operating at the high sampling rate of 176.4 kHz. Partly because of the fourfold oversampling and partly because of the feedback of the rounding-off errors in antiphase, rounding off to 14 bits does not result in a higher noise contribution in the audio band. This remains at the magnitude corresponding to a 16 bit quantization (signal-to-noise ratio about 96 dB), so that even though there is a 14 bit digital-to-analog converter it is still possible to think in terms of a 16 bit conversion system.

In comparison with direct 16 bit digital-to-analog conversion, which must be followed by a lowpass filter with a sharp cut-off to give sufficient suppression of signals at frequencies above 20 kHz, our conversion system has a number of advantages. The first is the linear phase characteristic, which can be obtained with a digital filter, but not with an analog filter; the second is a filter characteristic that varies with the clock rate and is therefore insensitive to variation in the speed of rotation of the disc. Finally, because the quantization steps are smaller, the maximum 'slew rate' that these circuits must be able to process is lower (the slew rate is the rate of variation of output voltage). There is therefore less chance of intermodulation distortion because the permitted slew rate has been exceeded.

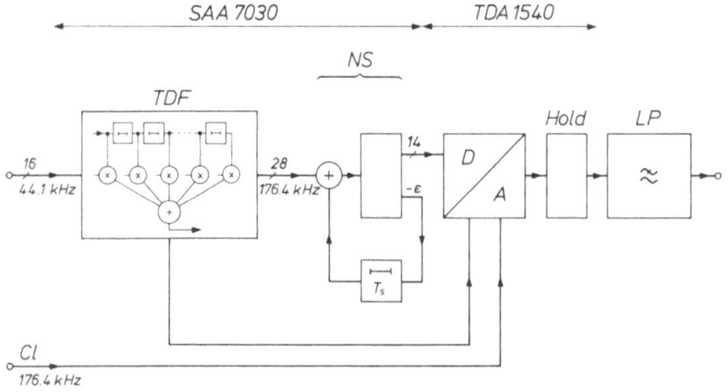

Fig. 2. Block diagram of the digital-to-analog conversion. TDF digital transversal filter which brings the sampling rate of 44.1 kHz to 176.4 kHz and attenuates signals in the bands around 44.1 kHz, 88.2 kHz and 132.3 kHz. NS noise shaper in which the rounding-off error is delayed by one period Ts after rounding-off to 14 bits and then fed back in the opposite sense. D/A 14 bit digital-to-analog converter. Hold hold circuit. Cl clock signal. LP lowpass 3rd-order Bessel filter.

The entire series of operations in the digital-to-analog conversion is shown as a block diagram in Fig. 2. The oversampling takes place in the digital filter TDF to which the input signal is fed. The filter output signal is then rounded off to 14 bits, and the rounding error is fed back in the opposite sense in the noise shaper NS. The digital filter and noise shaper are located in a single integrated circuit in NMOS technology (type SAA 7030). This IC processes both stereo channels. Then follow the digital-to-analog converter D/A and a hold circuit, combined in a single IC type (TDA 1540) in bipolar technology; for each stereo channel there is a separate IC. The analog signal finally passes through a lowpass filter.

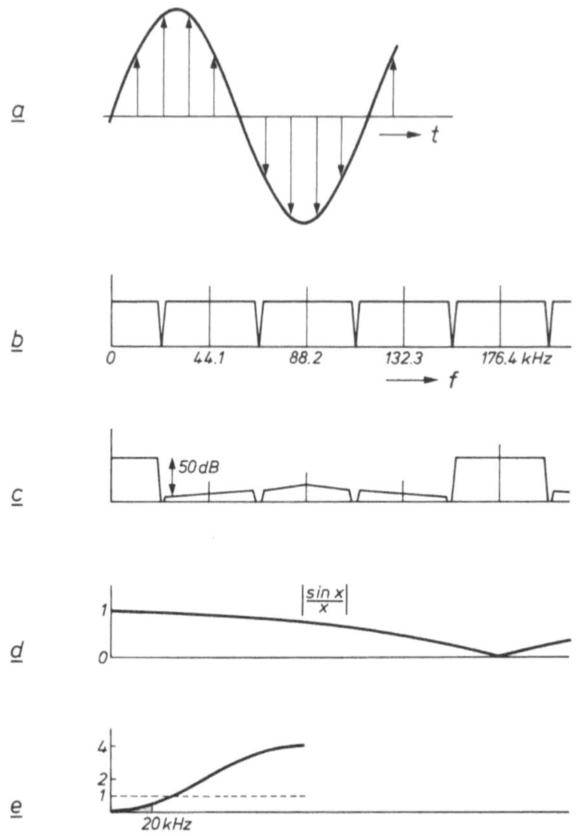

Fig. 3. a) A train of periodic pulses that sample an analog signal waveform. b) Frequency spectrum of such a pulse train. The pulse repetition frequency is 44.1 kHz, the sampled signal occupies the audio frequency band (0-20 kHz). c) Frequency spectrum for over-sampling and filtering of the same signal at 176.4 kHz. It is now much easier to filter out the frequencies above the audio band. d) A hold circuit after the digital-to-analog converter keeps a signal sample at the same value until the arrival of the next sample. The frequency spectrum in c is thus multiplied by the function |(sin x)/x| with a first zero at 176.4 kHz. e) Noise spectrum after the noise shaper. In the audio range of interest the noise is considerably attenuated compared with the flat noise spectrum (dashed line) that would be obtained without noise shaping.

3.5.2 Suppression of frequencies above the audio band

Direct digital-to-analog conversion of the presented signal provides a series of analog signal samples (Fig. 3a). These have the form of pulses that - in theory - are infinitely short, but have a content (duration times amplitude) corresponding to the sampled signal value. The repetition frequency is 44.1

kHz. The frequency spectrum of such a series is illustrated in Fig. 3b[1]. In theory it is infinite; above the baseband 0-20 kHz can be seen integral multiples of the sampling frequency with their left-hand and right-hand sidebands. Between these bands there are transition regions, the first for example being between 20 kHz and 24.1 kHz.

This entire spectrum must not be passed on to the player amplifier and loudspeaker. Even though the frequencies above 20 kHz are inaudible, they would overload the player amplifier and set up intermodulation products with the baseband frequencies or possibly with the high-frequency bias current of a tape recorder. Therefore all signals at frequencies above the baseband should be attenuated by at least 50 dB.

To produce such an attenuation, an analog filter after the digital-to-analog converter will inevitably have to contain a large number of elements and require trimming. In addition a linear phase characteristic is required in the passband so that the waveform of pulsed sound effects will not be impaired. In the Philips Compact Disc player these requirements are met in a different way, by means of:

- fourfold oversampling of the signal in the digital phase,
- a digital filter operation,
- a hold function after the digital-to-analog conversion,
- a third-order Bessel filter in the analog-signal path.

A digital transversal filter is used for the filtering after oversampling. To understand the operation of the filter, we can think of it as consisting of 96 elements (Fig. 4a), while the delay in each element is $(176.4 \times 10^3)^{-1}$ s, i.e. a quarter of the sampling period or ¼Ts. Four times in each period the filter takes up new data. At three of these four times the content of this data is zero, since the oversampling is done by the introduction of intermediate samples of value zero. This means that only 24 of the 96 elements are filled at any one time. The contents of each element are multiplied by a coefficient c. The filter provides data at a rate of 176.4 kHz; each number is the sum of 24 non-zero multiplications. In this way the filter always calculates three new sample values at the locations of the zero samples.

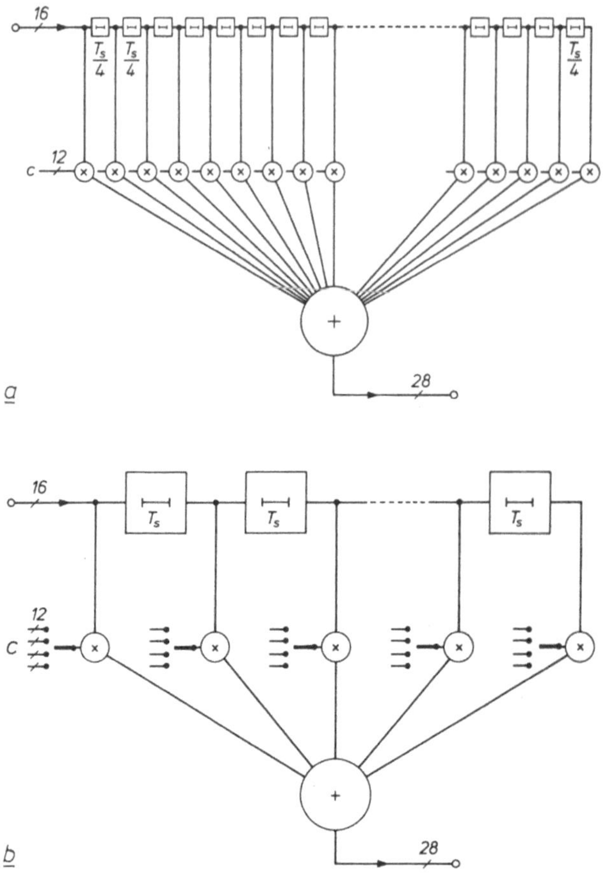

Fig. 4. Digital transversal filter. a) Filter consisting of 96 elements. A 16 bit word remains in each element for a quarter of the sampling period Ts. Since a new 16 bit word is only offered once per Ts, three-quarters of the elements are filled by the value zero. During the period Ts there are four multiplications by the 96 coefficients c; only 24 multiplications produce a product different from zero. These products are summed; in this way an output is provided four times in each sampling period, i.e. at a frequency of 4 × 44.1 kHz = 176.4 kHz. This means that there is a fourfold oversampling. b) An equivalent circuit that has been used in practice instead of (a) because it has 24 delay lines and multipliers instead of 96.

The practical version of the filter is in fact some-what different from the version referred to in the above explanation. In practice the filter consists of only 24 delay elements and a 16 bit word remains in each element for a time Ts (Fig. 4b). During this time Ts the word is multiplied four times by a coefficient c, which is different for each multiplication. The products are also summed four times during the time Ts and passed to the output. The frequency at which these summated values appear at the output is therefore 4/Ts = 176.4 kHz) again.

The coefficients are numbers with 12 bits. Each product has a length of 16 + 12 = 28 bits. The numbers have been chosen in such a way that the summation of 24 products does not introduce extra bits, so that the filter output consists of 28 bits with no rounding off.

The frequency spectrum of the oversampled and filtered signal is shown in fig. 3c. It can be seen that the bands in this spectrum around 1 ×, 2 × and 3 × 44.1 kHz are suppressed.

The digital-to-analog converter generates a current whose magnitude is proportional to the last digital word presented. This current is kept constant in a hold circuit until the next sample value is delivered, producing the staircase curve mentioned above. The signal samples have thus in theory changed from infinitely short pulses to pulses with the duration of a sampling period. This also has consequences for the frequency spectrum; the spectrum in Fig. 3c is multiplied by a curve of the form |(sin x)/x| that has a first zero at 176.4 kHz (see Fig. 3d). This gives an attenuation of signals in the 20 kHz sidebands on either side of 176.4 kHz by more than 18 dB. The hold effect causes no phase distortion.

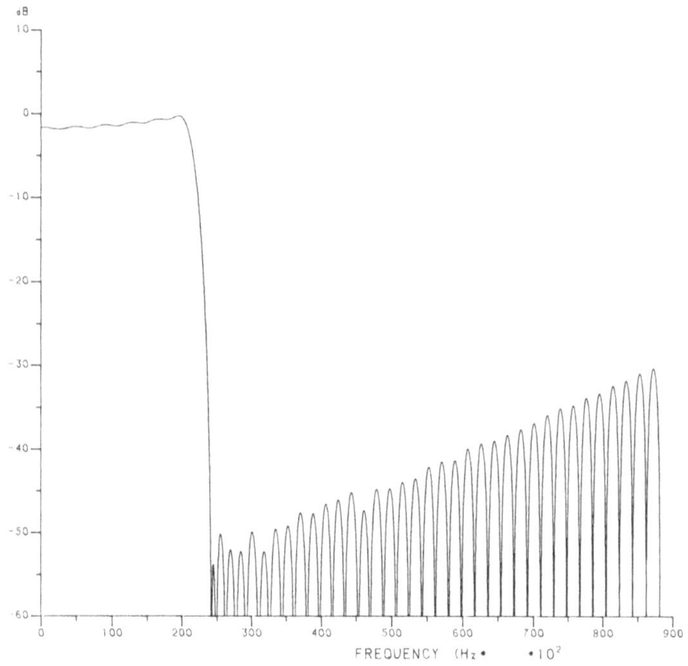

Fig. 5. Computer calculation of the detailed passband characteristic of the digital transversal filter. This has a small overshoot at the highest audio frequencies, which is used to compensate for the slight attenuation produced here by the curve in Fig. 3d and the analog Bessel filter. A very sharp lowpass cut-off of 50 dB is obtained. The irregularity in the suppressed band is caused by rounding-off the filter coefficients to 12 bits.

The attenuation is still not sufficient, however. As a supplement, a lowpass Bessel filter of the third order is used, which has its −3 dB point at 30 kHz. The Bessel type of filter has been selected because of its linear phase characteristic in the passband. This filter is simple and requires no highly accurate elements.

The hold function and the Bessel filter introduce some slight attenuation at the top of the passband. The digital filter is designed to correct this with a small overshoot (Fig. 5).

3.5.3 Suppression of the quantization noise

The presented signal, quantized to 16 bits, will contain some noise on conversion into an analog signal. This reproduces the errors due to the quantization in fixed steps. The root-mean-square value of the noise voltage in the sampled frequency band is $q/\sqrt{12}$, where q represents the magnitude of the quantization step. We see that when the quantization step is doubled, i.e. coding with one bit less, the noise voltage is also doubled, or, in other words, the noise level rises by 6 dB.

The samples that leave the filter at a repetition frequency of 176.4 kHz describe a signal with a band-width of 88.2 kHz. The quantization noise added due to the subsequent rounding off to 14 bits is spread over this band. With a signal of sufficient amplitude and a sufficiently broad frequency spectrum this distribution is uniform, since the quantization errors for successive samples are in principle uncorrelated; the quantization noise is 'white' noise. Only the band from 0 to 20 kHz is relevant; this is only about a fourth part of the sampled band, and the noise power in the band from 0 to 20 kHz is therefore only a fourth part of the total noise power. This means that because of the fourfold oversampling the signal-to-noise ratio in the relevant frequency band is 6 dB better than would be expected with 14 bit quantization. It is thus about 90 dB, which is what would have been obtained with a 15 bit system without over-sampling.

In rounding off from 28 to 14 bits it is useful to compare successive rounding-off errors. If the analog signal is a direct voltage, successive samples will have the same rounding-off error. The audio signal will not contain any direct current; it will however contain slowly varying signals that will resemble a direct current in a short time interval. If the error produced in the rounding-off from 28 to 14 bits is now changed in sign and added to the next sample to arrive (see Fig. 2), the average quantization error for slowly varying signals - i.e. low frequencies - can be reduced. This appears in the shape of the frequency spectrum of the quantization noise (see Fig. 3e); at low frequencies the noise level is lower, at high frequencies it becomes higher. With a sampling rate of 176.4 kHz, it follows that a 7 dB gain in signal-to-noise ratio is obtained in the audio band (0-20 kHz). The ratio of the maximum signal to the noise contributed by the entire digital-to-analog conversion system described above is thus brought to about 97 dB, i.e. the value corresponding to a 16 bit quantization.

3.5.4 The digital-to-analog converter

The 14 bit digital-to-analog converter has been dealt with in detail elsewhere[2]. Here we shall only indicate how it differs from other digital-to-analog converters.

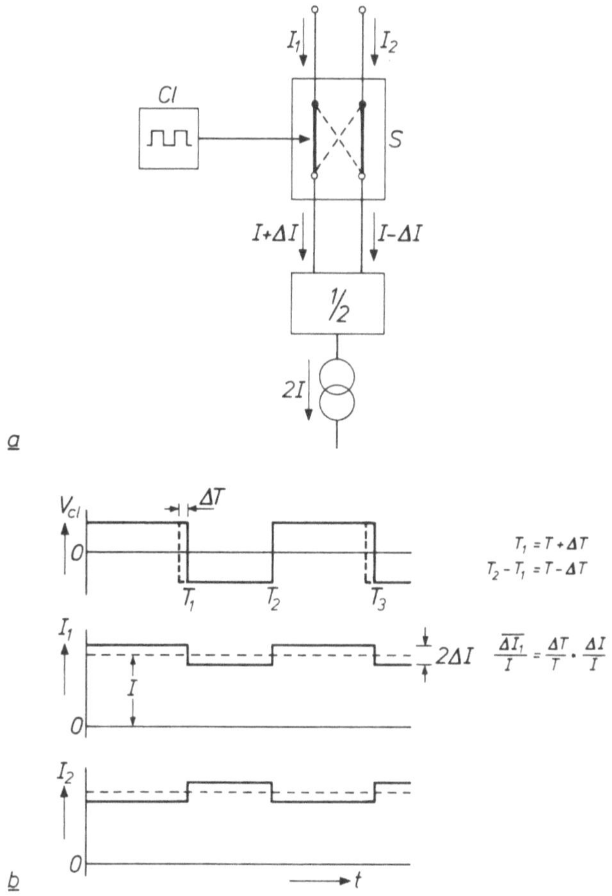

Fig. 6. a) Division of a current $2I$. *Cl* clock generator. *S* switches for periodically interchanging the two half-currents. *b*) The output currents I_1 and I_2 as a function of time t. Their mean value is the same. A difference between the mean output currents can be caused by an asymmetry ΔT of the clock signal V_{cl}. This difference is however an order of magnitude smaller than ΔI.

A characteristic feature is the way in which currents are generated that are accurately related by a factor of 2; a digital-to-analog converter requires a set of such currents. The exact ratio is obtained by periodically interchanging the currents that are derived by dividing down by two from a constant reference current (see Fig. 6), so that small differences are averaged out. This system is known as 'dynamic element matching'. Accurate division by four can be carried out with a slightly more complicated circuit, also based on periodic interchange. The full series of current dividers is shown in fig. 7. Here Cl is the clock signal that controls the periodic switching; only for the four least-significant bits are the currents obtained from a passive division by means of differences in emitter area.

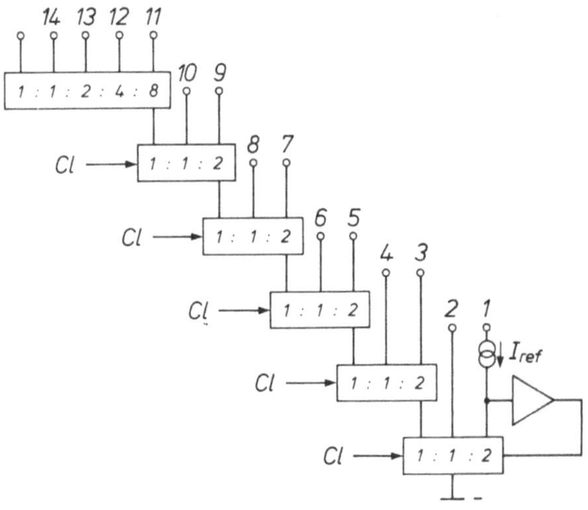

Fig. 7. Cascade of current dividers in the 14 bit digital-to-analog converter TDA 1540. The starting point is the reference current Iref. Currents that are accurately equal to a half and a quarter of the input current are obtained in the divider stages by periodic interchanges; the clock signal Cl controls these interchanges. Only the four least-significant bits 11 … 14 are obtained by passive division.

Fig. 8 shows the complete switching diagram of the 14 bit digital-to-analog converter. The cascade of divider stages can be seen in the figure. The ripple caused by the periodic switching is smoothed at the seven most significant bits by an RC filter; the seven capacitors (above in Fig. 8) are externally connected.

The nonlinearity of the digital-to-analog converter is extremely low: between −20°C and +70°C it is less than 3×10^{-5}, or half the least-significant bit. The TDA 1540 integrated circuit is followed by the low-pass Bessel filter of the third order, and the analog signal appears at the output.

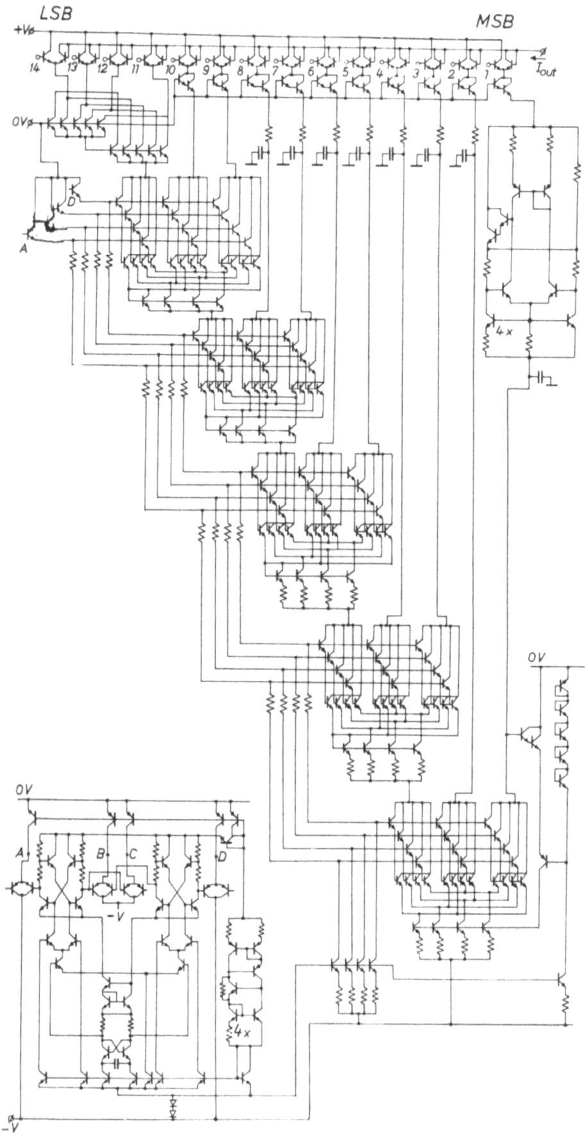

Fig. 8. Complete circuit diagram of the 14 bit digital-to-analog converter. The cascade of current dividers in Fig. 7 can be identified here. The capacitors (above), which smooth out the ripple on the divider-output currents, are external. Bottom left: The clock generator.

References

[1] The principles of the digital processing of audio signals are explained in a very readable account by B. A. Blesser in Digitization of audio: a comprehensive examination of theory, implementation, and current practice, J. Audio Engng Soc. **26**, 739-771 (1978).

[2] R. J. van de Plassche and D. Goedhart, A monolithic 14-bit D/A converter, IEEE J. SC-14, 552-556 (1979).

3.6 Compact Disc (CD) Mastering - An Industrial Process

W. Verkaik

Philips Electro-Acoustics Division, Optical Disc Mastering, Eindhoven, The Netherlands

Abstract
Compact Disc (CD) mastering is a process in which digital audio and subcode information is encoded into the standard CD format and recorded on a disk surface. The information is contained in pits of discretely varying lengths arranged in a spiral.
The disk-mastering process lies between tape mastering and replication. It involves the application of thin photoresistant layers onto glass substrates, encoding and recording the audio and subcode information, and developing and testing to generate the required pit dimensions (pit geometry).
The parameters influencing the pit geometry and other quality parameters of masters are many, and the process requires a specific philosophy and discipline to be performed industrially. This philosophy and the resulting equipment, operating requirements, quality control, and test methods are described.

3.6.1 Introduction

The introduction of Compact Disc (CD) digital audio signifies a new era in sound technology. The CD sets new standards in reproduction quality, impossible to achieve with traditional sound reproduction techniques. These standards, combined with the disk's compact size-both sides of a full LP on one side of a 120-mm diameter disk-make it a vital contribution to the future of commercial audio.

With CD digital audio, program origination and replication techniques show certain similarities to those for normal LPs. However, the mastering process is completely different. It is a process that Philips has developed against a considerable background of experience, gained with the LaserVision optical disk[1]. The LaserVision mastering technology has now led to the introduction of second-generation mastering equipment, specifically dedicated to the Philips Compact Disc.

Courtesy Audio Engineering Society (www.aes.org). Reprinted with permission from:
Digital Audio: AES Premiere Conference, Rye, New York, 1982, 189-195.

3.6.2 The production of CDs

The production chain from original sound recording to finished disks can be divided into the following stages (Fig. 1):

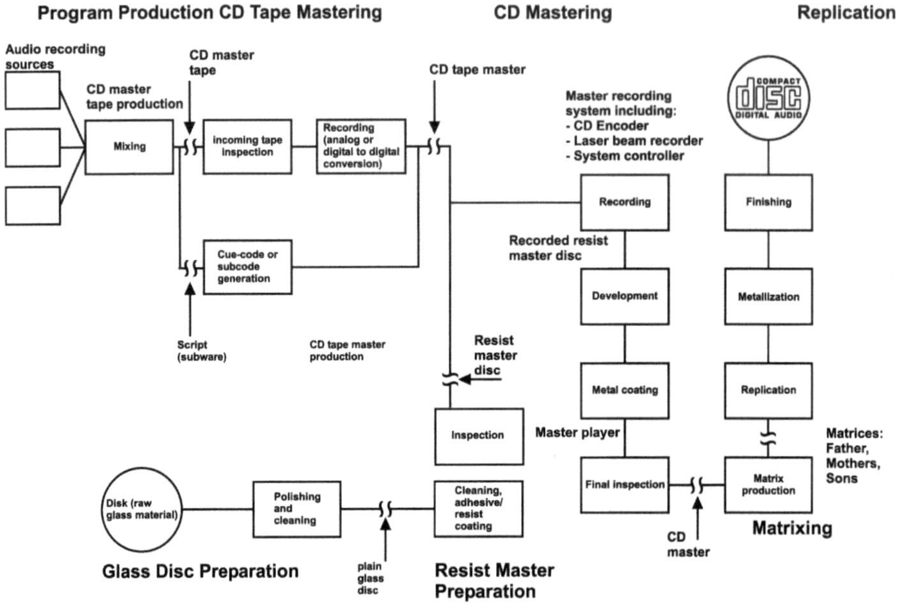

Fig. 1. Block diagram of CD production

1) *Program Production.* Here the original sound is recorded, mixed, and transcribed to generate the CD master tape, either analog or (preferably) digital, carrying the desired two stereo audio channels.
2) *CD Tape Mastering.* The Master tape at this stage is converted (analog to digital, or, if necessary, digital to digital), and the CD subcode information is generated and recorded (possibly in the form of cue codes) on the CD tape master. This tape master fulfills the requirements as specified in [2] and is the standard carrier of the CD digital audio information.
3) *CD Disk Mastering.* In this process the information from the CD tape master is encoded into the CD standard format and recorded (cut) on the surface of a photoresist-coated glass disk, the CD resist master disk. The result of this process is the CD disk master, the first disk-shaped carrier of the CD standard information contained in a vast number of pits arranged in a continuous spiral. This surface structure determines to a large extent the basic parameters of CDs and is optimized toward subsequent mass replication.

4) *Matrixing and Replication.* By galvanic processing the CD disk master surface is transferred onto a nickel shell (father) which, by the same process, can generate a number of positives (mothers). Each mother can generate a number of negatives (sons or stampers), which, after adequate processing, are used in replication. By compression or injection molding the stamper surface, information is pressed into a transparent plastic carrier, which after aluminum mirror coating (for reflection), protective lacquer coating, and label printing forms the final CD.

3.6.3 Disk mastering

The process steps involved in the disk-mastering process are the following:

1) *Glass Disk Preparation.* The glass substrate required to enter the mastering process is made by grinding, polishing, and cleaning. This substrate, which is standardized with regard to dimensions, clamping possibilities, and surface quality, is called "plain glass disk" and is effectively manufactured and distributed by a glass factory.
2) *Resist Master Preparation.* The plain glass disk enters the process area of the mastering facility. This is a clean room, climatically controlled, with a dustfiltering class of 10 000. In certain areas, where necessary, the equipment has dust filtering class 100 and facilities for the exhaust of chemical vapors. The disk is first visually checked for the minutest imperfection. It is then introduced into the resist master preparation system. Passing through the system, the disk is first thoroughly cleaned. It then receives an adhesive layer, followed by a coat of photoresist, after which careful inspection is carried out. The inspected disk is then placed in a special cassette, cured in an oven, and held in the store. The CD resist master disk, which has a shelf life of several weeks, is now ready for disk mastering.
3) *Recording and Developing.* Recording takes place in the recording area, which is very moderately controlled (class 100 000). With the CD tape master prepared, a CD resist master disk is taken from the store and passed to the CD master recording system. The system comprises a laser beam recorder with its own dust filtering of class 100, a system controller, encoder, and digital tape recorder. The signals from the CD tape master are recorded by the laser beam recorder, which exposes the CD resist master disk according to the CD tape master's content. A well planned facility will be designed for expansion, to include a second CD master recording system. This permits a doubling of output, without the need for extra facilities in the resist master preparation system. After recording, the exposed CD resist master disk is returned in its special cassette to the process area. There it passes through a developing

and evaporating stage. The latter imparts a silver coating, which permits inspection and subsequent galvanic processing prior to the replication process. The CD disk master is now ready for inspection and testing.

4) *Quality Control.* At every stage in the process, quality inspection on samples is undertaken with equipment installed in clean sections, which have dust class 100. Final testing of the CD disk master is carried out by the master player system, which permits playing the CD disk master. The readout signals are relayed to a silent room for audio assessment. The system also permits the testing of signals which determine other quality aspects of the recording and the status of the mastering process. Prior to passing on to the matrixing department for further processing, there is a visual and microscopic final inspection.

3.6.4 The readout mechanism

To understand the effects of the basic dimensions of the pits formed in the mastering process on the CD system quality; it is worthwhile to give an elementary model of the readout mechanism.

The information contained in the discretely varying length of the pits is read out by a focused laser beam in the CD player. The size and the energy distribution of the laser spot hitting a pit in the information surface are illustrated in Fig. 2.

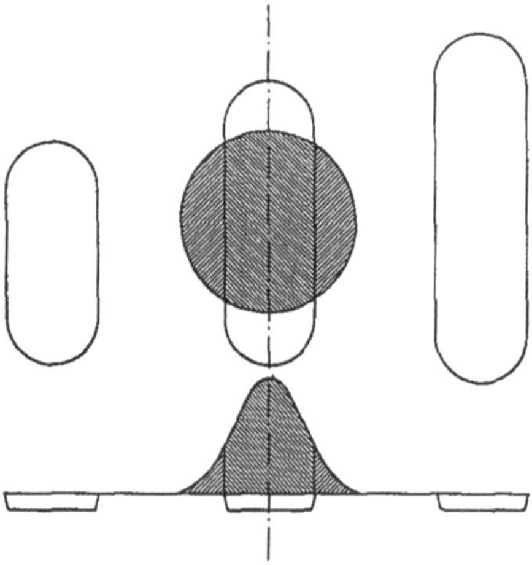

Fig. 2. Optical readout.

The light reflected from this surface is influenced by the presence of a pit and measured. This signal has to yield both the high-frequency signal containing all audio and subcode information (Fig. 3), and the radial tracking signal forming a servo signal for track following (Fig. 4).

The optimum high-frequency signal is achieved if the presence of a pit results in a total loss of reflected intensity. This situation occurs when the pit depth equals one-quarter of the apparent wavelength of the light, while the pit width is such that the intensity of the light reflected from the bottom of the pit equals the intensity of the light reflected from the surface (shaded areas in Fig. 2). In that case destructive interference will take place. Since the size and shape of the readout spot are standard in the CD system, there will be only one pit depth, and also one pit width, fulfilling the former requirement.

The optimum radial tracking signal unfortunately is not achieved at the same depth and width. On the contrary, an optimum signal is achieved when the pit depth equals one-eighth of the wavelength of the light.

Therefore a very carefully chosen compromise concerning the basic pit dimensions governs the mastering process, in which an optimum situation is specified yielding:

1) An acceptable high-frequency signal
2) An acceptable radial tracking signal
3) Mass-replicable structures
4) Minimum sensitivity to unwanted process parameters

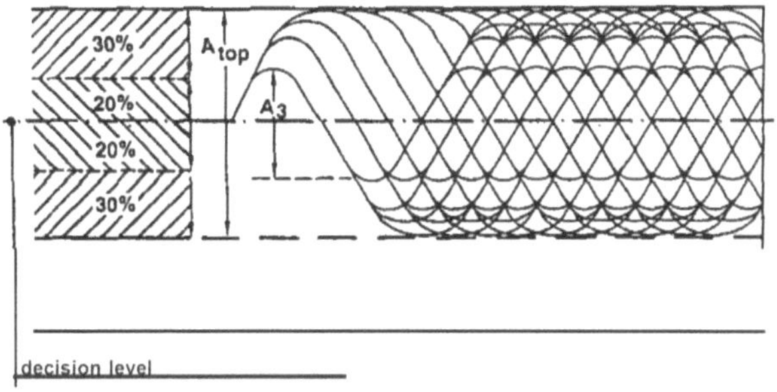

Fig. 3. High-frequency signal.

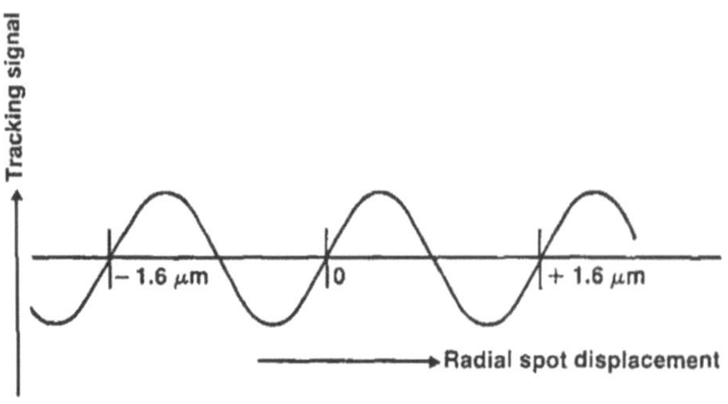

Fig. 4. Characteristic of radial differential signal.

3.6.5 Generation of pits

There are three steps involved in the generation of pits: 1) encoding, 2) recording, and 3) developing.

3.6.5.1 Encoding

The digital audio information read from the tape master and the subcode information generated by the subcode processor are fed into the professional CD encoder, the principle of which is shown in Fig. 5.

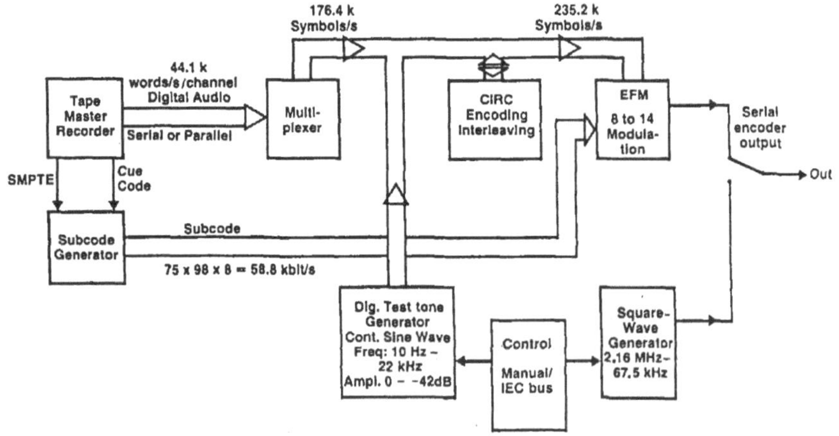

Fig. 5. CD encoder principle. Features: self-testing, decoding to digital audio optional. Serial encoder output: Audio-235.2 ksymb/s, 17 bits; Synchr-7.35 ksymb/s, 27 bits; Subcode-7.35 ksymb/s, 17 bits.

Apart from the multiplexing, coding according to the cross interleave Read-Solomon code (CIRC), and modulation according to the eight-to-fourteen modulation (EFM) principle, this encoder has facilities for the generation of test signals (sine waves and square waves adjustable in frequency and amplitude).

3.6.5.2 Recording

The serial encoder output (high-frequency signal) is connected to the driver of the acousto-optical (AO) modulator in the lightpath of the CD laser beam recorder. The optical configuration of this recorder is shown in Fig. 6. The light beam of the argon-ion laser is modulated by the AO modulator under control of the high frequency signal. The modulated laser beam, after passing various optical elements, is projected onto the objective lens, which focuses the laser beam on the surface of the resist master disk. By very accurately rotating the resist master disk and simultaneously translating the objective lens assembly, the focused recording spot will intermittently illuminate the photosensitive layer in a spiral fashion. The focusing of the objective lens on the moving resist master disk surface requires an active focusing servo system, comprising a primary focusing system using a separate diode laser beam and a secondary focusing system using part of the reflected light of the recording spot for fine tuning. The spot can be constantly monitored on a TV monitor. Alignment of the optical configuration after laser replacement, or maintenance, is greatly facilitated by the microcomputer-controlled beam-positioning facilities.

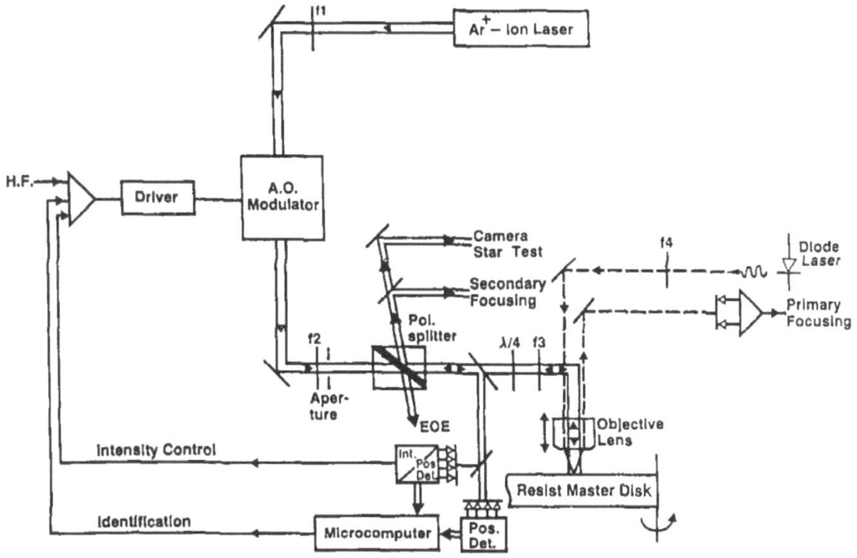

Fig. 6. Optical configuration of CD laser beam recorder.

3.6.5.3 Developing

The exposed resist master disk is developed in a developer system, where the rotating master is subjected to a flow of developing fluid, which is selectively etching away the illuminated portions of the photoresist. This etching process continues until the glass surface is reached and is terminated when the desired pit geometry is achieved. The progress of the pit formation is constantly monitored by measuring the zero-order and first-order diffracted intensities of the laser beam projected through the master.

3.6.6 Pit geometry

The following dimensions are related to the pit geometry: pit length, pit depth, pit width, slopes, and track pitch.

The pit length is dictated by the digital high-frequency signal to the optical modulator. Fig. 7 shows the correspondence between such a typical digital pattern and the resulting pit lengths. At a recording speed of 1.2 m/s this means that the pit length can have values between 0.833 and 3.054 μm, with minimum increments of 0.278 μm. The accuracy with which the pit length must be controlled must be one order of magnitude smaller than the smallest increment, such as ±30 nm. The pit depth depends on the thickness of the photoresist layer. During developing the exposed photoresist is etched away until the glass substrate is reached. The thickness and the homogeneity of the resist layer depend on the performance of the resist master preparation system and are crucial parameters in the mastering process.

Serial Input to Optical Modulator:

Audio:	235.2	ksymb/sec	17 bits
Subcode:	7.35	ksymb/sec	17 bits
Synch.:	7.35	ksymb/sec	27 bits

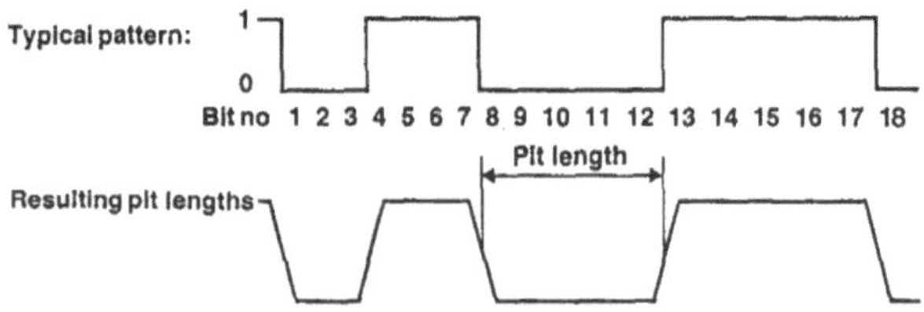

Fig. 7. CD pit length generation. Pit length increments at V_{lin} = 1.2 m/s-0.278 μm; minimum pit length (3 increments)-0.833 μm; maximum pit length (11 increments)-3.054 μm; tolerance-±30 nm. Serial input to optical modulator: Audio-235.2 ksymb/s, 17 bits; Subcode-7.35 ksymb/s, 17 bits; Synch-7.35 ksymb/s, 27 bits.

The pit width and slopes depend on the focused recording laser spot size and intensity distribution, together with the developing process. Depth, width, and slopes are carefully chosen to achieve the optimum as discussed in Sect. 3.6.4.

The track pitch is the distance between successive tracks on the disk and has the specified value of 1.6 μm. During mastering this track pitch depends on the rotational velocity of the resist master disk and the translational speed of the sledge carrying the focusing assembly. The specified track-pitch accuracy demands very stable and sophisticated control systems in the laser beam recorder.

3.6.7 Test parameters and methods

As has been shown in the previous paragraphs, the pit geometry is of basic importance for the quality of masters. Direct measurement of these dimensions is possible by means of electron microscopy, but this method is destructive for the test item, is very time consuming, and gives only a local indication. For routine measurements to ascertain master quality, the pit geometry is measured by playing the master on a master player and deriving test signals from the readout high-frequency signal. Track pitch and track-form stability are measured by monitoring the radial tracking signal during playback of the master.

Also a scan is made of information layer defects by counting appropriate indications and flags derived from the demodulating circuitry.

Finally (the "proof of the pudding is in the eating"), a master is released only after assessment of the total audio program quality and subcode integrity.

All these measurements are performed during the same real-time playback test session, using the specially designed CD master player system, which can also be used to perform similar measurements on stampers and replicas.

An additional aspect of the master quality is processability, which means the master's fitness to be processed in the subsequent matrixing department.

All test parameters and methods are summarized in Tables 1 and 2.

Table 1. Test parameters

Pit geometry
Carrier-to-noise ratio (CNR)
Surface noise
Symmetry
Phase depth
Track pitch
Track-form stability
Information layer defects
Block error rate (BLER)
C1,2 flags
Interpolations
Mutes
Overall program assessment
Audio signal quality
Ticks and clicks
Audio channel phase relation
Subcode
Processability
Metal coating
Scratches
Stains
Dust
Fibers

Table 2. Test method.

Master player with
Spectrum analyzer
Oscilloscope
Audio amplifier
Headphones
Loudspeakers in silent room
Subcode reader
Counters and chart recorders
Microscopes
Film viewer
Naked eye

3.6.8 Quality characteristic sourcing

Figure 8 shows how the important CD system performance parameters are influenced by the successive processes of disk making. The first column indicates the specified system performance characteristics. These characteristics are determined by the qualities of both the CD player and the disk. Since in this context our attention is focused on the disk production chain, in column 2 the corresponding disk parameters are indicated assuming an "ideal" readout spot.

Disk parameters can be generated entirely by the matrixing and replication process (indicated as source in column 3), or will be influenced by this process. For example, information layer defects of disks can stem from defects generated in the mastering process and magnified by matrixing and replication, but can also be generated in the latter process itself.

The pit geometry of the disk obviously stems from the pit geometry of the master, but will be influenced by the matrixing and replication process.

In general, if the effects on pit geometry in matrixing and replication are

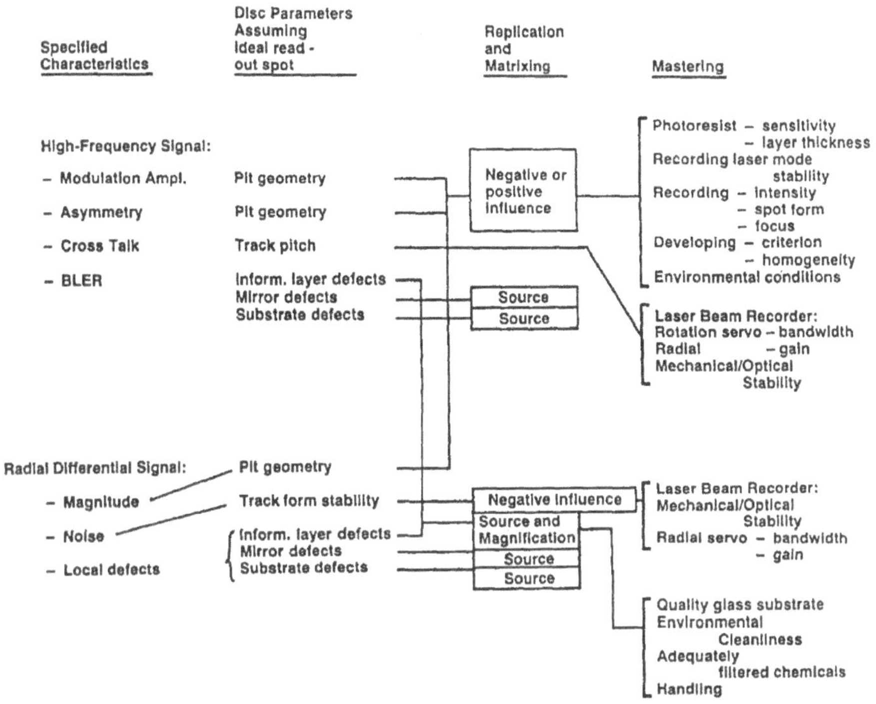

Fig. 8. Quality characteristics sourcing.

consistent, compensation of these effects during mastering can be executed. In those circumstances the replication process will have a "positive" influence on the pit geometry.

In general the mastering parameters in column 4 can be classified as quality of incoming materials, performance of mastering equipment, and environmental conditions and handling.

To arrive at an industrially acceptable situation concerning incoming materials, the Philips CD mastering process requires only commercially available chemicals and materials from several suppliers and standard items such as the tape master and the plain glass disk.

The CD mastering equipment is second-generation equipment specifically designed for long, trouble-free operation, with the help of well-defined quality control and maintenance procedures.

In order to make the mastering process less dependent on the environmental conditions and the skills of the operators, the CD mastering equipment was designed with built-in dust filtering (requiring much less investment in clean-room and air-conditioning facilities) and vastly automated handling. This not only has a direct positive effect on the costs of mastering, but also improves quality and yield.

3.6.9 Conclusions

From the previous paragraphs it can be concluded that the Philips CD mastering process

1) Is a process in which the basic parameters determining the CD system performance, as far as the disk is concerned, are well understood and under control
2) Makes use of equipment specifically designed for routine production
3) Is supported by a vast amount of basic and operational know-how
4) Is designed toward optimum quality and minimum cost of disk replication

May well be called "an industrial process."

References

[1] F. Olijkhoek, T. H. Peek, and C. A. Wesdorp, "Mastering Technology for the Philips Optical Disc Systems," Video Disc Technology Overview 25/2. Electro/81.

[2] "Specification of the 3/4-Inch Cassette Type CD Master Tape," Sony/Philips Publ.

3.7 Communications aspects of the Compact Disc digital audio system

Sophisticated coding and signal processing principles applied to a mass-marketed consumer product

J.B.H. Peek

3.7.1. Introduction

The compact disc digital audio system has already been introduced in a large number of countries. After an agreement between Philips and Sony in 1979, a common system standard was defined. This standard gradually became the world standard for this completely new system of storage and reproduction of audio signals. An extensive catalog of discs with various labels, and several brands of Compact Disc (CD) players are now available. Most people who have had the opportunity to listen to this new sound medium, not least performing artists, acknowledge that a more intense musical experience is achieved. The improvement in sound quality is in essence obtained by accurate waveform coding and decoding of the audio signals, and, in addition, the coded audio information is protected against disc errors.

From a systems point of view, the CD system was designed on the basis of communications concepts. The communications ideas that have been used will be described in this paper. The concepts applied in the CD player encompass demodulation, error correction and detection, interpolation, and bandwidth expansion to ease the D/A conversion. The paper describes an application of sophisticated communications coding and signal processing principles to a mass-marketed consumer product, and is therefore of general interest. It can be concluded that communications engineers can make valuable contributions in areas not traditionally part of the communications industry.

This restriction to communications concepts implies, however, that important and interesting aspects of the CD player such as the laser optical system, the tracking and focusing principles and control, and the integrated circuits designed for the player-will not be considered [1,2,3].

3.7.2 General System Description

As is usual in a communications system, some of the signal operations at the receiving end of the CD digital audio system are the inverse of those at the transmitting end. A block diagram showing the various signal operations is given in Fig. 1. Before a more detailed description of the successive signal operations in the CD player, we shall briefly describe the signal path from the studio to the optical readout in the CD player.

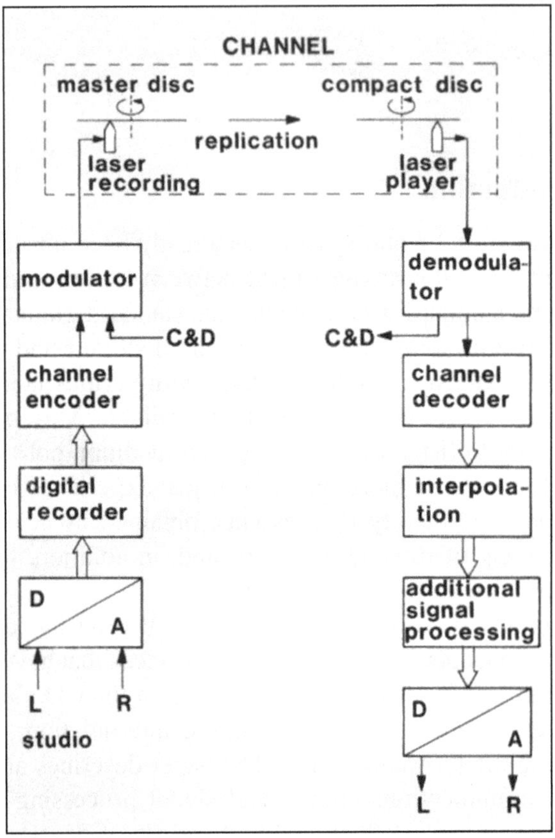

Fig. 1. The compact disc digital audio system, considered as a transmission system

Analog to Digital Conversion

Leaving aside any sound mixing, the two audio signals (left and right) that originate from the studio or concert hall are converted from analog to digital (A/D). The sampling frequency of the signals is quartz-crystal-controlled and is equal to 44.1 kHz. This sampling frequency of 44.1 kHz allows a recorded audio bandwidth of 20 kHz. The samples of both signals are uniformly quantized to 16 bits [4]. As a consequence of the quantization into 16-bit words and the crystal-controlled sampling, the conversion noise level is suppressed by more than 90 dB with respect to the peak signal level, and a total harmonic distortion of less then 0.005% can be achieved. The channel separation is more than 90 dB.

Recording

A video recorder is often used in combination with a PCM interface unit for digital recording of the audio signals on magnetic tape. It is because this video recorder uses the PAL television standard that the sampling frequency has been set at 44.1 kHz, which is $\frac{625-37}{625} \times 3 \times 15\,625 = 44.1$ kHz, where 625 is the number of lines in a PAL picture, 37 is the number of unused lines, 3 the number of audio samples recorded per line, and 15625 Hz the line frequency[5].

Channel encoding

Together with a subsequent modulation, channel encoding is part of the so-called disc mastering process. In this process, the information from the video tape recorder system is encoded into the standardized CD format.

In the channel encoding step, the digital information is protected against channel errors by adding parity bytes derived in two Reed-Solomon [6] error-correction encoders. Because the channel mainly has a burstlike error behavior, the wellknown communications technique of interleaving is used to spread the errors out over a longer time[6]. The data streams entering the first encoder, between the two encoders and leaving the second encoder, are scrambled by means of sets of delay lines. As a result of this the burst byte errors will, after deinterleaving, be spread over a longer time so that they can be more easily corrected. Those errors which cannot be corrected but are still detected, which would give corresponding unreliable samples, are restored by interpolation. This will be described in more detail later.

After the channel encoder, digital control and display (C&D) information is added to the encoded data. This information contains music-related data and a table of contents of the disc. With this table of contents, a CD player can be programmed so that only desired musical sections will be reproduced.

Modulation

Before the output data of the channel encoder can be conveyed to the master disc, a modulation operation, achieved by bit mapping, is necessary[7,8]. The reasons for modulation are the following:

- The frequency spectrum of the signal read from the Compact Disc should have low power at the lower frequencies such that the tracking control system is minimal1y disturbed. This requirement is similar to that encountered in digital magnetic recording.
- The binary signal transferred to the master disc must be such that the.bit clock frequency can be regenerated from the signal detected in the CD player. This requirement can be met by suitably mapping a block of n bits onto m (m > n) bits and by imposing an upper limit (say eleven) on the allowable length of a sequence of all ones or all zeros[8].
- Since the light spot with which the CD is scanned in the CD player has finite dimensions, intersymbol interference results which is compensated by processing a sequence of symbols. This imposes a lower limit on the length of a sequence of ones or zeros. A minimum run length of three turns out to be a good choice in practice. Thus, assuming for example m = 11, a sequence like 01010011010 is forbidden.

In the CD digital audio system, a modulation scheme called EFM (eight-to-fourteen modulation) is used which meets these requirements satisfactorily[8]. In EFM, a group of 8 bits (also called a byte or symbol) is mapped into 14 channel bits. It can be shown that there are 267 distinct 14-bit sequences that meet the run-length constraints. For a unique mapping of 8 bits, only 256 sequences are needed, so that 11 sequences can be discarded. At the receiver end, that is, the player end, the inverse operation can be obtained by a table look-up. The 14 bit sequences cannot, however, be run after the other without violating the constraints of at least 3 and at most 11 consecutive ones and zeros. By inserting 3 properly chosen merging bits between 14-bit blocks, the run-length requirements can again be satisfied while at the same time suppressing the lower signal frequencies.

In the section describing the error correction and detection systems, we use the concept of frame. A frame consists of 12 audio samples of 16 bits each. (This is equivalent to 24 bytes.) To such a frame, parity bytes and C&D bits are added and EFM is applied. After the addition of merging bits and a synchronization pattern, a final frame consisting of 588 channel bits results. Finally, a sequence of these frames is transferred to the master disc at a channel data rate of 4.32 Mb/ s.

The Channel

Next, the CD standardized format is optically recorded on the surface of a glass disc which is coated with photoresist[9]. Following development and evaporation, the result is the socalled master disc. By galvanic processing, the master disc surface is "transferred" into a nickel shell (or "father"). From this "father," "sons" or stampers are made, which are suitable for replication. By compression or injection molding, the information contained on the surface of the stamper is transferred in the form of about a billion minute pits to a transparent plastic disc. This CD has a diameter of 120 mm, a thickness of 1.2 mm, and a track pitch of 1.6 μm. Finally, after receiving a reflective aluminum coating, over which a protective lacquer is applied, the "Compact Disc" is ready for playing. In the CD player, the track on the disc is optically scanned by an AlGaAs laser (wavelength $\approx$ 0.8 μm) at a constant velocity of about 1.25 m/s. The speed of rotation of the disc therefore varies from 8 r/s when scanning the inner side of the disc to about 3.5 r/s when scanning the outer side. The maximum playing time is about 67 minutes (stereo, of course).

There are several sources of channel errors. First, small unwanted particles or air bubbles in the plastic material, or pit inaccuracies due to stamping and stamper errors, may be present in the replication process. This can cause errors when the information is optically read out. Second, fingerprints or scratches on the disc may occur when it is handled. Together with surface roughness, these disturbances cause additional channel errors. The channel mainly has a burstlike error behavior. As a consequence, a scratch or fingerprint will cause several 14-to-8 demodulated blocks to be in error, which in turn will result in several consecutive byte errors.

3.7.3 Some Error-Correcting Coding Principles

Before describing the error correction and detection that is used in the CD decoder (the channel decoder in Fig. 1), it might be useful to review some principles of error-correcting coding [6, 10].

Without any protective measures, channel errors would result in erroneous audio samples which in turn could cause considerable audible disturbances. It is the purpose of the channel code to reduce the errors at the output of the decoder to a sufficiently low level. In data communications systems, it is common practice, when retransmission is not practical, to use error-correcting codes to achieve such a goal. Since error-correcting block codes are used in the CD system, we will focus our attention solely on these codes. In a block code, a block of *k* information bits is encoded into *n* bits (Fig. 2).

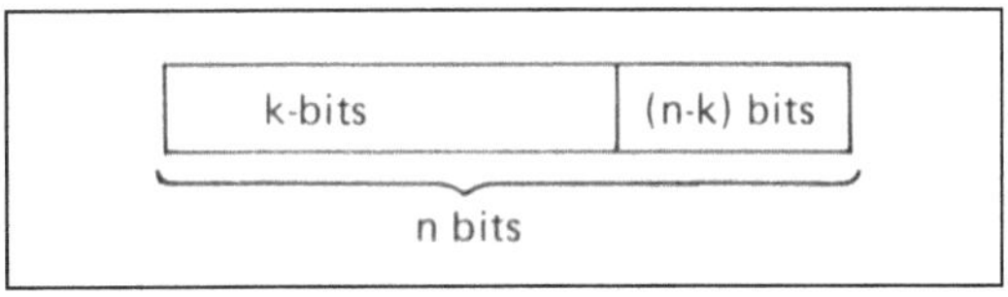

Fig. 2. A block code.

The $(n\text{-}k)$ bits which are computed from the k bits according to the mathematical structure of the code are called the parity bits. A block code is often specified by its (n,k) value.

Table I shows an illustrative example of a single-error correcting block code that is obtained by repeating the bit to be transmitted three times. The last two bits can be regarded as parity bits. If we assume that at the most one channel error can occur in a block of three bits, then it can be seen that if a zero were transmitted the number of zeros in a received block of three bits would still be in the majority. The same holds if a one were transmitted. This observation offers a simple single-error correction method based on a majority decision rule. If, however, at the most two errors can occur in a block of three bits, error correction is not always possible. Nevertheless, error detection is still possible, since any received code word other than 000 or 111 is detected as an error. In this simple example, correction and detection cannot be done simultaneously.

data bit	single error correcting code	channel outputs (max. one error)
0	0 0 0	0 0 0 0 0 1 0 1 0 1 0 0
1	1 1 1	1 1 1 1 1 0 1 0 1 0 1 1

Table I. Example of Single Error Correcting Code

At this point, it is useful to introduce the concept of "Hamming distance" between two code words. If two code words, each n bits long, differ in d ($d \leq n$) positions, then the Hamming distance between these code words is d. Hence, if d errors occur in a transmitted code word the distance between this word and the original code word becomes d.

The effect of applying our example of a triple-repeating, single-error-correcting code can now be clarified with the aid of the Hamming distance concept. Originally, the Hamming distance between the two data bits 0 and 1 is $d = 1$, which is too small to give protection against channel errors. Using the triple repeating code, the distance between the two code words 000 and 111 is increased to $d = 3$. A maximum of one channel error (in a block of three bits) will result in a distance of one (at most) between the received and transmitted code word. This distance is small enough to enable one to decide without doubt which word was transmitted. The principle of looking for the nearest neighbor is called maximum-likelihood decoding.

In general, if up to t errors in an arbitrary code word have to be corrected, then the minimum distance d_{min} min must satisfy the condition

$$d_{min} \geq 2t + 1.$$

This fact is visualized in Fig. 3, which is a two-dimensional representation of a multidimensional codeword space. The point z represents a received word, while x and y are code words. Furthermore, it can be seen that up to $2t$ errors can be detected in this case, provided correction is not attempted simultaneously.

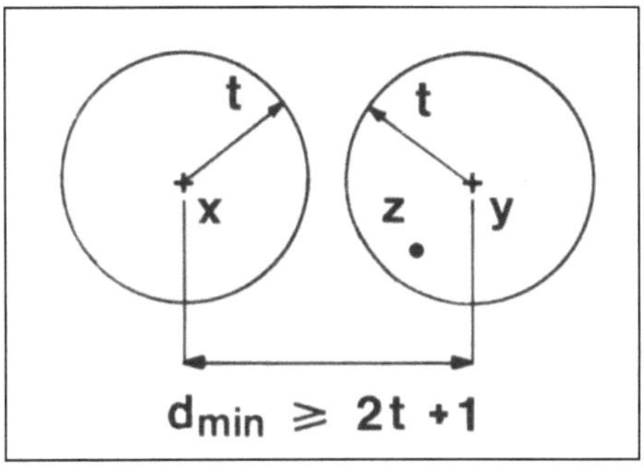

Fig. 3. Relation between minimum distance d_{min} and the maximum number of correctable errors t. The point z shows a received word, while x and y are code words.

In the theory of error-correcting codes, the concept of erasure decoding is of importance. The ith position in a block code, as given in Fig. 2, is called an erasure position if the bit value at that position is unreliable. How such an indication of unreliability can be obtained will become clear in the next section. It is the purpose of erasure correction to determine the correct bit values at a given number of erasure positions. Since, in the case of erasure correction, the positions of the unreliable bits are known, one can imagine that more bits can be corrected than when the positions are unknown.

This can be illustrated with the aid of the triple repeated code described previously. If the received word is unreliable at two arbitrarily chosen but known erased positions (and no further errors are present), then error correction is possible. By deciding on the nonerased bit as being the transmitted bit, simple error correction is obtained. In summary, the code given in Table I has only single-error-correction capability but double erasure correction capability. In general, for a code with minimum distance d_m, $(d_m - 1)$ erasures can be corrected at $(d_m - 1)$ given positions.

The principles of error-correcting block codes as described on a bit level can be extended to the symbol or byte level. Thus, from a block of k information symbols, $(n\text{-}k)$ parity symbols can be calculated and added so that a block of n symbols results. With symbols of s bits, only a small number, that is, 2^{ks} of the large number 2^{ns} of possible different words of n symbols become code words so that a large d_{min} can be created.

Reed-Solomon codes are particularly efficient since only $2t$ parity symbols have to be used to correct t symbol errors. In other words,

$$d_{min} = n\text{-}k + 1.$$

The decoding algorithm will not be described here. In the next section, attention will be given to the decoding strategy which is used by the CD decoder, and the way in which burst errors are treated.

3.7.4 The Compact Disc Decoder

Both audio channels (left and right) are sampled with a frequency of 44.1 kHz. Each sample is represented in 16 bits using uniform quantization. The audio samples are gathered in frames of 12 audio samples each, 6 samples from the left audio channel (L_n) and 6 samples from the right channel (R_n), as shown in Fig. 4. Now each sample of 16 bits consists of 2 bytes or symbols, so that each frame can also be viewed as consisting of 24 audio bytes.

L_{6n} ,R_{6n} ,L_{6n+1} , R_{6n+1} , ……,L_{6n+5} ,R_{6n+5}

Fig. 4. The nth frame. A frame contains 12 audio samples, 6 samples from the left audio channel (L_n) and 6 samples from the right channel (R_n).

In the CD encoder, the bytes of a number of consecutive frames are scrambled and parity bytes are added such that disc errors can be corrected (or detected if correction fails). The entire process of scrambling and adding parity bytes can best be explained with the help of the CD decoder scheme (Fig. 5) which is, of course, the inverse of the encoder scheme.

Roughly speaking, the CD decoder consists of two decoders (called C_1 and C_2) in series[11-14]. These two decoders have the same structure and are capable of correcting and detecting byte errors. Both codes are Reed-Solomon codes with (n,k) values (32,28) and (28,24) so that each uses four parity bytes. Thus, the minimum distance $d_{min} = n\text{-}k + 1 = 5$ and, since $2t+1 \leq d_{min}$, we have $2t \leq 4$ for each code. On the other hand, we have seen that a code with $d_{min}=5$ can correct $e=d_{min}\text{-}1=4$ erasures. Hence, it is plausible that each code can correct any number of errors (t) and erasures (e) simultaneously, provided

$$2t + e \leq 4 \quad (e,t \text{ in bytes}).$$

As has been mentioned earlier, an erasure is a byte in a known position of which the byte value is uncertain (It might be erroneous).

Error-detecting capabilities are dependent on the number of errors and erasures that simultaneously have to be corrected. In general, the larger the correcting capability used, the smaller the detecting capability. Hence there is a trade-off between error correction and detection. An undetected erroneous sample can give an annoying audible click, while for detected erroneous samples, interpolated sample values can be computed such that the result is inaudible. The decoders (C_1 and C_2) are separated from each other and from the demodulator by deinterleaving delay lines which are intended to scatter a burst of disc errors among many code words such that the number of errors per code word is minimized, which in turn maximizes the correction and detection probabilities. The first deinterleaving delay lines and the first decoder (C_1) are intended for the correction of most of the small random single byte errors and the detection of the larger burst errors. The second set of deinterleaving delay lines and the second decoder (C_2) are intended for the correction of burst errors and other error patterns which the C_1decoder could not correct. As will be described in more detail, the delay lines Δ after the C_2 decoder scramble uncorrectable but detected byte errors (which become unreliable samples) in such a way that these can often be interpolated between reliable neighbor samples.

The various parts of the CD decoder scheme (Fig. 5) will now be described in more detail.

The deinterleaving delay lines (D) before the C_1 decoder consist of one-symbol (byte) delays used in every even-numbered byte of the 32 byte codewords. The term "code word" will be used only for the full length n. By this procedure, two consecutive bytes on the disc will always end up in two different C_1 code words, thus ensuring that a relatively small disc error lying on the boundary of two bytes will not cause two byte errors in a single C_1 word.

In the currently available Philips CD players, the following strategy (Table II) is used in the C_1 decoder: First, try to correct at most one byte error; if this fails, detect a multiple byte error pattern (put erasure flags on all bytes of the outgoing C_1 word which is derived from a 24-byte frame, as explained earlier). From the mathematical properties of the code it can be proved that the C_1 decoder (given the strategy) will detect all double and triple byte errors with certainty, while error events leading from 4 up to a maximum of 32 error bytes per code word have a probability of not being detected equal to:

Pr (undetected error pattern in code word/ ≥ 4 erroneous bytes) $\approx 1.9 \times 10^{-6}$, where the symbol / denotes a conditional probability.

After the C_1 decoder, the 28 remaining bytes (the 4 parity bytes used in the C_1 decoder are no longer used) and the possible erasure flags are deinterleaved by a triangular shaped network of delay lines (Fig. 5).

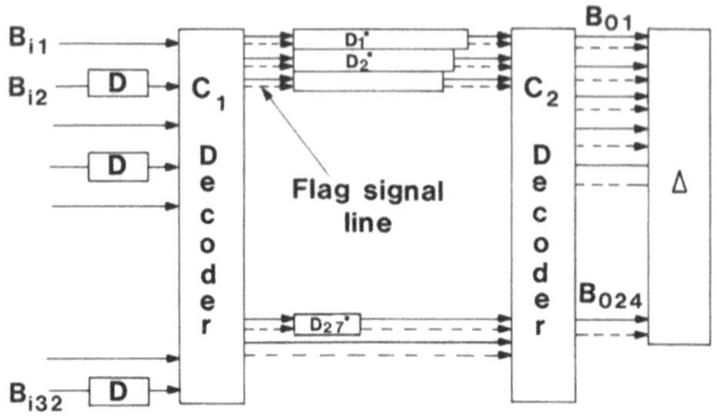

Fig. 5. Scheme of the CD decoder. The 32 bytes (B_{i1}, ... , B_{i32}) of a frame (24 audio samples and 8 parity bytes) are applied in parallel to the 32 inputs. The delay lines *D* have a delay equel to the duration of one byte, so that the information of the "even" bytes of a frame is cross-interleaved with that of the "odd" bytes of the next frame. The C_1 decoder is designed in accordance with the rules for a Reed-Solomon code with (n=32, k=28). It corrects one error, and if multiple errors occur passes them on unchanged, attaching to all 28 bytes an erasure flag, sent via the dashed lines. Due to the different lengths of the delay liner Di^* (i=1, ..., 27), errors that occur in one word at the output of the C_1 decoder are "spread" over a number of words at the input of the C_2 decoder. This result in reducing the number of errors per input word of the C_2 decoder. The second decoder C_2 is also designed to decode a Reed-Solomon code with (n=28, k=24). If the errors cannot be corrected, 24 bytes are passed on unchanged and the associated positions are given an erasure flag via the dashed output lines, B_{01}, ..., B_{024}. In most cases, the unreliable output samples (corresponding with the unreliable bytes) can still be restored by interpolation.

This network of delay lines (D_i*) is such that the length of the delay lines from bottom to top changes with increments of four bytes. Because of this network, the 28 symbols belonging to a single C_1 word and the possible attached erasure flags will be allocated to 28 different C_2 words which are equidistantly spaced. Fig. 6 illustrates how all the symbols of a single C_1 word which are given a flag (indicated by circles) arrive at the output of a triangular shaped delay network in distinct C_2 words, assuming for simplicity delay increments of one byte instead of the actual four bytes. This configuration of C_1 and C_2 code words explains the abbreviation CIRC (Cross Interleaved Reed-Solomon Code).

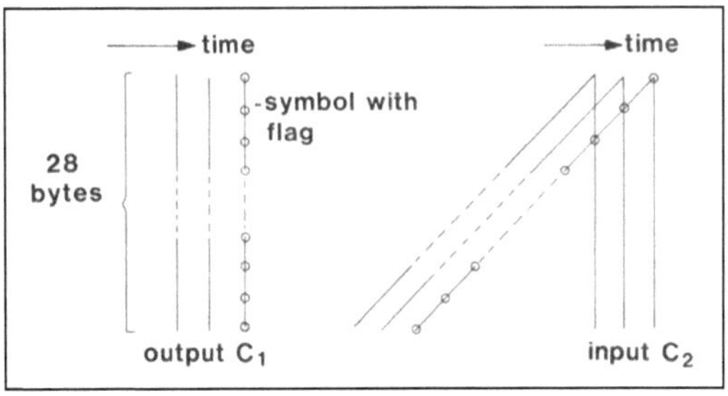

Fig. 6. Effect of deinterleaving: 28 bytes, with detected error flags, in a code word emerging from the C_1 decoder are distributed to 28 consecutive codewords which are then input to the C_2 decoder.

Suppose the increments in the delay lengths of the triangular network were indeed one byte. It would then be possible to correct a burst error encompassing four consecutive C_1 code words if four-erasure correction at the C_2 decoder was used (Fig. 7). In the actual CD system, the increment equals 4 bytes, thus offering a maximum burst-error-correcting capability of 16 consecutive uncorrectable C_1 words.

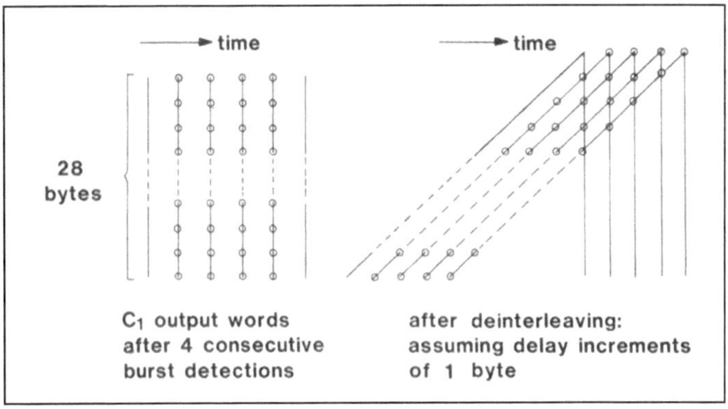

Fig. 7. Example: showing 4-erasure capability.

In current CD players, the possibility of correcting up to four erasures is not used since this would cause too high a probability of an undetected error (a "click"). The strategy adopted in these players allows up to two-erasure correction for the C_2 decoder (Table II). The translation of the C_2 correction strategy to the maximum correctable burst length on the disc is somewhat complicated because of the deinterleaving delay lines before the C_1 decoder. This is the reason for the unusual values of correction and interpolation length given in Table III. It must also be mentioned that the numbers given in Table 111" do not take error propagation (due, perhaps, to synch loss) into account.

For random symbol errors only, the probability of an interpolation or a click (see next section) can be calculated as a function of the disc random byte error probability. From measurements it appears that the average symbol error rate lies around 10^{-4} to 2×10^{-4}.

Starting from the currently used strategy as given in Table II, it can be calculated that the click probability for this case is negligible. The probability of an unreliable sample is 8.3×10^{-10} (once every 3-3/4 hours) if the random disc byte error is about 10^{-3}. For a random disc byte error rate of 10^{-4}, a realistic figure, the sample interpolation rate is about 10^{-15}.

C_1 decoder	C_2 decoder

if single- or zero-error is detected
then modify at most one symbol accordingly
else assign erasure flags to all symbols of the received word
	if single- or zero-error is detected
	then modify at most one symbol accordingly
	else if more than 2 flags
	  then copy C_2 erasure flags from C_1 erasure flags
	  else if two flags
	    then try 2 erasure decoding;
	    if less than two flags or **if** 2-erasure
	    decoding fails **then** assign erasure flags to
	    all symbols of the received word

Table II. Currently Used Error-Correction and Detection Strategy.

Up until now, decoder performance has been expressed in terms of the maximum correctable burst length and the interpolation and click rates for the case of random byte errors. The question, however, is: Do these quantities reflect the actual performance of the decoders? In practice it turns out that the interpolations can be attributed, in most cases, to clusters of small error bursts such as can be caused by fingerprints or scratches on the surface of the disc. In spite of the interleaving, such relatively small bursts can lead to errors which will meet at the input of the C_2 decoder (Fig. 8) if they fall within the constraint length ($\approx$ 1.8 cm on the disc). Hence there will be C_2 code words which cannot be corrected and will thus cause interpolations.

From the above, it can be seen that it is worthwhile to increase the correction capabilities of the decoders. In order, however, not to increase the click probability at the same time, it is necessary to introduce multiple-level reliability information (that is, distinction in flag qualities such as certainly in error, and less probable in error) at the entrance of both decoders. Current IC technology offers the possibility to implement these more-complex decoders, and they may be provided in future generations of CD players.

A final observation on the subject of error correction and detection in CD players is that all error control procedures are in vain if track loss occurs either through an improper design of the optical tracking servo system or because of excessive disc damage.

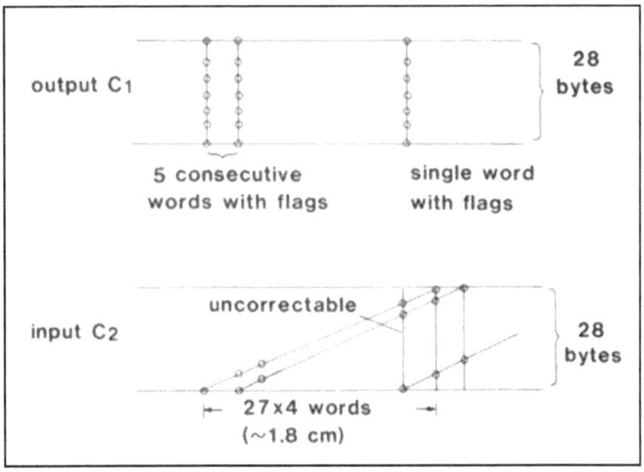

Fig. 8. Uncorrectable situation due to two smaller bursts. If, at the end of the C_1 decoder, 5 consecutive words are attached with flags and if, in addition, a single word attached with flags follows within a distance of 27x4 words (constraint length), an uncorrectable situation can occur. In that case, the input of the C_2 decoder can consist of three erroneous bytes which the present decoder cannot correct.

3.7.5 Interpolation and Muting

As has been mentioned earlier, those byte errors which cannot be corrected by thc C_2 decoder can still be detected. Without any further signal processing these unreliable samples could cause large audible disturbances. It is the purpose of interpolation to insert new samples instead of the unreliable ones [13]. Of course, the interpolated samples should be such that the final result gives no audible disturbance.

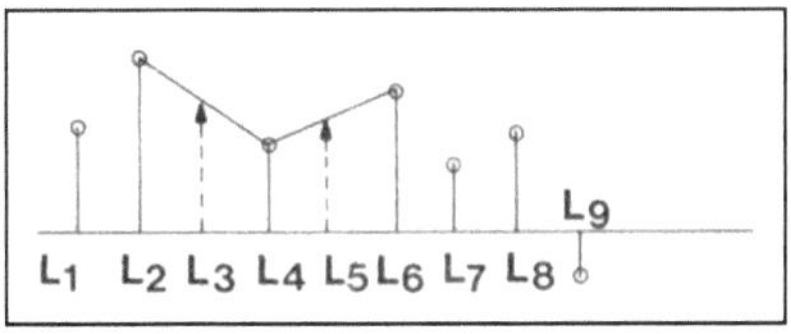

Fig. 9. First-order linear interpolation.

If two reliable neighbor samples are present, an interpolated sample can be obtained from a linear (straight line) interpolation (Fig. 9). Listening tests indicate that the result of this interpolation method in CD systems gives inaudible effects. If an entire C_2 word is detected as unreliable, this would, without taking precautions, make it impossible to apply the suggested interpolation method since both the even and odd numbered samples are declared unreliable. This situation arises if the C_1 decoder fails to detect an error but the C_2 decoder detects it. It is the purpose of the deinterleaving delay lines (Δ) in Figs. 5 and 10 to obtain a pattern, in such a situation, where the unreliable even-numbered samples can be interpolated from the reliable odd-numbered samples or vice versa. Two successive unreliable words consisting of 12 sample pairs are indicated in Fig. 10. A sample pair consists of a sample from the right and a sample from the left audio channel. After the delay lines Δ (length= two frames) the pattern is suitable for interpolation.

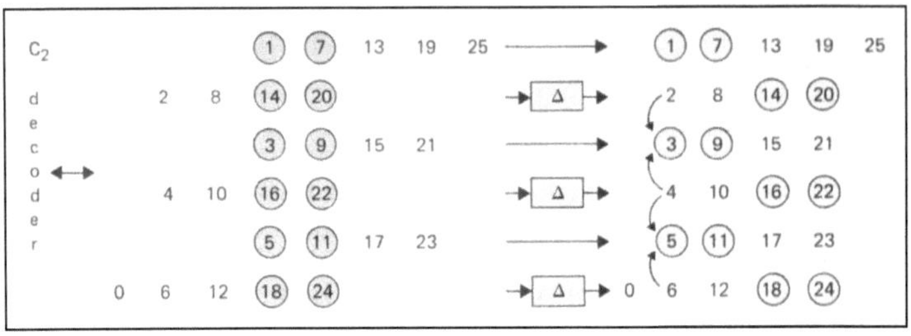

Fig. 10. The effect of delay lines Δ (2 frame times) on sets of samples. The numbers indicate the ordering of the sets of samples. An encircled sample set denotes an erasure flag. After the delay lines, the unreliable samples shown in the figure can be estimated by a first-order linear interpolation.

Because of the various deinterleaving operations and the shuffle of the samples at the input of the deinterleaving delay lines Δ, it is again somewhat complicated to determine the maximum burst length (on the disc) that can be dealt with using first-order linear interpolation. This maximum burst length turns out to be 48 frames (Table III).

C_2 decoder	correction length	interpolation length
1-symbol correction	4 frames 0.68 mm (track length) on disc	48 frames 8.16 mm
2-symbol correction	8 frames 1.36 mm	48 frames 8.16 mm
4-symbol erasure correction	15 frames 2.55 mm	48 frames 8.16 mm

Table III. Maximum Burst Correction and Interpolation Length.

In current CD players, a last remedy is provided in case a burst length of 48 frames is exceeded and two or more consecutive unreliable samples result. In this case, a gradually increasing attenuation of the reliable samples before the burst, then an insertion of zero-valued samples instead of the unreliable samples, and finally a decreasing attenuation of the reliable samples after the burst is applied. This muting of the signal is inaudible provided the muting time does not exceed a few milliseconds and the muting is only incidental.

Digital audio signals can be processed with a digital computer and listened to in a specially designed listening room. Various interpolation methods for the case of two or more consecutive unreliable samples have been tested using such a digital audio computer facility. Since the Compact Disc turns about 10 times a second when the inner side of the disc is read out, error patterns that occur every 0.1 seconds were used in the computer simulations. From these tests it can be concluded that simple straight-line interpolation performs satisfactorily if the number of consecutive unreliable samples is less than eight.

Further research revealed that if 16 consecutive samples are unreliable, restoration is always possible by using adaptive interpolation[15]; the word adaptive indicates an interpolation that uses the statistical properties of the music before and after the burst. Although adaptive interpolating is not used in current players, it is a future possibility.

3.7.6 Additional Signal Processing and D/ A Conversion

As has been mentioned earlier, the two audio signals (left and right) are uniformly quantized in 16 bits at a sampling rate of 44.1 kHz. After the interpolation or muting, the digital signal is in principle ready for conversion to the analog domain. The implementation of a 16-bit D/A converter at an acceptable price level is not, however, an easy task. Besides, as will be explained later, the analog filter following the D / A converter would be complex and expensive if a direct conversion were used.

It will be shown that a 16-bit D/ A performance is obtained from a 14-bit D/ A converter together with additional signal processing. A 14-bit D/A converter is easier to realize, but the 16-bit accuracy would be lost and a dynamic range of more than 90 dB would no longer be maintained. The operations on a 16-bit signal $s(n)$ that is first rounded off to 14 bits and D / A converted are depicted in Fig. 11. Rounding introduces an error $e(n)$ that can be considered as an independent (white) noise sample added to the signal $s(n)$. We have

$$-\frac{q}{2} < e(n) \leq +\frac{q}{2}$$

where q is the step size of the least significant bit (LSB), in our case the 14th bit. The mean square error $\overline{e^2(n)}$ is approximately

$$\overline{e^2(n)} \approx \frac{1}{12} q^2.$$

A rounded digital signal, with roundoff noise, can be modeled as a signal which has passed through a noisy communications channel. From communications theory it is known that a signal can be protected against noise by introducing redundancy, which requires bandwidth expansion at the transmitter end. This bandwidth expansion idea can be used to ease the D/ A conversion[16,17].

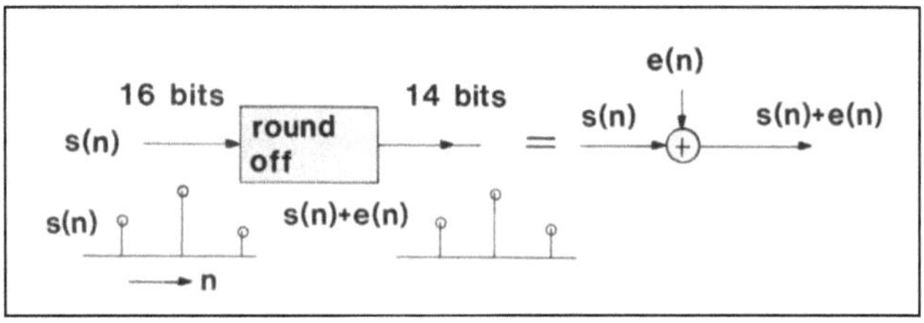

Fig. 11. Rounding a digital signal *s(n)* can be regarded as if the signal has passed through a communications channel.

In the D/ A conversion system, a bandwidth expansion by a factor of four is realized by what is called interpolation in the area of digital signal processing. The interpolated samples are obtained in a way that differs from that described in the previous section. Here the interpolated signal values are obtained by first a fourfold increase of the sample frequency through insertion of three zero signal values between every two input samples, and next by lowpass filtering this signal with a finite impulse response (FIR) digital filter. This digital filter has 96 taps, and the 96 coefficients are each represented in 12 bits, so that an attenuation in the stop band (above 24 kHz) of about 50 dB results (Fig. 12). In a digital system, only the signal frequencies in the band from zero to half the sampling frequency are relevant and consequently only these frequency bands are indicated in Fig. 12.

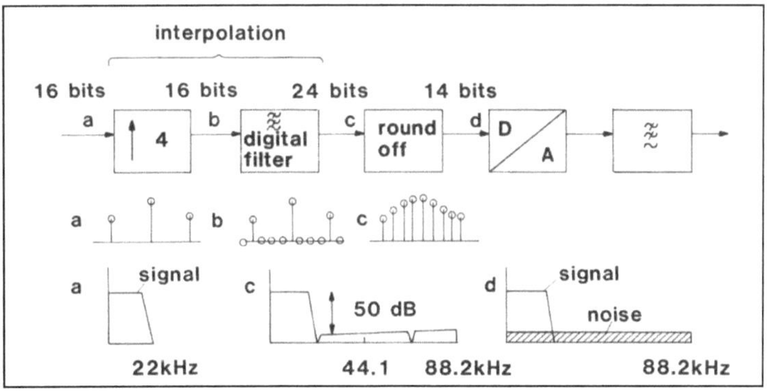

Fig. 12. D/A conversion based on interpolation.

Because of the digital lowpass filter, the signal at the filter output has acquired a word length of 16 + 12 = 28 bits. Reducing this word length by rounding to 14 bits gives a mean square error of $\frac{1}{12}q^2$ (where q is the step size of the LSB of a 14-bit D/A converter). This noise power, however, is now evenly distributed over a four-times-larger interval (Fig. 12). For reasons of simplicity, the attenuated signal components around 44.1 kHz and near 88.2 kHz are no longer indicated in the picture that shows the round-off noise spectrum. The noise power in the 0-22 kHz bandwidth, however, is four times less than in the case of a direct round-off from 16 to 14 bits. A factor of four in noise power (6 dB) corresponds with a factor of two in amplitude, and thus with one bit.

In the Philips D/A conversion system, a noise-shaping filter is used after the digital filter in Fig. 12 which redistributes the noise power in such a way that the noise power in the audio bandwidth 0-20 kHz is reduced at the expense of an increase in noise power outside this bandwidth. Since the ear responds only to frequencies up to 20 kHz, the 7-dB gain in signal-to-noise ratio obtained with the noise-shaping filter in this bandwidth can directly be translated as an extra bit gained. Thus the combination of interpolation (factor of four), the noise-shaping filter, and a 14-bit D/A gives about the same performance as a straight 16-bit D/A converter.

As mentioned earlier, the digital lowpass filter attenuates the frequencies above 24 kHz by about 50 dB. Consequently, the analog filter following the D/A converter is rather simple. Such an analog filter is necessary in order to prevent signals around multiples of the sampling frequency from overloading the power amplifier or from mixing with other signals (such as the bias signal from a tape recorder) and thus causing audible distortion. If, however, a direct 16-bit D/A is used, the analog filter after this converter would have to be quite complex. For this filter the transition band would have to be only a few kHz without affecting the flatness of the passband, while giving an attenuation above 24 kHz of at least 50 dB.

Acknowledgment

I would like to thank S. Baggen for valuable discussions and comments.

References

[1] M. G. Carasso, J. B. H. Peek, and J. P. Sinjou, "The compact disc digital audio' system," Philips Tech. Rev., (Special Issue), **40**, 151-156 (1982).

[2] S. Miyaoka, "Digital audio is compact and rugged," IEEE Spectrum, pp. 35-39, March 1984.

[3] "Draft Compact Disc Digital Audio System," now under discussion in the International Electrotechnical Commission as document 60A (Central Office) 82. For text, reference is made to doc. 60A (Secretariat 96).

[4] B. A. Blesser, "Digitization of audio," J. Audio Eng. Soc., **26**, 739-771 (1978).

[5] T. Doi, Y. Tsuchiga, and A. Iga, "On several standards for converting PCM signals into video signals," J. Audio Eng. Soc., **26**, 641-649 (1978).

[6] G. C. Clark and J. B. Cain, Error-Correction Coding for Digital Communications, Plenum Press, 1981.

[7] K. A. Immink, "Modulation systems for digital audio discs with optical readout," IEEE Int. Conf. on Acoust., Speech, and Sig. Process., pp. 587-589, Atlanta, GA (1981).

[8] J. P. J. Heemskerk and K. A. Schouhamer Immink, "Compact disc: system aspects and modulation," Philips Tech. Rev., (Special Issue), **40**, 157-164 (1982).

[9] W. Verkaik, "Compact Disc (CD) mastering-an industrial process," The AES Premier Conference- The New World of Digital Audio, New York, (1982).

[10] A. B. Carlson, Communication Systems, McGraw-Hill, 1975.

[11] L. B. Vries and K. Odaka, "CIRC-the error correcting code for the Compact Disc," The AES Premier conference-The New World of Digital Audio, New York (1982).

[12] L. M. H. E. Driessen and L. B. Vries, "Performance calculations of the compact disc error correcting code on a memoryless channel," Int. Conf. Video and Data Recording, University of Southampton (1982).

[13] H. Hoeve, J. Timmermans, and L. B. Vries, "Error correction and concealment in the compact disc system," Philips Tech. Rev., (Special Issue), **40**, 166-172 (1982).

[14] T. Doi, "Error correction for digital audio recordings," The AES Premier Conference- The New World of Digital Audio, New York (1982).

[15] A. J. E. M. Janssen, R. N. J. Veldhuis, and L. B. Vries, "Adaptive interpolation of time-discrete signals that can be modelled as autoregressive processes," submitted for publication to IEEE Trans. Acoust., Speech, and Sig. Process.

[16] D. Goedhart, R. J. van de Plassche, and E. F. Stikvoort, "Digital-to- analog conversion in playing a compact disc," Philips Tech. Rev., (Special Issue), **40**, 174-179 (1982).

[17] T. A. C. M. Claasen, W. F. G. Mecklenbräuker, J. B. H. Peek, and N. van Hurck, "Signal processing method for improving the dynamic range of A/D and D/A converters," IEEE Trans. Acoust., Speech, and Sig. Process., ASSP-28, no. 5, Oct. 1980.

Chapter 4

COMPACT DISC STANDARDS AND FORMATS

Sorin G. Stan
Philips Consumer Lifestyle

4.1 Introduction

Soon after the market introduction of the audio CD it became clear that computer data could also benefit heavily from the advantages offered by optical disc storage. The audio CD entered the consumer electronics arena shortly after the first personal computers made their debut. This coincidence of events turned out to be very prolific and led to many new applications that have culminated with complex combinations of text, sound, graphics, and video addressed nowadays by the term multimedia.

The coming years after 1982 would therefore witness the appearance of several types of compact discs, most of them emerging from the seminal physical format standardized for audio playback. New standards came forth to cover the increasing number of utilization areas. A diagram showing the evolution of the CD family and the related standards is depicted in Fig. 4.1.

Initially, the compact disc was defined as a read-only medium with a specific manufacturing technology. The recordable and rewritable CDs appeared somewhat later, about 8 years after the introduction of the digital audio system. Since then, all CD recordable drives and the recording computer software have been designed to format and write data in compliance to the previously defined standards for read-only discs. This leads to written discs that are backward compatible with their read-only counterparts in terms of both media physical parameters and the logical structure of the recorded data.

This chapter addresses the various compact disc specifications and emphasizes the essential differences between them. Other CD formats that do not belong officially to the CD family but have either tried to or penetrated already the consumer electronics market will be discussed as well herein, toward the end of the chapter.

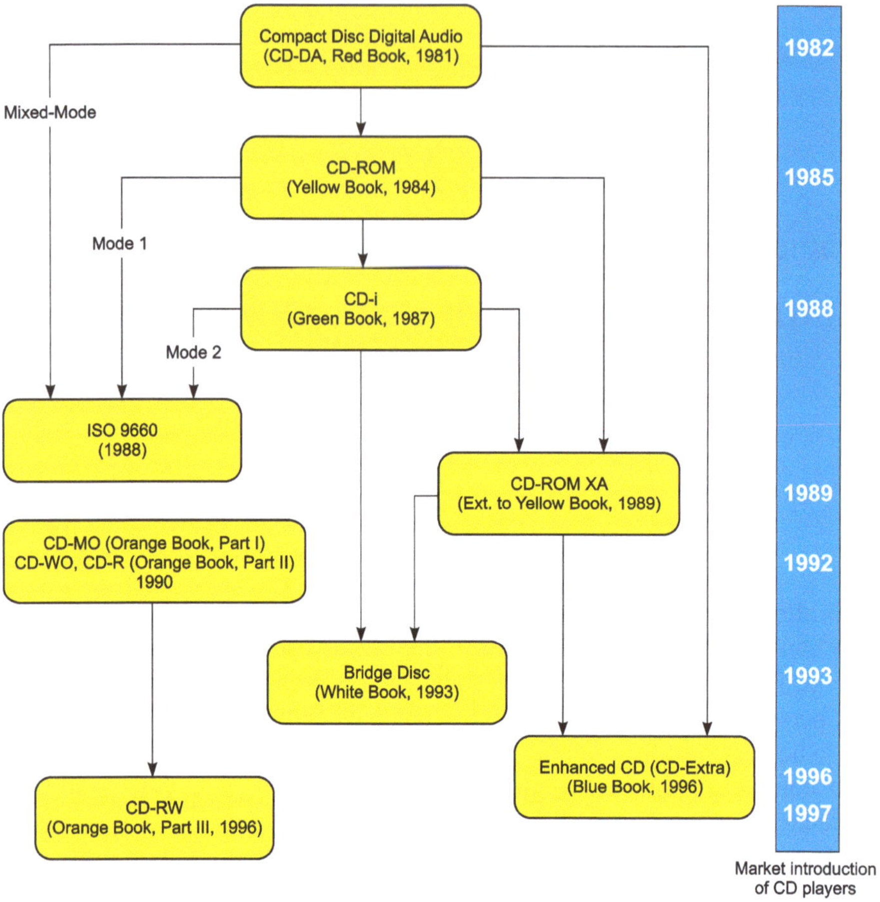

Fig. 4.1. An overview of the compact disc standards and their related standardization and market introduction years.

4.2. Read-only CDs

The audio CD system proposals[92] of Philips and Sony prepared for the technical community in 1981 have been amended slightly several times during the years that followed, with the latest version dating from 1999. These system specifications are known as the **Red Book** and define both the physical layout of the **Compact Disc Digital Audio (CD-DA)** and the logical structure of all data recorded on disc. The International Electrotechnical Commission (IEC) has recompiled the Red Book into an international standard, of which the first edition[23] was published in 1987. This document too was updated at a later stage and also changed its catalog number from IEC-908 into IEC-60908 [17].

The CD-DA players were introduced on the market in 1982, one year after Philips and Sony made available their first version of the Red Book. The two electronics companies worked closely together to define a set of consistent media and system specifications. These definitions allowed any disc manufactured accordingly to be read out in any player produced under responsible licensing agreement terms. A licensing program offered by Philips and Sony would represent in the years to come the basis for cross-compatibility between discs and playback devices yet to be produced in many flavors by so many other companies.

For further understanding of the Red Book derivatives it will very be useful to summarize in this section the main CD-DA characteristics. Before being recorded on disc, the Red Book demands the digitization of the stereo analog audio by two analog-to-digital converters (ADCs) sampling in parallel at 44.1 kHz, with each ADC producing 16-bit samples represented in pulse code modulation format. This means that, at any sampling instant, four bytes would become available as user data for subsequent digital signal processing operations. As indicated already in the original articles preceding this chapter, the user data is arranged in frames. The CD-DA standard specifies a data frame consisting of 24 bytes, that is, carrying six PCM samples per audio channel. The fixed sampling rate of 44.1 kHz leads to a user data rate of 44100/6 = 7350 frames/second or $7350 \times 24 \times 8 = 1.4112$ Mbit/s.* Expressed in kilobytes, the CD-DA media delivers 172.3 kB/s** toward the two digital-to-analog converters (DACs) used to restore the original 2-channel audio stream.

Further along the data path, the user data must be accompanied by error detection and correction information. Two sets of four parity bytes are added to each frame, with each set calculated independently by one Reed-Solomon (RS) code. The two codes work in cooperation on a two-dimensional array of user data and will provide during readout a combined straightforward error and erasure correction. The complete two-dimensional data structure makes also use of interleaving, whereby frames following each other in time are spread across the error correction matrix according to fixed delays. The entire construction is called Cross-Interleaved Reed-Solomon Code (CIRC) and has the ability to restore the information erroneously retrieved from a maximum length of a damaged track equal to 2.3 mm. To this performance, the cooperating RS codes contribute with at most two erroneous bytes that can be corrected straightforward along one matrix direction and another maximum four bytes that can be corrected using the erasure method along the second matrix direction. Statistically, one erroneous byte among one billion of correct ones are expected at the output of the CIRC error correction circuitry.

* One megabit per second is equal to 10^6 bits/s.

**In computer terms 1 kB = 2^{10} = 1024 bytes and 1 MB = 2^{20} bytes.

The 32-byte frame containing user data and RS parities, as explained in the articles from the previous chapters, is given a preceding 8-bit symbol that carries well-defined control information. Each bit within such a symbol is part of a so-called subcode channel and is designated by one of the uppercase letters between P and W. There are, hence, eight subcode channels that collect bit-wise information from consecutive subcode or control symbols. It is like arranging all frames with their preceding control bytes below each other in a matrix and reading vertically, at once, the information contained within a bit-oriented column. The subcode channels P through W are obtained by reading along the first eight columns, with 98 consecutive frames being needed to form one unit of subcode data. The P channel indicates the start and stop positions of each audio track while the Q channel contains addressing information in the form of total playback time elapsed from the beginning of the data spiral on disc and relative playback time from the beginning of an audio track. It becomes therefore possible to locate any group of 98 user data frames on disc and search for a particular audio title, a given musical passage, etc. Taking into account the rate of 7350 frames per second, the resolution with which audio data can be addressed is equal to 98/7350 = 1/75 seconds. The playback time is usually displayed in minutes and seconds on the little screens of the audio equipment.

Finally, each frame containing 33 bytes undergoes the channel modulation before being recorded on disc. The particular technique employed in audio CDs is called Eight-to-Fourteen Modulation (EFM) and converts each byte into a symbol of 14 bits. To this symbol three more bits are appended to control the DC level of the resulted signal toward zero and to ensure that this signal fulfills the modulation rules of minimum and maximum number of digital ones and zeros (see the details revealed throughout the preceding original articles). A synchronization pattern consisting of 27 bits is attached at the forepart of the modulated frame and this ultimately leads to a total of 588 bits per frame. The resulting channel bit rate can then be calculated as equal to $588 \times 7350 =$ 4.3218 Mbit/s.

The audio compact disc can hold in its standardized format between 606 and 807 MB of digital audio samples on media with outer diameters of 12 cm, where 1 MB = 2^{20} bytes according to the convention already explained. The range of storage capacities does not represent something that a manufacturer can really choose, but is due to the allowed tolerances with which the disc may be manufactured according to the Red Book. However, by controlling very precisely the fabrication process, it is possible to consistently produce 800-MB media while still fulfilling the international CD-DA standard. The total playback time of an audio CD ranges from 60 minutes and 2 seconds to 79 minutes and 57 seconds, corresponding to the above limits of the storage capacity. Obviously, a disc does not have to be recorded to its full storage capacity. In addition to the 12-cm media, the Red Book also specifies a disc having an outer diameter of 8 cm. Mainly intended for the release of single audio titles, these smaller discs can play back at most for 24 minutes and 6 seconds, which is equivalent to a maximum of 248 MB of digital audio. The typical playback time for these CD-DA discs, which are sometimes called **CD Single**, is 20 minutes. Quite confusing, note that a similar name was also given in the Japanese market to an 8-cm audio compact disc on which data was arranged according to a computer file structure. The Red Book also specifies the **CD Audio Maxi-Single**, which is nothing but a CD with a maximum playback time of 30 minutes but still with a geometrical outer diameter equal to 12 centimeters.

Several very important physical parameters standardized by the Red Book define the type of light to be used for optical readout and the characteristics of this readout process itself. We shall mention here only that a semiconductor laser emitting in the infrared spectrum and having the wavelength of 780 nm is needed for the CD-DA playback as well as for all other compact disc formats derived from the Red Book. An overview of the compact disc system parameters is given in Tables 4.1 and 4.2 in this section.

During the years which followed the introduction of the audio compact disc, the Red Book specifications led to several other formats as illustrated in Fig. 4.2. The **CD-Graphics**, sometimes denoted by **CD+G**, makes use of the six out of eight defined subcode channels. As previously mentioned, the audio discs were designed to use two of these channels, namely P and Q, to store track-related information but the other six were left empty (i.e., filled with digital zeros). The CD+G format distributes pieces of graphic bitmaps within these empty fields and, while playing back the audio CD, static images can be restored and displayed, for example, on a TV screen. Some karaoke CDs make use of this feature to store song lyrics. Three CD+G modes are defined for displaying text in two colors (line-graphics), 16 colors (TV-graphics), and 256 colors (enhanced TV-graphics), respectively. The latter is sometimes referred to as **CD+EG**. A fourth mode, which has led to the name **CD-MIDI**, allows the use of the six subcode channels previously mentioned to store information according to the Musical Instrument Digital Interface (MIDI) specifications [24]. Finally, a user mode for professional applications does not specify any particular graphical structure to be recorded in the available subcode channels, but it rather leaves this choice to the user.

The latest variation of the compact disc digital audio format, called **CD-Text**, is standardized by an add-on chapter[102] of the Red Book that became available separately in 1996. This document allows the usage of the six empty subcode channels from CD-DA to store text information, e.g. related to each recorded song, that can be displayed on small screens during playback. The audio players should be equipped with corresponding screens and must be able to

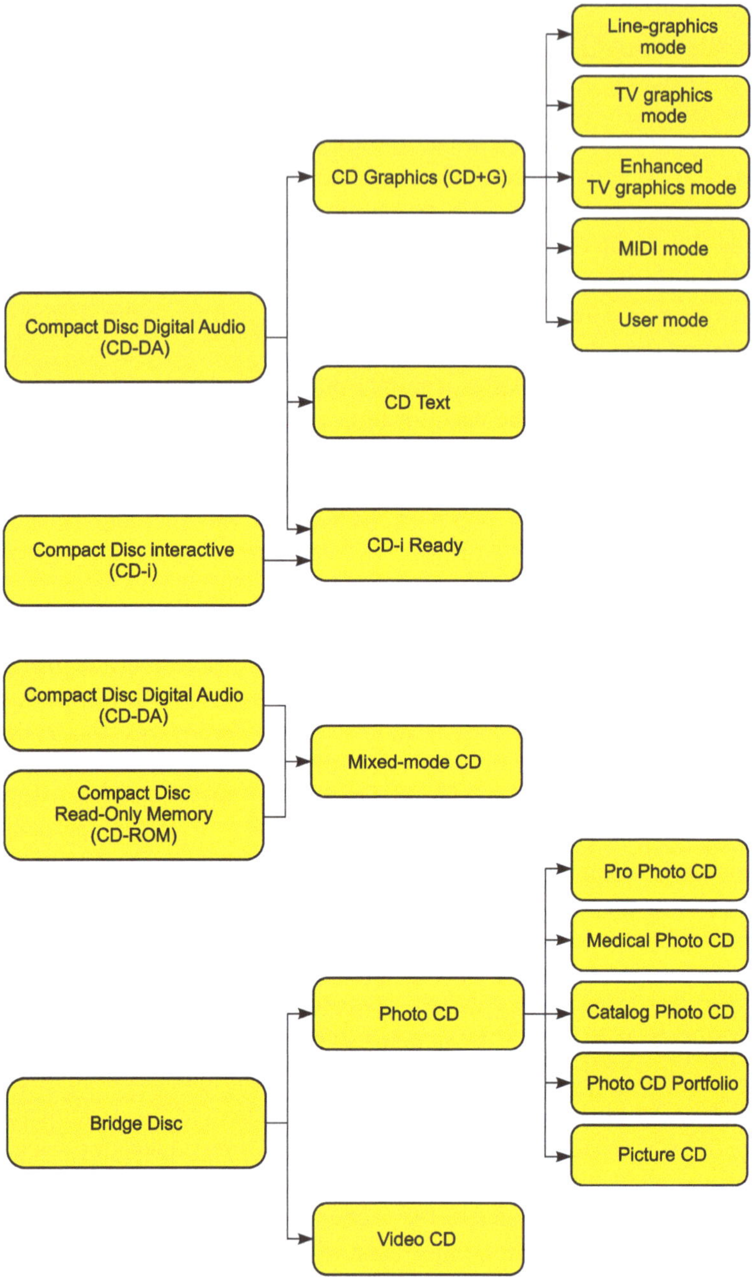

Fig. 4.2. Compact disc formats emerging either from the same standard or from a combination of already established standards.

decode the text embedded in the control (subcode) bytes. It is also possible to use this information for implementing menu-based features, like language or artist selection. One of the most important aspects of the CD-Text format is its compatibility with the Interactive Text Transmission System (ITTS), commonly referred to as teletext[21], which allows the display of up to 21 lines of 40 colour alphanumeric or graphic characters each.

The **Compact Disc Read-Only Memory (CD-ROM)** format was also elaborated by Philips and Sony, who extended the work and expertise they already had on CD-DA to an optical disc system for computer applications. The original CD-ROM standard[94] was completed and officially submitted to international organizations in 1984. The first CD-ROM players hit the market one year later as peripherals for large computer systems. The International Standard Organization (ISO) and the International Electrotechnical Commission (IEC) adopted the CD-ROM format as a standard in 1985, the corresponding and now updated document[33] being known as the **Yellow Book**.

The most important upgrade from CD-DA to CD-ROM was the addition of a second pair of cooperating Reed-Solomon codes to improve the reliability of the readout information. This modification can better be understood when first having a look at a feature specific to audio compact disc systems. Called data concealment, this feature complements the error detection and correction functions when they fail to deliver correct data. An erroneous audio sample is then approximated through linear interpolation between its preceding and succeeding neighbors. Should the interpolation also fail because many consecutive samples are declared in error, the concealment circuitry will hold the last correct audio sample for several clock cycles. In either situation of sample interpolation or hold, the audio degradation perceived by an average listener is not significant. The data concealment can therefore be regarded as a trade-off between unnecessarily muting the audio stream and a slight degradation of the sound quality. In the case of computer data, however, data concealment is clearly not allowed. At the beginning of 1980s, the computer industry was already requesting bit error rates below 10^{-13} when measured at the host interface level. This data reliability requirement was equivalent to one erroneous bit delivered during the playback of more than 180 fully-recorded compact discs. It was this reason for which the data bytes collected from 98 consecutive frames in a manner similar to collecting the subcode information were set to form yet another error correction matrix. Two RS codes with their corresponding parity symbols could then operate upon this matrix and perform additional error detection and correction functions, increasing therefore the data reliability. In addition, 2048 bytes out of $98 \times 24 = 2352$ formed a sector (the rest being used for sector synchronization, parity symbols, etc.), which represented a convenient unit to operate with in computer data storage. Obviously, only the 2048-byte sector would have to be transferred through the host interface, which

Parameter		Value	Unit
Outer diameter of the disc	8-cm	80 ± 0.2	mm
	12-cm	120 ± 0.3	
Diameter of the center hole		$15^{+0.1}_{0}$	mm
Disc thickness		$1.2^{+0.3}_{-0.1}$	mm
Thickness of the transparent substrate		1.2 ± 0.1	mm
Disc weight	8-cm	6 . . . 16	g
	12-cm	14 . . . 33	
Maximum disc unbalance	CD-DA	10	g·mm
	CD-ROM	7	
Wavelength of the laser light		780 ± 10	nm
Numerical aperture of the objective lens		0.45 ± 0.01	–
Refractive index of the transparent substrate		1.55 ± 0.1	–
Maximum substrate birefringence		100	nm
Minimum disc reflectivity		70	%
Track pitch		1.6 ± 0.1	μm
Maximum track eccentricity		± 70	μm
Starting diameter of the program area		$50^{0}_{-0.4}$	mm
Maximum diameter of the program area	8-cm	75	mm
	12-cm	116	
Reference scanning velocity		1.3 ± 0.1	m/s
Channel bit length		278 . . . 324	nm
Typical pit depth		140	nm
Recording density		$207.8 \cdot 10^6$	bits / cm^2
Recording efficiency (audio applications)		32.65	%
Recording efficiency (CD-ROM, Mode 1)		28.43	%

(continued on the next page)

Table 4.1. System parameters of the read-only compacy discs.

sets the user data rate in a CD-ROM system to $2048/1024 \times 7350/98 = 150$ kB/s at the reference constant linear velocity.

Although still spinning at the speed of an audio disc at the end of the 1980s, the CD-ROM would later be required to deliver its data much faster in computer environments. One of the essential parameters of an optical disc system is the data rate at which the user information is retrieved from the disc. This parameter can be exactly specified if the recorded information is correlated with a reference linear velocity v_0 at which the laser beam should scan the spiral track. Both v_0 and the data rate are defined by the disc standards and depend on the requirements for continuously streaming digital information in

(continued from the previous page)

Parameter		Value	Unit
Stoarge capacity (audio applications)	8-cm	171 . . . 232	MB
	12-cm	606 . . . 807	
Stoarge capacity (CD-ROM, Mode 1)	8-cm	149 . . . 202	MB
	12-cm	528 . . . 703	
Maximum playback time at 1X CLV	8-cm	17 . . . 24	min
	12-cm	60 . . . 80	
Channel bit rate at 1X CLV		4.3218	Mbit / s
User data rate at 1X CLV (audio applications)		172.3	kB / s
User data rate at 1X CLV (CD-ROM, Mode 1)		150.0	kB / s
Channel clock period		231.4	ns
Maximum jitter during readout		35	ns
Maximum length of a correctable defect on disc		2.29	mm
Maximum block error rate before CIRC error correction		0.03	–
Maximum bit error rate after third-layer error correction		10^{-12}	–

Table 4.2. System parameters of the read-only compact discs.

various applications. In computer environments, however, variable data rates can easily be handled and it is very well possible to increase the data throughput by spinning the disc faster if possible. The ratio between the linear velocity v at which the spiralled data track is scanned in practice and the reference velocity v_0 at which the optical disc is specified is usually called overspeed or X-factor[126–128]. Tables 4.1 and 4.2 display several parameters at the reference velocity commonly denoted by 1X. Later specifications produced for other CD types would purposely introduce higher overspeeds like, for example, 8X or 32X, and these will be addressed later in this chapter.

The particular manner of organizing data on CD-ROM media in 2-kB sectors protected by Reed-Solomon codes is called **Mode 1** and represents the most used format in computer applications. Accordingly, between 528 and 703 MB of reliable computer data can fit on a CD-ROM disc, with 650 MB and 74 minutes being commonly promoted as typical storage capacity and playback time, respectively. Two additional formats, namely **Mode 0** and **Mode 2**, are also defined to indicate the unused areas on the disc and to provide 2336 user bytes, respectively. The former contains only digital zeros preceded by a synchronization pattern and a header. Mode 2 allows the user to fill in the available 2336 bytes in a convenient manner, without having any obligation to protect them by means of error detection and correction codes. It is the number of $2336-2048=288$ bytes that makes the difference between Mode 1 and Mode 2 CD-ROM media. Combinations of these these two formats on one disc also exist, especially for those applications that need a mixture of graphic information and reliable computer data.

Nevertheless, a player or drive capable of only optically reading out standardized CD-ROMs did not suffice for computer applications. The corresponding user data was also required to support specific identification techniques and to be organized on disc according to well-defined rules. Many CD-ROM drive manufacturers started in the early 1980s to introduce file descriptions similar to those currently based on directory trees. Not all these descriptions, technically called file systems, were compatible with each other. The situation degenerated to such an extent that computers had to be restarted and loaded with another file system description when not the default CD-ROM player but a different one, already connected to the same computer, had to be used. The proliferation of proprietary file systems determined several industry representatives to adopt common definitions that are nowadays referred to as the High Sierra format. This denomination simply recalls the High Sierra Hotel in Nevada, U.S.A. where the discussions took place in 1986. Two years later, an updated version of the initial proposals from 1986 was converted by the International Standards Organization into a standalone document addressing the **ISO 9660** CD-ROM file interchange format[71]. This standard describes a file system which does not depend on the application itself. Accordingly, reliable computer data recorded in Mode 1 format as well as data containing some special graphical information (Mode 2) can be found on disc by means of a root directory and a path table containing the addresses of all files. The file structure was originally developed for personal computers running MS-DOS and it failed to support other operating systems, like UNIX. The solution was found very soon and consisted in adding a so-called CD-ROM extension to the operating system, which led to the successful interfacing between any computer and a CD-ROM drive playing back ISO 9660 media. The CD-ROM extension for UNIX is known as **Rock Ridge Interchange Protocol (RRIP)** and its counterpart for the Microsoft Windows operating systems bears the name **Joliet**.

Another compact disc format, called **Compact Disc interactive (CD-i)**, was originally proposed by Philips and Sony in 1984 but the first CD-i players were introduced on the market only in 1987. The corresponding standard[93], commonly designated as the **Green Book**, addresses an optical medium carrying digital audio, static text and images, as well digitized video information. The Green Book also defines a complete hardware system, which is built around the microprocessor 68000 developed by Motorola Inc. and is rigorously needed to play back the disc independently from any other readout equipment. The description of the OS-9/68000 real-time operating system[81] developed by Microware Systems Corp. is also part of the Green Book. The only additional electronics which would also be required during the CD-i operation is a TV set to display the static and moving images. As for the second half of the 1980s, the CD-i contents represented a very good trade-off between the quantity and the quality of mixed digital audio and video information stored on an optical disc. Quite significantly, it is considered nowadays that CD-i media opened the way toward multimedia applications with their specific user-friendly interfaces and a high degree of interactive functions.

The CD-i defined a complete interactive environment that included both the application and the user data, own file and directory structures, choices for pointing devices and keyboard, etc. The stand-alone hardware specified by the Green Book differed from other compact disc players because it used a built-in computer and a dedicated, also standardized, operating system. They provided real-time operation, a requirement that is essential for many multimedia functions. In a sense, the CD-i represented the predecessor of the many current game consoles based on optical media. It is also important to mention that full-motion video data was encoded on a CD-i disc according to the MPEG-1 standard[38–42] defined by the Motion Picture Experts Group (MPEG). At the optical channel level the same EFM and CIRC techniques were used and the physical parameters of the optical readout, including the disc itself, did not differ from those standardized by the Red and Yellow Books. A modification with respcct to the audio CD, however, was the choice of using either PCM to convert the analog audio signal into digital values or the adaptive differential pulse code modulation (ADPCM). By coding only the magnitude difference between successive samples and adapting the code to accommodate the magnitude changes, the latter is able to store more digitized information within a given disc area. The number of stereo audio channels could thereby be increased to eight, while also reducing the number of bits per sample and the sampling frequency. Accordingly, the Green Book defined three audio quality levels to be chosen from during encoding. Due to the compression techniques used for encoding the audio and video streams, the CD-i storage capacity can be considered remarkable for the end of the 1980s. A CD-i disc could hold up to 19 hours of monophonic sound, about 7500 still images or, alternatively, more than 70 minutes of full-motion video. The strength of CD-i, though, was given by the interactive combinations of audio and video multimedia files and by the possibility offered to commercialize these combinations, including games, on separate discs. Notwithstanding, in a consumer market that was increasingly dominated by personal computers, the CD-i system had difficulties to establish itself as a desired product and had practically ceased to exist by the end of the past century. Soon after the Green Book was standardized, some discs containing both CD-i and CD-DA data became available. On these media known as **CD-i Ready**, the digital audio was located near the central hub and started at the logical track one. A number of tracks were therefore complying with the Red Book and could be played back on the existing audio equipment. However, in order for the CD-i operating system to recognize and start playing the disc, the CD-i Ready format made use of the so-called Index 0 area of Track 1, where Green Book information was recorded. According to the CD-DA standard, each track begins with a pause, which is designated as digital silence and is marked with the index 0. The digital audio data then starts within the same track at index 1. Most audio players manufactured in the 1980s skipped the digital silence of the first track and, hence, did not see the special information replacing the Red Book digital silence at Index 0 on the first track. Under these circumstances, the same disc could be used in CD-i multimedia systems as well as in legacy audio players, providing a reasonable merge between these two categories of products.

With the proliferation of both personal computers and the associated multimedia software, it became theoretically possible to play back CD-i media on CD-ROM drives and run the corresponding applications on PCs. To solve this issue, Philips and Sony proposed in 1988 the extended architecture format for CD-ROM. Microsoft Corporation rapidly adhered to these proposals as they pioneered already multimedia applications on personal computers. The adopted standard[98] is currently known as an **extension to Yellow Book**. It basically specifies a sort of boot program on the **CD-ROM Extended Architecture (CD-ROM XA)** disc, which runs on the computer to which the CDROM player is attached and universally provides the software interface toward CD-i applications. Note that the computer itself is responsible for the decoding in software and/or hardware of the digital audio and video information.

The introduction of the CD-ROM XA standard, however, did not guarantee the backward compatibility of these discs with the already existing CD-i format. Quite funny, these two compact disc formats were dragged into a sort of chicken-and-egg question to be answered that led to the impossibility to make use of CD-ROM XA interactive applications on the installed base of CD-i players. Recall that the former were standardized to allow CD-i applications being run on personal computers. Since the CD-ROM XA media were seen as important carriers for many multimedia applications on PC, the compatibility with the existing stand-alone CD-i equipment, also used for entertainment, was regarded very much as a marketing strategy. A new standard[97] defining the so-called **CD-i Bridge Disc** was introduced by Philips and Sony in 1991, with the latest version[99] being published in 1995. Essentially, the bridge disc was defined to hold a small CD-i program that would be executed when the disc is first mounted on a CD-i player. A similar role was fulfilled by the boot program previously added to the original CD-i format, which increased for a while the users' confusion but solved the mutual compatibility between CD-ROM XA and the CD-i boxes in living rooms.

By solving the mentioned compatibility between two CD formats, the CD-i Bridge Disc introduced a sort of convergence between computer-based applications and stand-alone entertainment solutions. For this reason, and because the CD-i has practically disappeared as a multimedia system, the standard[97] set in 1991 and known as the **White Book** is very often said to simply define a **Bridge Disc**. The White Book promoted a certain amount of freedom toward users and allowed them to define their own application within the existing bridge disc specifications. Eastman Kodak Company of the U.S.A., for example, proposed together with Philips in 1992 the **Photo CD**, an interactive disc on which high-quality photographic images can be stored in six resolution levels ranging from 128×192 to 4096×6144 pixels. The two companies elaborated a set of specifications[4] that provide means to electronically scan a photographic film, digitally process the pictures, and subsequently depositing the resulting data on read-only or recordable media. As suggested already in Fig. 4.2, many professional applications make use of Photo CDs. The **Picture CD**, on the other hand, contains photographic images at only one resolution, namely 1024×1536 pixels, and is intended for the average user. The computer software needed to view the read-only Picture CD data is included on disc, as opposed to a stand-alone software package that is required for professional applications. Another bridge disc was proposed in 1993 by JVC and Philips Electronics and has become very popular in many Asian Countries. Known as **Video CD**, it holds more than 70 minutes of full-motion video with accompanying sound complying with the MPEG-1 audio/video encoding standard[38–42]. JVC and Philips were joined one year later by the Japanese companies Matsushita Electric Industrial Co., Ltd.* and Sony Corp. and finalized together an improved Video CD specification[78]. In a sense, the Video CD featuring the MPEG-1 parameters listed in Table 4.3 is regarded nowadays as the predecessor of the DVD-Video. Apart from the MPEG-1 files, the data structure on Video CDs also allows combinations of tracks recorded in CD-DA, CD-ROM XA, and CD-i formats and provides the possibility to store dedicated karaoke information.

A disc format that borrowed characteristics from both the Red Book and the Yellow Book also exists and is designated as **mixed-mode CD**. This disc contains one audio track located at the inner diameter and complying with the CD-DA standard, followed by any number of tracks between 2 and 99 holding computer data. Although lacking in sophistication, this mixed configuration has caused many headaches to the owners of audio players. Because most audio equipment was not designed to cope with CD-ROM data, the corresponding tracks could not be detected and consequently attempted to be played back. This led not only to annoying sounds but even to the possibility of destroying the loudspeakers. The so-called "track-one" problem was solved in 1995 when Philips and Sony released together with Microsoft Corp. the newest member of the read-only compact disc standards commonly designated as the **Blue**

*Matsushita Electric Industrial Co., Ltd. changed its name in October 2008 to Panasonic Corporation.

<table>
<tr><td rowspan="7">Video stream</td><td colspan="2">Encoding system</td><td colspan="2">MPEG-1</td></tr>
<tr><td colspan="2">Data rate</td><td colspan="2">fixed (1.15 Mbit/s)</td></tr>
<tr><td colspan="2">Packet size</td><td colspan="2">2324 bytes</td></tr>
<tr><td colspan="2">Television system</td><td>NTSC</td><td>PAL</td></tr>
<tr><td colspan="2">Frame rate [Hz]</td><td>29.97</td><td>25.00</td></tr>
<tr><td rowspan="2">Resolution [pixels]</td><td>Still picture</td><td>352×240
704×480</td><td>352×288
704×576</td></tr>
<tr><td>Full motion</td><td>352×240</td><td>352×288</td></tr>
<tr><td rowspan="6">Audio streams</td><td colspan="2">Encoding system</td><td colspan="2">MPEG-1, Layer II</td></tr>
<tr><td colspan="2">Number of streams</td><td colspan="2">2 mono or 1 stereo</td></tr>
<tr><td colspan="2">Surround sound</td><td colspan="2">Dolby ProLogic</td></tr>
<tr><td colspan="2">Sampling frequency</td><td colspan="2">44.1 kHz</td></tr>
<tr><td colspan="2">Data rate</td><td colspan="2">fixed (224 kbit/s)</td></tr>
<tr><td colspan="2">Packet size</td><td colspan="2">2324 bytes</td></tr>
</table>

Table 4.3. Characteristics of the data streams recorded on Video CD media.

Book[80]. This specification explicitly addressed the issue of mixed-mode CDs by defining multiple sessions on disc, with each session representing a distinct recording that can be played back individually. As will be seen later, the idea of having multiple sessions on a disc originates from the Compact Disc Recordable format introduced at the beginning of the 1990s. The Blue Book specifies the **Enhanced Music CD**, also known as **CD-Extra** or **CD-Plus**, which contains one session recorded with digital audio and a second session formatted with CD-ROM Extended Architecture data. The Enhanced Music CD can, hence, store graphics, music, full-motion video, etc. and provide thereby support for multimedia applications on computers. The audio players, on the other hand, only recognize the first CD-DA session and will not proceed any further during playback.

4.3. Recordable and Rewritable CDs

The read-only audio and data compact discs were already established as very successful products by the end of 1980s. By contrast, the technologies needed to record and erase the digital information on optical discs were still being studied by several research laboratories, but a breakthrough toward cheap consumer products was not yet expected. However, the general feeling was that the market introduction of a recordable compact disc system could not lie far ahead into the future.

Among the technologies suitable for implementation in the next generations of compact disc drives, magneto-optical (MO) recording was considered to have a significant development lead. The MO media could be manufactured at a sufficiently low price and without considerable production efforts. The recorder electromechanics and optics appeared to have a reasonable complexity, which could be compared with their counterparts in read-only compact disc systems. In fact, several manufacturers of computers and dedicated peripherals at that time were already aligning their efforts toward producing MO data drives. Still available at present in some countries, magneto-optical media can be written by a laser beam that increases locally the temperature of a dedicated recording layer while a magnetic field realigns the molecular structure of that particular heated spot. A similar procedure is employed to erase the written data. The readout process takes places only optically, as the molecular structure previously mentioned modifies the electromagnetic properties, namely the polarization, of an incident laser beam. In 1990, Philips and Sony proposed this MO technology for writing, erasing, and reading of 12-cm discs formatted according to the ISO 9660 specification. The new **Compact Disc Magneto-Optic (CD-MO)** was submitted for standardization in 1990 and became the **Orange Book, Part I**. This standard defined two types of media: one containing both a read-only and an MO area, and another type that only featured rewritable magneto-optical fields[95].

Obviously, the physics of a CD-MO disc was incompatible with the existing read-only compact disc formats. Because of this incompatibility, the Orange Book was consequently extended with its **Part II**[96] that defined a write-once optical medium based on a recording technology also developed at the end of the 1980s but totally different from MO. Independent of these developments, Sony Corporation pursued the CD-MO standard toward introducing in 1992 a successful product called **MiniDisc (MD)**. While still using the same laser wavelength for readout, the same channel modulation code (EFM), and the same CIRC error detection and correction technique as in audio CDs, the MiniDisc employs an advanced audio compression method. Called Adaptive Transform Acoustic Coding (ATRAC), the 16-bit PCM samples also obtained at 44.1 kHz sampling rate are processed by separating first the audible spectrum in 52 frequency bands according to psychoacoustic principles[129, 131]. This mechanism provides means to dynamically reduce the number of bits by which the audio samples are represented and finally stored on disc, while hardly affecting the quality of the reproduced sound. As a result, 80 minutes of stereo music can fit on a 64-mm rewritable disc. The ATRAC technique has been continuously improved by Sony and, currently, its third and fourth generations are implemented in the MiniDisc systems still on the market. It is also possible nowadays to record in the so-called MDLP (that is, long play) mode, which compresses up to 320 minutes of music on a conventional disc.

The Orange Book, Part II specification[96] currently in use defines a write-once (WO) technology based on recordable but not erasable media. After being written, the discs can be played back directly and practically indefinitely in the compact disc equipment already existing on the market. The media were first designated as CD-WO but this name was soon superseded by **Compact Disc Recordable (CD-R)**. During the recording process, a high-power laser determines irreversible modifications in the structure of an organic layer stacked behind the transparent substrate. At the incident positions the absorbed light energy produces marks that alternate with the non-exposed land areas along the circumferential data spiral. During playback, the areas previously illuminated by the high-power laser spot (that is, the marks) appear like pits for the readout beam and modulate both the intensity and the phase of the reflected light.

The Orange Book does not restrict the usage of CD-Rs to particular applications but provides the compatibility means of a user-written disc with the existent read-only formats. The user data desired to be recorded must then be processed according to the CIRC and EFM rules, control and synchronization symbols should be added, and the channel bit clock should be equal to 4.3218 Mbit/s. The parameters of the recordable media and of the optical recording process are defined in such a way that, in principle, a compact disc player does not see any essential physical difference between recorded CD-Rs and read-only counterparts. At the logical level, however, the CD-R format will disclose its identity through several bit settings included in the control data stream. By using adequate computer software it is possible for the user to record any compact disc format.

Besides the physical parameters that guarantee the compatibility with read-only discs, the Orange Book also defines the configuration of a 3-dimensional structure called groove. The groove extends in a helical fashion from the inner to the outer diameter of the disc and features a slight sinusoidal deviation from its geometrical middle axis as well. The presence of the spiralled-undulation engraved on the blank disc helps the laser beam stay on track during recording, facilitates the generation of the write clock used to handle the data to be written at regular intervals, and contributes to the identification of the empty spots on the blank disc where data must be written. The common denomination for this relief structure on disc is wobbled groove.

A concern that had been expressed before the standardization of the CD-R format took place was related to the protection of copyrighted material recorded on the already available read-only discs. To prevent the illegal duplication of prerecorded content distributed on audio CDs, the music industry requested the implementation of specific copy protection measures to be obeyed by media as well as drive manufacturers. However, sophisticated digital technologies to satisfy the content providers were not readily available. Taking also into account the desired CD-R backward compatibility with read-only compact discs, there was not too much room left in the new physical and logical format to accommodate extremely efficient copy protection measures. Nor had anyone envisaged the magnitude of today's illegal, commercially-oriented duplication activities. A relatively simple copyright management system was consequently adopted and included in the Orange Book, but it was believed at that time to fulfill the requirements of the music industry.

Yet another significant feature introduced by the Orange Book Part II is called multisession recording. Previously standardized read-only CDs could only have one session of digital audio or data, flanked by some descriptor areas called lead-in and lead-out. A table of contents (TOC) describing the location of each recording unit (or track) was always embossed in the leadin area. By analogy with read-only media, the single session recording, also called disc-at-once (DAO), implies writing the entire contents of the disc (i.e., lead-in, data, and lead-out) exclusively during one uninterrupted process carried on by the CD-R drive without any pause. At the end of this process, the disc is said to become finalized and receivcs a table of contents that indexes the cumulated tracks. It is this final step that turns the CD-R into a media compliant with one of the read-only standards. However, a disc recorded at once may end up with a lot of unused storage capacity while the user would probably like to append more data at a later stage. This potential efficiency problem is addressed by another recording strategy, called session-at-once (SAO), that allows the user to write separate sessions and finalize the disc at a convenient moment later in time. This technique also permits the user to remove the disc before being finalized and continue the recording process, in principle, on another CD-R drive. Finally, a third recording method is called track-at-once (TAO) and supports the sequential recording of separate data units organized in tracks. A session does not have to be closed immediately, which means that more tracks can be added at a later stage. Combinations of track-at-once and session-at-once recordings are also allowed. Once the user decides to close the last open session and finalize the entire disc, the information about all written tracks is collected from temporary locations and compiled into a definitive table of contents. The disc becomes readable in legacy non-recordable players and computer drives only after having received this final TOC. By contrast, recordable drives are needed to decipher the unfinished sessions although some players may have built-in capabilities to read not-finalized CD-R discs.

As a matter of fact, because the concept of multisession recording did not exist before the release of the Orange Book, the installed base of read-only players and drives in the early 1980s could not handle the newly introduced

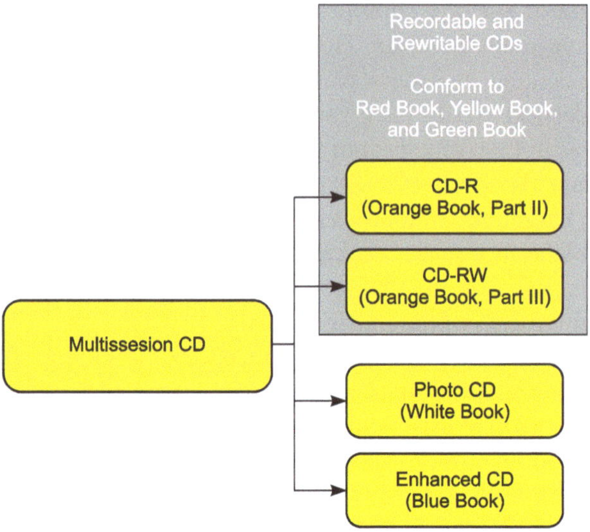

Fig. 4.3. CD formats which may contain data arranged in more than one session.

media. The old compact disc equipment was programmed to search only for one lead-in and the corresponding lead-out and, hence, only identified the first session on disc. At a later stage, however, most manufacturers of stand-alone players and computer drives implemented firmware changes that finally led to the compatibility between their released hardware and the already existent recordable multisession media. The optical storage community also realized that having more than one session on a compact disc, irrespective of its read-only or recordable format, would represent a general feature. The Photo CD and Enhanced Music CD previously discussed are multisession read-only discs as indicated in Fig. 4.3. From this perspective, a unified approach of this feature that would not depend on the compact disc type appeared as necessary. Philips and Sony proposed in 1995 a separate standard[100] for the **Multisession CD**, which was adopted immediately by the optical storage community and has served ever since to guarantee the full compatibility between CD players or CD-ROM drives and the new multisession media. It is also worthwhile to mention that the effective storage capacity of multisession discs decreases with about 13 MB per session due to the additional space allocated to individual lead-in and lead-out areas.

Another recordable CD system was proposed by Philips and Sony at the end of 1996 but the corresponding optical disc drives were only introduced one year later by Ricoh Company, Ltd. of Japan. The standard, commonly designated as the **Orange Book Part III**[101], defines a **Compact Disc ReWritable (CD-RW)** which is similar to CD-R in many respects. The essential difference is made by a recording layer whose internal structure can be frozen locally to either a crystalline or an amorphous state. The transitions between these two states take place by irradiating and consequently heating the layer with well-defined energy levels of the incident laser beam, followed by a controlled cooling. The corresponding technology is called phase-change recording. The amorphous areas behave like the pits embossed on read-only discs and return an insignificant amount of light to the photodetectors during the optical readout. By contrast, the crystalline regions return sufficient incident light to be assimilated with the highly-reflective lands on read-only media. Either aggregation state can be reversed, which allows the user to record and erase data or directly overwrite the previously written areas on CD-RW discs. One may safely erase and rewrite these discs several hundred times, with a theoretical upper limit of about 1000 direct overwrite (DOW) cycles. From the standpoint of the hardware manufacturers, note that not all CD stand-alone players and computer drives could straightforwardly read CD-RW discs when the latter became available. The reason for this incompatibility was the reduced reflectivity of these discs as compared with all other read-only CDs or with CD-Rs. The latter reflect 50-75% more incident light than their rewritable counterparts. Minor modifications that assumed the scaling of some signals and additional gains introduced in the servo and read channels, as well as more sensitive photodetectors, have provided the playback compatibility of CD-RW media with practically all current CD equipment.

One very important and also useful feature introduced by the Orange Book Part III was the so-called packet writing. Thereby have the restrictions to record data in disc-, session-, or track-at-once modes been superseded be the possibility to record smaller units of data called packets. The length of a packet may either remain fixed or vary depending on the amount of data to be written, with the smallest packet size being equal to 64 kilobytes (1 kB = 1024 bytes). Note that a data track must contain at least one packet and mandatory run-in and run-out blocks are required to link the consecutive packets to each other. A pair of run-in and run-out blocks takes up 14 kB of user space on disc, which may reduce the total storage capacity on CD-RWmedia from 650 MB to about 533 MB in the worst case, when the entire disc is recorded with fixed packets of 64 kB. A track may remain continuously open, providing thereby means for a host computer to directly add and erase data on CD-RW media in a manner similar to writing and overwriting the information on floppy and hard-disks. For consistency reasons, the packet writing technique was added to the last versions of the Orange Book Part II, although most of the users record CD-R discs only in disc-, session-, or track-at-once modes.

A particular aspect of writing recordable and rewritable compact discs is related to the compliance of the various CD-R/RW format features with the rapidly evolving personal computers. A host system must be able not only to recognize the attached drive, but to write multiple sessions, format the rewritable media, etc., operations that essentially require a dedicated file system. For these purposes, a standard[43–47] has been developed and updated repeatedly since the introduction of the CD-R recorders. A subset of the file structures specified by this international standard has been redefined by the Optical Storage Technology Association (OSTA). The new proposal, called **Universal Disk Format (UDF)**, has been developed to maximize the data interchange and minimize the cost and complexity associated with the file system implementation. The UDF provides translation algorithms for many operating systems running on small or large computers and has evolved from its version 1.0 introduced in 1995 for CD-R/RW to subsequent versions able to handle other optical media[91].

After the introduction of the CD-R and CD-RW, the series of related standards has further been extended with several additions to the Orange Book. Driven by the fierce market competition, the manufacturers of recordable and rewritable drives did not cease to increase the recording speed. In a manner similar to playing back CD-ROM media at overspeeds between 1X and 48-56X on current drives, the today's CD-R/RW systems have achieved the performance of finalizing a complete CD-R disc within less than 2 minutes and a CD-RW disc within 3-4 minutes. These figures correspond to recording speeds that exceed the reference velocity by 52 and 32 times, respectively.

The Orange Book was first updated for write-once media because the dye materials appeared extremely suitable for recording at high speeds and the drive vendors did not hesitate to use the available technologies to release new, faster drives about twice a year. Version 3.1 of the Orange Book Part II provides standardized write strategies for CD-R systems operating in constant linear velocity mode at the overspeeds 1X, 2X, and 4X[103]. A subsequent addition to this document was released in 2001 and pushed the standardized recording overspeed to 16X and was labeled as **Volume 2: Multi-Speed**. For the sake of correctness, version 3.1 of Part II has also received the additional title **Volume 1: 1x/2x/4x** and became in 2005 version 3.2[111]. Note that a second power calibration area has been defined for optional use at the outer disc diameter, which allows the drive to better calculate the laser power required during writing at different speeds throughout the disc.

By the end of 2001, Volume 2 was upgraded to cover also the specifications for recording at 20X, 24X, and 32X for both 8- and 12-cm media. The most recent version of this volume[113] standardizes the **Compact Disc Recordable Multi-Speed** up to 48X while preserving the backward compatibility with Volume 1. It is very important to mention that all high-speed CD-R media are backward compatible with the installed base of low-speed recorders. It is therefore possible to write any disc at a user's convenient constant linear velocity (CLV) that does not exceed the maximum one for which the disc was certified. For this reason, many recorders are also capable to write in constant angular velocity (CAV) mode by continuously adapting the write strategy while the laser spot scans the groove from the inner to the outer radius. Note, however, that no official CAV specification exists for CD-Rs mainly because it is relatively difficult to guarantee the compatibility of the finalized discs with the installed base of legacy players and CAV recording was consequently seen in the beginning only as a drive feature. The standardization of the CAV mode was solved for the first time in an upgrade of the CD-RW specifications to be discussed shortly. Philips and Sony announced in 2002 that no further standardization would take place for higher CD-R recording speeds despite the availability on the market of drives and CD-R media capable to write at 52X and even at 56X. Stressing the system beyond 48X does not pay off in terms of customer satisfaction (i.e., significantly shortening the recording time), while relatively unsafe operating electromechanical conditions are approached by increasing the spinning rate toward 200 Hz.

The upgrade of the Orange Book Part III followed similar milestones as previously described for the revision of Part II. The CD-RW discs typically used at 1X were first provided with the correct recording and erasing parameters for two more overspeeds: 2X and 4X. This upgrade took place in 1998 by releasing the version 2.0 of the Orange Book Part III. Further efforts to increase the recording speed on thc existing CD-RW media showed that different write strategies than adopted up to 4X would be needed. This was an important and very decisive discovery since new media would have been rendered obsolete by the installed base of low-speed CD-RW recorders. Nevertheless, this conclusion did not hamper the upgrade of the Orange Book Part III and **Volume 2: High-Speed**[108] was published in 2000 to cover the recording and erasing between 4X and 10X in CLV mode. This standard has also been updated later and finally arrived in 2006 at version 1.2. Although not explicitly stated in the standard, the CAV operation could also be used and was adopted by the drive manufacturers when the appropriate hardware became available. With the advent of the second volume dedicated to higher recording speeds, version 2.0 of the initial Part III specifications from 1998 received an additional title: **Volume 1: 1x/2x/4x**[114]. The prospects for increasing the writing performance beyond 10X appeared also in sight since the 4X barrier was now surmounted by adopting a write strategy different from the one specified in Volume 1. In 2002, Philips and Sony released the **Volume 3: Ultra-Speed**[109] of the Orange Book Part III and pushed the writing and erasing capabilities on phase-change compact discs up to 24X. The new standard also stated explicitly that ultra-speed CD-RW media can be written in CAV mode while remaining compatible with the write strategies specified in Volume 2. However, the incompatibility of 4X-24X discs with the specifications[104] that were available prior to the introduction of Volume 2 could not be solved at that particular time due to technical problems related to phase-change materials. A later format upgrade added an offset to the embossed, logical addressing scheme of these discs to prevent the old drives from attempting to write on high-speed and ultra-speed CD-RWs. The new media, that also bear specific logos, have become thereby unrecognizable in old recorders and are protected thus against inappropriate use. The final specifications[117] were published in 2006.

In addition to the recordable and rewritable CD standards previously discussed, yet another important document was prepared by a group of four companies to facilitate the use of CD-RW media in computer applications. Entitled **Compact Disc Mount Rainier Rewritable (CD-MRW)**, or shortly Mt. Rainier specifications, this document[3] had been worked out since the end of 1999 by Compaq Computer Corp. and Microsoft Corp., both of the U.S.A., together with Philips and Sony and released in 2001. More than 35 other companies expressed their immediate support for the Mt. Rainier specifications, which aimed at replacing the floppy disk by CD-RW media starting from 2002. This goal in itself was considered in some sense very ambitious, but when reformulated it meant nothing but seamlessly writing CD-RW media in a drag-and-drop fashion and without the use of additional software. The certified implementation of the Mt. Rainier specifications can be recognized by the "EasyWrite" logo emblazoned on the front panel of CD-RW drives, but nowadays many such devices do not even display the logo anymore since it has become a de facto implementation for quite some time. The key improvements with respect to the Orange Book Part III emerge from a set of physical formatting requirements that, if fulfilled by the manufacturers of CD-RW peripherals, provide defect management capabilities and increase thereby the data reliability. Note that compliant changes are also needed in the recording software and in the operating system running on the host computer.

To begin with, after the CD-RW introduction most operating systems could not handle directly the operations performed on these media, which was not the case at that time when writing and erasing a hard-disk or a floppy disk. For this reason it was necessary to install a dedicated application (software) to translate the read/write requests into UDF-based commands interpretable by the CD-RW drive. It was the Mt. Rainier document which requested native operating system support for all CD-RW functions, which is a default feature at present on practically any computer. A second very important requirement compelled the CD-RW drive manufacturers to build defect management into their products. Thereby all defect areas on the disc should be bypassed in a transparent mode for the operating system, which had always been a mandatory feature in hard-disk drives. A third proposal aimed at reducing the shortest data slice to be addressed from 64 kB (equal at that time to the minimum size of a written CD-RW packet) to 2 kB. The proposal brought the rewritable CDs yet another step closer to hard-disks. The fourth important Mt. Rainier requirement obliged the CD-RW drives to format the media in the background, that is, without receiving any particular command from the application. The user would clearly benefit from this feature by being able to record immediately on newly purchased discs, while the drive itself would proceed with formatting during the idle periods. Finally, it was also desired to standardize both the Mt. Rainier set of commands issued by the operating system toward the CD-RW drive and the disc physical layout.

4.4. Miscellaneous CD Versions and Formats

During the relatively long evolution of the CD family, many other companies and organizations proposed and manufactured their own versions of compact discs. Obstacles like the incompatibility with the existing and successfully established CD standards, the reduced interest from the optical storage community, insufficient added value when compared with existing compact discs, etc., could often not be surmounted and many of the proposed media failed to become successful products. Several notable exceptions, however, have been accepted worldwide as complementary solutions to the internationally standardized CDs. Note also that several companies attempted and sometimes succeeded to obtain the international recognition of their disc format from a standardization forum.

By promoting compact disc formats different from the widely accepted ones, their supporters also hoped in some cases to convince the legal owners of CD standards about new features missed by consumers, which could eventually also lead to the co-participation in a new licensing program. The standards owners proved, many times, impervious to the arguments brought in favor of the new proposals and avoided further cooperation. In some cases, the standards owners even proceeded toward legally banning those products that could violate the consumer trust in the established specifications. A notorious example in this sense is the **Mono CD**, which made use of the two stereo channels defined in the Red Book to hold monophonic digital audio. Although featuring an increased playback duration, the Mono CD positioned the consumers back in time, before the stereo sound was even invented. Another example is given by a category of CDs produced in ingenious shapes sometimes not even resembling a standardized discoidal medium. Such media have been banned because their dynamic unbalance owing to the non-discoidal shape gives rise to heavy vibrations and can easily damage the players or drives operating at high rotational velocities.

The **Commodore Dynamic Total Vision (CDTV)** was introduced in 1991 by Commodore Business Machines of Canada as a multimedia system similar to CD-i. Despite some initial marketing success, the standardization efforts of Commodore eventually failed and the CDTV has not really become a consumer product. The proprietary file system used in this application was blamed for the failure, because it rendered CDTV media unreadable by IBM-compatible and Macintosh personal computers. It was these two types of PCs that drew a lot of attention toward the end of the 1980s, and adding the right interface to your own device was about to become the rule of success in the growing market of personal computers. It will be shown throughout this section that several other optical disc formats failed or needed to be modified because of not being readable in established computer environments.

A rewritable CD format known as **Tandy High-intensity Optical Recording (THOR)** was proposed in 1988 by Tandy Corp. of the U.S.A. but the technology was ahead of its time and never reached the consumer market. A significant drawback of the THOR-CD was its limited number of erasure cycles, about 100, but this could still be considered an achievement for the 1980s. Yet another failure was the **Compact Disc Read-Only Data Exchange (CD-RDx)** standard proposed at the beginning of the 1990s by the Central Intelligence Agency (CIA) of the U.S.A. From a file system viewpoint, this format was meant to become a common denominator for many computers and their operating systems, superseding the ISO 9660 specifications. The CD-ROM, however, was sufficiently well established at that time and the CD-RDx could not prevail by simply replacing an existing product.

A relatively successful compact disc format was the **hybrid CD-ROM**. These optical medium contained both ISO 9660 files for the operating systems developed by Microsoft Corp. (MS-DOS and the various Windows versions) and for the Macintosh Hierarchical File System (HFS) supplied by Apple Computer Inc.* of the U.S.A. with its computers. The international optical disc community showed some interest and almost accepted the hybrid CD-ROMs because they could provide important savings for companies delivering software solutions to various computing platforms. This argument failed to stay true during the years because all operating systems started to deploy software extensions that allowed ISO 9660 files to be universally handled. It is worth to note at this point that the nomenclature "hybrid" was first used for the type of CD-ROM media holding ISO 9660 as well as HFS files. Later, during the many years of optical disc history, several other discs were also called hybrid without having any relation whatsoever to the hybrid CD-ROM. Although no official body introduced a specific definition, the present consensus seems to be that a hybrid disc either provides more than one function on a single platter without obeying entirely a given CD/DVD standard or has a physical structure that usually belongs to more than one conventional media.

A step further was taken in 1995 when the **Bootable CD-ROM** format was defined by two engineers of the American companies Phoenix Technologies

*Apple Computer Inc. changed its name at the beginning of 2007 in Apple Inc. to emphasize its market strategy covering also consumer electronics products.

Ltd. and IBM, respectively. The specification is called **El Torito**[118] and provides means for a computer to start its operating system from a suitable CD-ROM disc, bypassing thereby other internal bootable devices. It follows that a CD-ROM drive must be first recognized without any prior loading of a dedicated driver and that special information must be located on disc to allow the automatic running of a bootable software sequence. The drive recognition can essentially be arranged to take place while executing the basic input-output system (BIOS) program. This program becomes then also responsible for issuing the commands needed for booting. However, the logical format promoted by the ISO 9660 standard lacked the bootable records on disc and did not suffice to respond the BIOS commands. It was the El Torito specification that provided the specific start-up support while still maintaining the backward compatibility with the High Sierra file interchange format. Multiple boot configurations can also been defined to allow the user to start one of the operating systems residing on the CD-ROM disc and ensure, in addition, the compatibility of one disc with IBM-compatible and Macintosh computers.

A compact disc format that emerged from the international standards but did not fully obey any one of them was known as the **CD Single**. It was introduced on the Japanese market in the early '90s as an 8-cm audio disc. The digital audio information recorded on a CD Single was in line with the Red Book, but it was organized at a higher level according to the ISO 9660 file structure.* The users of CD-ROM drives could, hence, easily handle their favorite audio titles just like working with any other computer data file. This feature became obsolete within several years, namely as soon as some support for recognizing and handling the multi-megabyte CD-DA tracks started to be incorporated into various operating systems. The subsequent development of small applications or even of complex software packages that turned the computer and its CD-ROM drive into an audio system eradicated definitively the need for CD Single media. At present, the difference between CD-ROM and CD-DA files is completely transparent for unaware users.

After the inception of the CD-i, several audio studios realized that more data could be stored on 12-cm digital audio discs when replacing the standardized PCM coding by ADPCM already adopted by the CD-i. Essentially, the ADPCM compression technique reduces the CD-DA data rate by 50-85% and allows thereby longer playback times than stipulated in the Red Book. Depending on the required reproduction quality, between five and ten hours of digital stereo sound can be provided at the expense of a dedicated ADPCM audio equipment. The corresponding discs, known since their introduction as **CD-Background Music**, have never convinced the consumers to spend additional money on audio sets for playing back more hours of music than CD-DA at no increase in

*Some people designate those 8-cm CD-DA discs which hold only one song as CD Single. These discs conform entirely with the Red Book and are not the subject of the current discussion.

the reproduction quality.

Another compact disc format that failed to establish itself as a successful product was the **CD-Video**. The format was brought forward by Philips Electronics in 1986 and represented a hybrid between CD-DA and LaserVision. At that time, the already existing Video CDs could not offer sufficient video quality, an attribute that was already associated with the analog recordings stored on LaserVision media. The audio quality, on the other hand, was related to the digitized sound stored on compact discs. In this context, the CD-Video was thought to achieve a compromise and was designed to accommodate 5-6 minutes of full-motion analog video for either PAL or NTSC television systems[15, 16] along with up to 20 minutes of digital audio complying with the Red Book. Originally proposed under the name Blue Book, the CD-Video standard did not gain market acceptance and disappeared completely. The name Blue Book was reused nine years later when the Enhanced Music CD specifications, having nothing in common with CD-Video, were endorsed by many companies.

A relatively successful disc format bearing the name **Phase-change Dual (PD)** was announced in 1990 by Matsushita Electric Industrial Co., Ltd. of Japan. Some sources[25] refer to the abbreviation PD as standing for "Powerful optical Disk system." Although similar to CD-RW from the recording technology standpoint, neither the physical parameters nor the logical data format were compatible with the established compact disc standards. A particular aspect of the proprietary data format on PD media was the use of embossed structures (i.e., sequences of pits and lands) for both addressing and synchronization purposes. The pits were formed along the phase-change grooves but these particular areas were forbidden for recording. Note that both the PD and CD-RW systems emerged from the same pool of research ideas, with Matsushita betting on being first on the recordable optical disc market and the CD-RW champions delaying their product for several years, until solving the backward compatibility issue. It was Panasonic, a brand name of the parent company Matsushita, which commercialized in 1994 the very first optical disc drive capable of playing back standardized compact discs as well as of writing and erasing 650-MB PD media. Significantly, the latter could not be used without a protective cartridge. The physical and logical PD formats together with the required cartridge are described in two documents[6, 65] approved in 1996 and 1997 by international organizations for standardization. Dual CD-ROM/PD drives are still being used in Japan but have hardly penetrated other world markets. The corresponding technology, however, has served Matsushita toward the development of a rewritable DVD format, called DVD-RAM. The high-density compact discs are optical media with deep roots in the CD history. Among them, the **80-minute CD** that holds 700 MB of data emerged from the existing standards by stretching several disc physical parameters to their lowest accepted margins. The 80-minute disc did not represent a new format by itself, but proved the capability of the media manufacturers to control the production process in a consistent and very accurate manner. By reducing the track pitch by 5-6%, reducing the start diameter of the program area by 0.3-0.4 mm, and decreasing the reference velocity (which determines the length of the channel bit) by about 7.5%, all considered with respect to the nominal values, more data could be fit on disc while still obeying the international standards. These media were initially manufactured as CD-DA and CD-ROM to prevent their complete copying on CD-Rs, with the latter being only capable to hold up to 70-74 minutes of data. This situation did not persist too long because the improvements made to the CD-R manufacturing process led to the market introduction of the **80-minute recordable disc**, an optical medium that was not disapproved and eventually would even be endorsed by the owners of the Orange Book Part II.

In an attempt to increase the storage capacity above 1 GB, several companies had tried already in the 1980s to substantially reduce the physical dimensions of the embossed patterns. The resulting discs could thereby hold more data but, unfortunately, the technologies needed for their mass production only became available during the past decade when these research activities also contributed to the development of the DVD. Several versions of high-density compact discs with playback times exceeding 80 minutes were promoted to consumers before the end of the 1990s. Note that only the 80-minute media discussed previously obeyed the international standards, though marginally, while any other type of long-play disc remained for a while unsupported by such documents. From a practical point of view, most CD players and drives were able to play back compact discs versions holding up to 100 minutes and featuring a track pitch slightly narrower than standardized by the colored books. Their recordable counterparts also existed in the form of **90-** and **99-minute CD-Rs** but only a few recordable drives on the market could write on such media. The main incompatibility reasons were given by the difficulties arising when tracking the tight grooves and by the existing hardware and software that could not identify tracks and addresses beyond the standardized 79 minutes and 59 seconds. In some cases, if written, the long-play CD-Rs also exhibited negative sector addresses or played back with a poor performance in read-only drives. Besides, such media could practically be recorded only at very low and safe overspeeds, while most data recorders on the market operated above 32X. For a while, the high-capacity CDs could not catch the consumers' attention, but an eventual rework of the existing standards and the opportunity to really satisfy the users was certainly retained by Philips and Sony.

A somewhat special category of high-density compact discs were the Video CDs that stored between 100 and 150 minutes of MPEG-1 video and were relatively wide spread in many Asian countries. These media had a reduced track pitch and shorter pit/land structures than allowed by the White Book, which led to storage capacities between 880 MB and 1.3 GB. However, because most of the players and drives manufactured according to the CD standards could cope with such discs, the 100-minute versions became a sort of threshold reference. In order to remain active in the Asian markets although at the expense of some additional costs, the hardware manufacturers have hardly had any choice but to guarantee the playback of at least some Video CDs, mainly of those not exceeding 100 minutes of video playback. Media versions with storage capacities above this threshold required dedicated optical disc engines for playback and have gradually disappeared.

Because the introduction of the digital versatile disc did not disturb at all the Asian businesses operating with cheap high-density CDs, the latter market continued to grow and remained very profitable. Before the end of the 1990s, several companies recognized the need of standardizing a CD format with a storage capacity exceeding 1 GB and, if possible, create read-only as well as recordable versions. Philips and Sony took the lead once more and proposed the **Purple Book**, a standard which defines the **Double Density Compact Disc (DDCD)**. Three separate specifications belonging to this new standard describe the read-only **DD-ROM**[121] and its **DD-R**[122] and **DD-RW**[123] recordable and rewritable versions, respectively. All three formats were derived from their 650-MB CD counterparts by reducing the track pitch with 31.25%, reducing the start diameter of the program area with 1 mm, and decreasing the reference velocity by 30%. At the channel clock level, however, both CD and DDCD media featured the same data rate equal to 4.3218 Mbit/s at the nominal 1X readout overspeed. The storage capacity of the 12-cm DDCD format lies between 1.24 and 1.36 gigabytes (1 GB = 1024^3 bytes) of reliable computer data. Accordingly, a DD-ROM disc can be played back for at least 147 minutes and 27 seconds and up to 154 minutes and 7 seconds. The user data is organized in 2048-kB sectors protected by an error correction matrix, which is very similar to the Mode 1 of the ubiquitous CD-ROM format. The physical conversion between user data and channel bits also resembles the EFM and CIRC schemes described in the Yellow Book, with one noticeable exception: the consecutive frames are spread across the CIRC error correction matrix with larger delays than was previously the case. More precisely, the new CIRC7 scheme was designed with an interleaving length of seven frames as opposed to four in the CD-ROM format. This modification, when correlated with a shorter channel bit length on DDCD media, led to a total length of a damaged track that could be completely corrected equal to 2.7 mm (versus 2.3 mm on compact discs). Sony was the first company to commercialize read-only DDCD drives at the end of 2000 and these devices were followed about half a year later by their recordable versions. The advent of the DVD, however, quickly rendered the DDCD media and computer drives systems obsolete despite the recognized international standardization support.

Almost simultaneously with the standardization of the double density compact disc, Philips and Sony also proposed a hybrid read-only media designated as **CD2**[124]. This disc featured one single-density audio session at the inner diameter to obey the Red Book, and one double-density session complying with the Purple Book. The two distinct spirals were separated by a narrow unwritten annular region. Significantly, the CD2 format contained provisions for an elaborated security scheme meant to protect the data encrypted in the second session. The decryption key, called physical disc mark (PDM), was embossed in the lead-in of the first session where the pits wobbled radially while also exhibiting sudden phase changes. Each individual sign inversion that occurs when the sinusoidal pit sequence changes its phase can be seen as a carrier of one bit of information. A possible application of the CD2 format was for distribution of copyrighted audio content. A dedicated audio player would then be needed to retrieve the secret key and consequently decrypt the information stored along the high-density second session. As the digital information is not output by the CD2 player, a potential attempt to replicate the copyrighted audio would have to rely only on a bit-by-bit copy, i.e., making an exact image of the original disc with a CD or DDCD recorder. However, illegally making such a copy was discouraged by the impossibility to replicate the wobbled pit sequence with consumer recorders. To increase the security scheme, part of the decryption key was stored in a small area on disc where the channel clock exhibits well-defined frequency variations instead of having the fixed frequency of 4.3218 MHz. These frequency variations encode the decryption information and are designated by the CD2 standard as the hidden channel. For identification purposes, the CD2 format also allowed the manufacturer to optionally burn a unique code in the unwritten annular region between tracks, which would have to be replicated using a very high-power laser (unavailable for hobby purposes) when attempting to illegally copy the disc. The identification code was anticipated to be used, for example, when distributing copyrighted content by means of electronic transfer via the Internet. The CD2 was well thought out technically, but never took off as a product.

An optical media format relatively similar to CD2 was introduced by Sega Enterprises, Ltd., of Japan in 1998. Prior to that date, Sega had already been working for several years toward its new game console, called Dreamcast, that needed more storage capacity than a simple CD-ROM could offer. As DVD had not gained sufficient acceptance by that time, Sega launched the double-session **Gigabyte Disk Read-Only Memory (GD-ROM)** that could hold up to 1.2 GB of digital data and was produced by several selected optical media manufacturers. The disc became immediately quite popular because the Dreamcast games pioneered new and very powerful graphics while Sega was already an established player in this market. The first session of a GD-ROM complied with the international compact disc standards and contained about 35 MB of computer data and raw digital audio. This part of the disc could be played back in any CD-ROM player, including Sega's old CD-based game machines, and allowed a potential user to enjoy for approximately 4 minutes a limited-content version of the main application. By contrast, the second session was recorded in a high-density physical format that could only be read out optically by dedicated Dreamcast players. This session could hold more than 110 minutes of Mode 1 computer data (a complete game as an executable file) and the accompanying Red Book digital sound arranged in separate tracks. It is worth noting that a GD-ROM disc was bootable in Dreamcast drives but, unlike the CD2 format, there was no provision for protecting the copyrighted information. In fact, illegally copying the disc was practically hampered by the large size of the main application that could not fit on one standardized CD-ROM and by the inherent difficulty of playing back Dreamcast media in legacy CD-ROM drives.

Yet another sort of high-capacity compact disc was developed by Sanyo Electric Co., Ltd. using the physical formats of blank CD-R and CD-RW media. Known as **HD-BURN**, Sanyo's discs could store 1.4 GB of user data when written on dedicated data drives commercialized from 2002 onward. To achieve a higher storage density, the lengths of all written marks were reduced by 25% with respect to their counterparts on CD-R/RW. In addition, the HD-BURN technology replaced the EFM channel modulation code and the CIRC error correction strategy, both standardized for CD systems, by more efficient counterparts specified at that time already for the DVD media. The HD-BURN technology aimed from its introduction at recording digital video streams in a format similar to that of DVD-Video (to be addressed later), but on cheap CD-R and CD-RW discs. The recorded media would be basically compatible with DVD players upgraded by firmware, but the interest shown for this technology by manufacturers of data drives, DVD-Video players, and users alike remained rather confined to equipment mainly commercialized by Sanyo itself and disappeared within a few years.

An interesting version of the audio CD format was introduced in 1995 by Pacific Microsonics, Inc. of U.S.A. under the appellative of **High Definition Compatible Digital (HDCD)**. The signal processing principles behind this format were described in several publications[11, 75] at that time, but the disc has never been standardized until now despite the current availability of more than 5,000 HDCD audio titles. It was claimed that by digitizing the audio signal with 24 bits at a high sampling rate (192 or 176.4 kHz versus 44.1 kHz in CD-DA) and consequently reducing these parameters through digital signal processing to 20 bits and 44.1 kHz, respectively, a superior audio fidelity could be obtained. These operations took into account the critical frequency bands specific to the human auditory system. The digital audio stream could then further be converted into the CD-DA 16-bit format by intelligently spreading the additional four bits throughout the least significant positions in the CD-DA samples. When played back on common compact disc players, no perceivable sound degradation was caused by the modified least significant bits. When playing back the disc in HDCD players, however, these bits permitted the reconstruction of the original signal. It was claimed that a much higher reproduction fidelity was obtained because previously misunderstood or unknown sources of distortion in digital audio could be identified and corrected. As one might expect, the quality improvement comes at the price of an advanced digital signal processor used to recover the digital data. HDCD recordings used to be popular in U.S.A. and Japan, although sales until now have been many orders of magnitude lower than those of standardized audio CDs.

The concept of a hybrid disc containing both an embossed read-only surface and an adjacent recordable area has always challenged the optical recording community. In response, Eastman Kodak Company of U.S.A. introduced the **Compact Disc Programmable Read-Only Memory (CD-PROM)**. After the inception of the CD-R and CD-RW formats, the Orange Book standards had theoretically and thus legally allowed the use of such hybrid discs and several companies had attempted to manufacture them[79]. Notwithstanding the efforts made by many media manufacturers, hybrid read-only/recordable discs have always been and still remain difficult to produce. Apart from some incompatibility issues between the various required technologies, the main challenge was set by obtaining a seamless transition between the addresses embedded inside the read-only data stream as subcode information and the addresses contained in the recordable wobbled groove. The solution found in 1999 by Kodak was based on a first session embossed with an interrupted groove, followed by the recordable area where the same groove extended in a continuous fashion. Although slightly undulated, the groove interruptions resembled the dull pits and shiny marks of read-only media and the detected RF signal remained fairly undistorted by the low-frequency wobble. Note also that a perfect synchronization could be obtained between the limited set of read-only subcode addresses and the entire sequence of addresses embedded into the helical wobble. Because the CD-PROMs are multisession discs, they are fully compliant with the existing data and audio players. The primary advantage, of course, originates in the possibility of customizing a purchased application supplied on CD-PROM by appending personal information written by users on legacy CD-R/RW recorders.

Following Kodak's example, Optical Disc Corp. (ODC) of U.S.A. also released a hybrid compact disc that obeyed the Orange Book, Part II specifications. The disc is known as **CDR-ROM** and became available to content providers at the beginning of 2003. Just like with CD-PROM media, the user may append own data to the prerecorded content commercialized on a CDR-ROM. ODC made use of a nonconventional manufacturing technology known as dye polymer mastering to produce stampers with significantly different pit and groove geometries. Like Kodak's hybrid disc addressed above, CDR-ROM media can be geared toward enabling content owners and users to better organize the information, but also toward safeguarding proprietary material from illegal use and/or distribution.

In the digital audio market, the MPEG-1, Layer III encoding has started to drive new trends within a few years from the beginning of the new century. Commonly called MP3, this technology makes use of sophisticated digital signal processing techniques to model the human auditory system, while achieving remarkable compression ratios of up to 12:1 (plain speech can be compressed up to 24:1). CD-DA tracks that would typically need several tens of megabytes can thereby be stored within 2-5 MB of an MP3 file. The audio community feels comfortable about archiving hundreds of audio titles on one compact disc, especially because not too much quality loss takes place to any but the most discerning listeners. The exchange and download of MP3 files through the Internet followed by their recording on CD-R/RWmedia (and later on solid-state removable memories) has also boosted the interest for this compression technology. These developments have finally led to the introduction of stand-alone CD/MP3 audio players. Several companies have also started to commercialize read-only optical media recorded with MP3 audio. One example is the **Digital Automatic Music (DAM) CD**, which contains CD-DA tracks complying with the Red Book along with their MP3 counterparts and a software program to play back the MP3 files on a personal computer. Other developments include the market introduction during the early 2000s of CD players capable to handle digital audio compressed in Windows Media Audio (WMA) or Sony's ATRAC formats. An international standard for optical media that hold compressed sound tracks has not been published and remains unlikely to be proposed because the recording industry is very much concerned about the enormous amount of illegally exchanged sound tracks via the worldwide computer networks.

One of the attempts that aimed at and also succeeded in the standardization of a CD format had led to the introduction of the **Super Video Compact Disc (SVCD)** on the Asian markets. The history of this format is rather complicated, being strongly biased by the Chinese user requirements. Due to the enormous expansion of the Chinese economy that started after 1990, the demand for video products grew very rapidly. For some reasons, the highly-priced Video Cassette Recorders (VCRs) at that time hardly caught the attention of the consumers. Instead, they chose for Video CD, a new product at the beginning of the 1990s, sufficiently cheap and with tremendous possibilities (think, for instance, at interactive features). The market grew from 1 million hardware

Video stream	Encoding system		MPEG-1	
	Data rate		fixed (1.15 Mbit/s)	
	Packet size		2324 bytes	
	Television system		NTSC	PAL
	Frame rate [Hz]		29.97	25.00
	Resolution [pixels]	Still picture	352×240 704×480	352×288 704×576
		Full motion	352×240	352×288
Audio streams	Encoding system		MPEG-1, Layer II	
	Number of streams		2 mono or 1 stereo	
	Surround sound		Dolby ProLogic	
	Sampling frequency		44.1 kHz	
	Data rate		fixed (224 kbit/s)	
	Packet size		2324 bytes	

Table 4.4. Overview of the Super Video CD features.

units in 1995 to an estimated 6.5 million players in 1996, and further to about 20 million players in 1997.

Very much aware of the business opportunities in China, C-Cube Microsystems of U.S.A. proposed an enhanced Video CD system with increased picture quality (480 lines horizontal resolution versus 352 featured by the standardized Video CD), overlay graphics, and provisions for either 4-channel mono sound or two stereo channels. The format was launched in 1998 under the name **China Video Disc (CVD)**. At the same time, the China Recording Standards Committee proposed their own **Video CD 2.0** standard, which featured 480-line horizontal resolution, Java-based interactive multimedia, and also supported the high-density discs (up to 150 minutes) commonly known on the local market as **Super-Video CDs (S-VCDs)**. Because CVD, Video CD 2.0, and S-VCD were incompatible with each other, a compromise was forced by the Chinese authorities and a new proposal emerged: the **Chao Ji Video CD**. Unfortunately, the latter did not represent a disc but a player standard, which was intended to read out CDV, S-VCD, VCD 2.0, and CD-DA. The incompatibility between the various optical media available only in China remained, hence, unsolved and practically created an opportunity for other companies to think about a viable solution.

The third contribution to the improvement of the Chinese video disc has been collectively brought by the owners of original Video CD standard: Victor Company of Japan (JVC), Matsushita Electric Industrial Co., Ltd., Sony Corp., and Royal Philips Electronics. They also worked with the China Recording Standards Committee to develop the **High-Quality Video Compact Disc (HQ-VCD)**. This format was eventually renamed **Super Video Compact Disc (SVCD)**, was thoroughly documented[77] by its proponents, and has also been standardized by the International Electrotechnical Commission[22]. The SVCD format uses the MPEG-2 video compression technology[50–58] to enhance the image quality. An overview of the Super Video CD features is given in Table 4.4. Note also that the SVCD physical format is based on the CD-ROM XA specifications, which means that the user data occupies 2324 bytes out of 2352 that form a complete CD-ROM sector. In order to reach the bit rate needed for MPEG-2 streaming, the SVCD media has to rotate two times faster (i.e., at 2X) than it was previously the case with Video CDs. Other features included a high level of interactive functions which allow the user to select a particular video passage, graphic overlay that provides up to four selectable movie subtitles or karaoke lyrics, and multilingual sound. Significantly, the SVCD format can be flawlessly played back on DVD systems provided that it is recognized by the player's firmware. As a last remark, note that pure 5.1-surround sound (to be also discussed throughout the next chapter) can only be conveyed by the MPEG-2 audio compression. The SVCD, however, offers the option to compress these six audio channels using the MPEG-2 backward-compatible mode[53], which means that an MPEG-1 decoder will suffice to extract the audio information.

References

Some documents listed below are not publicly available and can only be obtained from the licensers of the standard in question.

[1] ATSC. *Digital Audio Compression (AC-3) Standard*, 1995.

[2] CIE 1978. *International Commission on Illumination, Recommendations on Uniform Color Spaces, Color-Difference Equations, Psychometric Color Terms. Suplement No. 2 to CIE Publication No. 15 (E-1.3.1)*, 1971.

[3] Compaq Computer Corp., Microsoft Corp., Royal Philips Electronics, and Sony Corp. *CD-MRW. Defect Management & Physical Formatting. Version 1.2*, 2004.

[4] Eastman Kodak Co. and Philips Electronics N.V. *System Description Photo CD*, 1994.

[5] EBU Tech 3258. *Specification of the Systems of the MAC/packet Family*, 1991.

[6] ECMA-240. *Data Interchange on 120 mm Optical Disk Cartridges using Phase Change PD Format - Capacity: 650 Mbytes per Cartridge*, 1996

[7] ECMA-287. *Safety of electronic equipment*, 1999.

[8] ECMA-43. *8-Bit Coded Character Set Structure and Rules. Third Edition*, 1991.

[9] EIA 608. *Recommended Practice for Line 21 Data Service*, 1994.

[10] ETSI. *ETS 300 294: 625-Line Television Wide Screen Signaling (WSS) Specification for Terrestrial TV Sevices (PAL, SECAM, PALplus); Wide Screen Letterbox*, 1995.

[11] Harley, R. "HDCD". *Fi: The Magazine of Music and Sound*, December, 1997.

[12] IEC 60068-2-3. *Environmental testing–Part 2: Tests. Tests B: Dry heat*, 1974.

[13] IEC 60068-2-30. *Environmental testing–Part 2: Tests. Test Db and guidance: Damp heat, cyclic (12 + 12-hour cycle)*, 1980.

[14] IEC 60721-3-5. *Classification of environmental conditions–Part 3: Classification of groups of environmental parameters and their severities–Section 5: Ground vehicle installations*, 1997.

[15] IEC 60856. *Pre-recorded Optical Reflective Videodisk System. Laser Vision 50 Hz/655 lines – PAL*, 1986.

[16] IEC 60857. *Pre-recorded Optical Reflective Videodisk System. Laser Vision 60 Hz/525 lines – M/NTSC*, 1986.

[17] IEC 60908. *Audio Recording–Compact Disc Digital Audio System*, 1999.

[18] IEC 60958. *Digital Audio Interface*, 1998.

[19] IEC 61096. *Methods of Measuring the Characteristics of Reproducing Equipment of Digital Audio Compact Discs*, 1992.

[20] IEC 61104. *Compact Disc Video System–12 cm CD-V*, 1992.

[21] IEC 61866. *Audiovisual systems–Interactive text transmission system (ITTS)*, 1997.

[22] IEC 62107. *Super video compact disc–Disc-interchange systemspecification*, 2000.

[23] IEC 908. *Audio Recording–Compact Disc Digital Audio System. First Edition*, 1987.

[24] IMA. *General MIDI Specification: Musical Instrument Digital Interface (MIDI) File Format*, 1992.

[25] Imanaka, R., Okazaki, Y., Saimi, T., Kawamura, I., Ohta, T., and Nishino, S. "PD (Powerful Optical Disk System) for Multimedia". *Japanese Journal of Applied Physics*, Vol. 35, Part 1, No. 1B, pp. 490-494, 1996.

[26] ISO 2375. *Data processing–Procedure for registration of escape sequences*, 1985.

[27] ISO 3166-1. *Codes for the representation of names of countries and their subdivisions–Part 1: Country codes*, 1997.

[28] ISO 3166-2. *Codes for the representation of names of countries and their subdivisions–Part 2: Country subdivision code*, 1998.

[29] ISO 3166-3. *Codes for the representation of names of countries and their subdivisions–Part 3: Code for formerly used names of countries*, 1999.

[30] ISO 3901. *Information and documentation–International Standard Recording Code (ISRC)*, 2001.

[31] ISO 5800. *Photography–Colour negative films for still photography–Determination of ISO speed*, 1987.

[32] ISO 639. *Code for the representation of names of languages*, 1988.

[33] ISO/IEC 10149. *Information technology–Data interchange on readonly 120 mm optical data disks (CD-ROM). Second Edition*, 1995.

[34] ISO/IEC 10918-1. *Information technology–Digital compression and coding of continuous-tone still images: Requirements and guidelines*, 1994.

[35] ISO/IEC 10918-2. *Information technology–Digital compression and coding of continuous-tone still images: Compliance testing*, 1995.

[36] ISO/IEC 10918-3. *Information technology–Digital compression and coding of continuous-tone still images: Extensions*, 1997.

[37] ISO/IEC 10918-4. *Information technology–Digital compression and coding of continuous-tone still images: Registration of JPEG profiles, SPIFF profiles, SPIFF tags, SPIFF colour spaces, APPn markers, SPIFF compression types and Registration Authorities (REGAUT)*, 1999.

[38] ISO/IEC 11172-1. *Information Technology–Coding of Moving Pictures and Associated Audio for Digital Storage Media at up to about 1.5 Mbit/s–Part 1: Systems*, 1993.

[39] ISO/IEC 11172-2. *Information Technology–Coding of Moving Pictures and Associated Audio for Digital Storage Media at up to about 1.5 Mbit/s–Part 2: Video*, 1993.

[40] ISO/IEC 11172-3. *Information Technology–Coding of Moving Pictures and Associated Audio for Digital Storage Media at up to about 1.5 Mbit/s–Part 3: Audio*, 1993.

[41] ISO/IEC 11172-4. *Information Technology–Coding of Moving Pictures and Associated Audio for Digital Storage Media at up to about 1.5 Mbit/s–Part 4: Compliance Testing*, 1995.

[42] ISO/IEC 11172-5. *Information Technology–Coding of Moving Pictures and Associated Audio for Digital Storage Media at up to about 1.5 Mbit/s–Part 5: Software Simulation*, 1998.

[43] ISO/IEC 13346-1. *Information technology–Volume and file structure of write-once and rewritable media using non-sequential recording for information interchange. Part 1: General*, 1995.

[44] ISO/IEC 13346-2. *Information technology–Volume and file structure of write-once and rewritable media using non-sequential recording for information interchange–Part 2: Volume and boot block*, 1999.

[45] ISO/IEC 13346-3. *Information technology–Volume and file structure of write-once and rewritable media using non-sequential recording for information interchange–Part 3: Volume structure*, 1999.

[46] ISO/IEC 13346-4. *Information technology–Volume and file structure of write-once and rewritable media using non-sequential recording for information interchange – Part 4: File structure*, 1999.

[47] ISO/IEC 13346-5. *Information technology–Volume and file structure of write-once and rewritable media using non-sequential recording for information interchange–Part 5: Record structure*, 1995.

[48] ISO/IEC 13490-1. *Information technology–Volume and file structure of read-only and write-once compact disk media for information interchange–Part 1: General*, 1995.

[49] ISO/IEC 13490-2. *Information technology–Volume and file structure of read-only and write-once compact disk media for information interchange–Part 2: Volume and file structure*, 1995.

[50] ISO/IEC 13818-1. *Information technology–Generic coding of moving pictures and associated audio information: Systems*, 2000.

[51] ISO/IEC 13818-10. *Information technology–Generic coding of moving pictures and associated audio information–Part 10: Conformance extensions for Digital Storage Media Command and Control (DSM-CC)*, 1999.

[52] ISO/IEC 13818-2. *Information technology–Generic coding of moving pictures and associated audio information: Video*, 2000.

[53] ISO/IEC 13818-3. *Information technology–Generic coding of moving pictures and associated audio information–Part 3: Audio*, 1998.

[54] ISO/IEC 13818-4. *Information technology–Generic coding of moving pictures and associated audio information–Part 4: Conformance testing*, 1998.

[55] ISO/IEC 13818-5. *Information technology–Generic coding of moving pictures and associated audio information – Part 5: Software simulation*, 1997.

[56] ISO/IEC 13818-6. *Information technology–Generic coding of moving pictures and associated audio information – Part 6: Extensions for DSM-CC*, 1998.

[57] ISO/IEC 13818-7. *Information technology–Generic coding of moving pictures and associated audio information–Part 7: Advanced Audio Coding (AAC)*, 1997.

[58] ISO/IEC 13818-9. *Information technology–Generic coding of moving pictures and associated audio information–Part 9: Extension for real time interface for systems decoders*, 1996.

[59] ISO/IEC 14496-1. *Information technology–Coding of audio-visual objects–Part 1: Systems*, 2001.

[60] ISO/IEC 14496-2. *Information technology–Coding of audio-visual objects–Part 2: Visual*, 2001.

[61] ISO/IEC 14496-3. *Information technology–Coding of audio-visual objects–Part 3: Audio*, 2001.

[62] ISO/IEC 14496-4. *Information technology–Coding of audio-visual objects–Part 4: Conformance testing*, 2000.

[63] ISO/IEC 14496-5. *Information technology–Coding of audio-visual objects–Part 5: Reference software*, 2001.

[64] ISO/IEC 14496-6. *Information technology–Coding of audio-visual objects–Part 6: Delivery Multimedia Integration Framework (DMIF)*, 2000.

[65] ISO/IEC 15485. *Information technology–Data interchange on 120 mm optical disk cartridges using phase change PD format –Capacity: 650 Mbytes per cartridge*, 1997.

[66] ISO/IEC 16969. *Information technology–Data interchange on 120 mm optical disk cartridges using +RW format– Capacity: 3,0 Gbytes and 6,0 Gbytes*, 1999.

[67] ISO/IEC 2022. *Information technology–Character code structure and extension techniques*, 1994.

[68] ISO/IEC 4873. *Information technology–ISO 8-bit code for information interchange–Structure and rules for implementation. Third edition*, 1991.

[69] ISO/IEC 646. *Information technology–ISO 7-bit coded character set for information interchange*, 1991.

[70] ISO/IEC 8859-1. *Information technology–8-bit single-byte coded graphic character sets–Part 1: Latin alphabet No. 1*, 1998.

[71] ISO/IEC 9660. Information processing–Volume and file structure of CD-ROM for information interchange, 1988.

[72] ISO/IEC 9945-1. *Information technology–Portable Operating System Interface (POSIX)–Part 1: System Application Program Interface (API)*, 1996.

[73] ITU/CCIR Recommendation 601-2. *Encoding Parameters of Digital Television for Studios*, 1990.

[74] ITU/CCIR Recommendation 709. *Basic Parameter Values for the HDTV Standard for the Studio and for International Program Exchange*, 1990.

[75] Johnson, K.O. and Pflaumer, M.W. "Compatible Resolution Enhancement in Digital Audio Systems". *101st Convention of the Audio Engineering Society*, November 8-11, 1996, Los Angeles, U.S.A.

[76] JVC, Matsushita Electronics, Philips Electronics N.V., and Sony Corp. *Video CD Specifications. Version 2.0*, 1994.

[77] JVC, Matsushita Electronics, Philips Electronics N.V., and Sony Corp. *Super Video CD Specifications. Version 1.0*, 1999.

[78] JVC, Matsushita, Philips Electronics N.V., and Sony Corp. *Video CD Specification. Version 2.0*, 1994.

[79] Kawanishi, T. Nakagawa, E., and Itonaga, M. "Development of Compact Disk-Write-Once Hybrid Disk". *Japanese Journal of Applied Physics*, Vol. 31, Part 1, No. 2B , pp. 688-690, 1992.

[80] Microsoft Corp., Philips Electronics N.V., and Sony Corp. *Enhanced Music CD Specification. Version 1.0*, 1995.

[81] Microware Systems Corporation. *OS-9/68000 Technical Manual. Version 2.4*, 1983.

[82] OSTA. *MultiRead Specifications for CD-ROM, CD-R, CD-R/RW and DVD-ROM Devices. Revision 1.11*, 1997.

[83] OSTA. *MultiPlay. Playback Compatibility Specification for CD and DVD Consumer Devices. Revision 1.1*, 2001.

[84] OSTA. *MultiAudio Specification. Revision 1.1*, 2002.

[85] OSTA. *Dublin Core-Normalized Metadata Format Profile Specification. Revision 1.01*, 2003.

[86] OSTA. *MPV Basic Profile Specification. Revision 1.01*, 2003.

[87] OSTA. *MPV Core Specification. Revision 1.01*, 2003.

[88] OSTA. *MPV Presentation Profile Specification. Revision 1.01*, 2003.

[89] OSTA. *Normalized Metadata Format Specification. Revision 1.01*, 2003.

[90] OSTA. *MPV Music Profile Specification. Revision 1.0*, 2004.

[91] OSTA. *Universal Disk Format Specification. Revision 2.60*, 2005.

[92] Philips Electronics N.V. and Sony Corp. *System Description. Compact Disc Digital Audio*, 1981.

[93] Philips Electronics N.V. and Sony Corp. *Compact Disc Interactive. Full Functional Specification*, 1984.

[94] Philips Electronics N.V. and Sony Corp. *Compact Disc Read-Only Memory. System Description*, 1984.

[95] Philips Electronics N.V. and Sony Corp. *Recordable Compact Disc Systems. System Description. Part I: CD-MO*, 1990.

[96] Philips Electronics N.V. and Sony Corp. *Recordable Compact Disc Systems. System Description. Part II: CD-WO*, 1990.

[97] Philips Electronics N.V. and Sony Corp. *An Introduction to CD-i Bridge Disc*, 1991.

[98] Philips Electronics N.V. and Sony Corp. *System Description CD-ROM XA*, 1991.

[99] Philips Electronics N.V. and Sony Corp. *CD-i Bridge Specification. Version 1.0*, 1995.

[100] Philips Electronics N.V. and Sony Corp. *Multisession Compact Disc Specification*, 1995.

[101] Philips Electronics N.V. and Sony Corp. *Recordable Compact Disc Systems. System Description. Part III: CD-RW. Version 1.0*, 1996.

[102] Philips Electronics N.V. and Sony Corp. *System Description. Compact Disc Digital Audio. Sucode/Control and Display System, Channels R..W. Chapter 5.8. The CD TEXT Mode. Version 1.0*, 1996.

[103] Philips Electronics N.V. and Sony Corp. *Recordable Compact Disc Systems. System Description. Part II: CD-R. Version 3.1*, 1998.

[104] Philips Electronics N.V. and Sony Corp. *Recordable Compact Disc Systems. System Description. Part III: CD-RW. Version 2.0*, 1998.

[105] Philips Electronics N.V. and Sony Corp. *CD Reference Measuring Methods. Specification Guidelines. Version 1.0*, 1999.

[106] Philips Electronics N.V. and Sony Corp. *CD-ROM. Compact Disc Read-Only Memory. System Description*, 1999.

[107] Philips Electronics N.V. and Sony Corp. *Compact Disc Digital Audio. System Description*, 1999.

[108] Philips Electronics N.V. and Sony Corp. *Recordable Compact Disc Systems. System Description. Part III: CD-RW. Volume 2: High-Speed. Version 1.0*, 2000.

[109] Philips Electronics N.V. and Sony Corp. *Recordable Compact Disc Systems. System Description. Part III: CD-RW. Volume 3: Ultra-Speed. Version 1.0*, 2002.

[110] Philips Electronics N.V. and Sony Corp. *CD-R and HC-R Disc Capacity Guidelines*, 2003.

[111] Philips Electronics N.V. and Sony Corp. *Recordable Compact Disc Systems. System Description. Part II: CD-R. Volume 1: 1X/2X/4X. Version 3.2*, 2005.

[112] Philips Electronics N.V. and Sony Corp. *High Capacity Recordable Compact Disc Systems. System Description. Version 1.1*, 2006.

[113] Philips Electronics N.V. and Sony Corp. *Recordable Compact Disc Systems. System Description. Part II: CD-R. Volume 2: Multi-Speed. Version 1.3*, 2006.

[114] Philips Electronics N.V. and Sony Corp. *Recordable Compact Disc Systems. System Description. Part III: CD-RW. Volume 1: 1X, 2X, 4X. Version 2.1*, 2006.

[115] Philips Electronics N.V. and Sony Corp. *Recordable Compact Disc Systems. System Description. Part III: CD-RW. Volume 1: 2X. Version 1.1*, 2006.

[116] Philips Electronics N.V. and Sony Corp. *Recordable Compact Disc Systems. System Description. Part III: CD-RW. Volume 2: High-Speed. Version 1.2*, 2006.

[117] Philips Electronics N.V. and Sony Corp. *Recordable Compact Disc Systems. System Description. Part III: CD-RW. Volume 3: Ultra-Speed. Version 1.2*, 2006.

[118] Phoenix Technologies Ltd. and IBM Corp. *El Torito. Bootable CDROM Format Specification. Version 1.0*, 1995.

[119] RIAJ-RIS506. *Music Shift JIS Kanji Character Set*, 1992.

[120] Royal Philips Electronics. *CD Logo Guide*, 1999.

[121] Royal Philips Electronics and Sony Corp. *1.3GB Read-Only Compact Disc Systems. Double Density CD Read-Only. Version 1.0*, 2001.

[122] Royal Philips Electronics and Sony Corp. *1.3GB Recordable Compact Disc Systems. Double Density CD Recordable. Version 1.0*, 2001.

[123] Royal Philips Electronics and Sony Corp. *1.3GB ReWritable Compact Disc Systems. Double Density CD ReWritable. Version 1.0*, 2001.

[124] Royal Philips Electronics and Sony Corp. *CD2. Physical Format. Version 0.40*, 2001.

[125] Sony Corporation. *SDDS Specification for Disc. Version 1.0–Digital Audio Multi-channel Coding.*

[126] Stan, S.G. *The CD-ROM Drive. A Brief System Description.* Kluwer Academic Publishers, 1998.

[127] Stan, S.G. and Bakx, J.L. "Adaptive-Speed Algorithms for CD-ROM Systems". *IEEE Transactions on Consumer Electronics*, Vol. 42, No. 1, pp. 43-51, 1996.

[128] Stan, S.G., van Kempen, H., Lin, C.C.S., Yen, M.S.M., andWang,W.W. "High-Performance Adaptive-Speed/CAV CD-ROM Drive". IEEE Transactions on Consumer Electronics, Vol. 43, No. 4, pp. 1034-1044, 1997.

[129] Tsutsui, K., Suzuki, H., Shimoyoshi, O., Sonohara, M., Akagiri, K., and Heddle, R.M. ATRAC: *Adaptive Transform Acoustic Coding for Mini-Disc*, 93rd AES Convention, San Francisco, October 1-4, 1992.

[130] Unicode, Inc. *The Unicode Standard–Worldwide Character Encoding. Version 1.1*, 1991

[131] Yoshida, T. "The Rewritable MiniDisc System". *Proceedings of the IEEE*, Vol. 82, No. 10, pp. 1492-1500, 1994.

Chapter 5
DIGITAL VERSATILE DISCS

Sorin G. Stan
Philips Consumer Lifestyle

5.1. From Compact to Digital Versatile Discs

The history of the digital versatile disc (DVD) is relatively short when compared to the 20-year time period that was needed for so many CD standards to become established. Nevertheless, setting up the DVD specifications took place after many struggles, with even more companies playing an essential role than it was ever the case for the compact disc and with many technical proposals to be carefully weighted against each other. New for such a standardization process was the fact that quite some audio and video content providers joined the efforts of the technical companies from the very beginning and influenced significantly all subsequent decisions.

At the end of the twentieth century optical storage was already a well-established research and development field. The path toward DVD had been consistently underlined during the 1990s by challenges taken up worldwide to increase the amount of information stored on optical discs and improve the quality of the readout and recording, as well as to improve the quality of the media itself and of the manufacturing processes. By contrast, the pre-CD efforts were aimed at the *creation* of those data storage technologies for optical discs that would ultimately lead to the replacement of the existing analog audio and video recordings by digital counterparts.

Among the very first proposals made to increase the amount of recorded information on optical discs was a technology that employed two sets of data sequences imprinted along the flanks of a V-shaped groove[161, 162]. Due to its numerous disadvantages, this technology was very soon considered impractical. It obviously required a special disc manufacturing process that would have made the grooved read-only media less attractive from a financial perspective. In addition, a potential recordable disc compatible with its read-only counterpart would have increased the complexity of the system. Another proposed technology considered an average depth of the pits that could be made to vary from track to track[1, 2]. This would allow the encoding of more than the two signal levels conventionally specified by a pit/land sequence and would consequently lead to an increase of the storage capacity.

The true adventure of high-density* read-only optical media began after the successful market introduction of the compact disc. Many research laboratories around the world were convinced that the CD physical format, as it was already defined, could be stretched by reducing the system margins and tolerances to increase the storage density. Several attempts were made to obtain CD-like media with even smaller track pitches and shorter channel bits (in fact, as discussed in the Sect. 4.4, these attempts had led to the introduction of CDs with playback times up to 150 minutes and to the DDCD standardization). Another basic idea was to shrink the physical dimensions of the disc relief structures while using a laser beam of shorter wavelength than employed in the CD systems for readout. Appropriate optics designed to reduce the laser spot diameter by increasing the focusing power of the objective lens would then be required. Unfortunately, semiconductor lasers emitting in the visible spectrum had not left the research laboratories by the end of the 1980s. In search for other solutions, several companies also exploited the generation of the second harmonic of the fundamental laser wavelength in conventional solid-state (semiconductor) lasers[164, 179] or used external resonant cavities to multiply the light frequency[106, 155]. Note that many such experiments were carried out with green light and did not employ small-size emitting devices but laboratory equipment, which was mainly intended to demonstrate the feasibility of optical storage densities higher than on compact discs.

An ingenious alternative to the short-wavelength laser was to continue using the available CD infrared light sources. Two satellite spots were employed already for tracking in many CD systems, while the central spot was meant to retrieve the recorded information. It was proven that the optical crosstalk induced into the central spot by the adjacent tracks could be cancelled by appropriately combining three RF signals[152]. This technique would then allow discs with a track pitch significantly smaller than standardized for CDs to be read out correctly at the expense of some additional electronics. Optical methods for canceling the distortions induced by too close adjacent tracks had also been suggested[5, 183].

Yet another method proposed to increase the storage density on optical discs was the optical superresolution[156, 186]. Thereby it was possible already at

*The term "high-density" should be understood throughout the entire book in the historical context of increasing the storage density on optical media from previously established norms to the next successful set of standards (for example from CD to DVD).

the end of the 1980s to read out topographical patterns whose dimensions were below the theoretical limit imposed by the diffraction theory. This technology required not only additional optical components, but also the assembly of the superresolution optical heads was to be carried out with high precision. The final products were foreseen to become expensive and unsuitable for the types of markets already conquered by the CD systems by the mid-1990s. An alternative to these expensive optical heads was proposed in 1993 and employed a read-only disc specially designed to increase the resolution of the optical readout[188]. The key advantage was that the readout could take place with conventional CD optoelectronics.

A possibility to increase the amount of information recorded on optical discs was introduced by Sony Corp. under the name single carrier independent pit edge recording (SCIPER). Sony proposed a method to encode the information in the edges of the embossed patterns on disc, by varying the position of these edges in very small discrete steps while spacing equally the centers of the patterns along the track. The method was first demonstrated with standard CD optics and infrared lasers[153] and was eventually extended to read-only systems operating with red laser light[154]. In parallel with the various optical approaches used to increase the storage density, intensive research concentrated also on improving the efficiency of the channel modulation techniques and of the error detection and correction codes.

Although experimentally demonstrated at the end of the 1980s, the optical disc systems based on media with denser pits and lands than on CDs were still far from being ready for commercial applications. A number of key issues held back the enthusiasm shown by researchers until solutions were found toward the end of the past century. Among these issues, the commercial availability of new laser types and the immaturity of the cheap replication technologies needed to produce high-density discs played an important role in delaying the introduction of the next generation of read-only optical media. The mass production in a cost-effective manner of many components, such as the newly required optical heads, represented another challenge toward the commercial introduction of a CD successor.

It became very clear for many companies that a new optical disc system could only be successful if it remained affordable just like the CD system. This would imply that the race toward increasing the storage capacity could only be carried on if semiconductor lasers emitting in the visible spectrum and generating smaller readout spots than their infrared counterparts would become available. Following several years of research, a few laboratories succeeded before the end of the 1980s to produce novel semiconductor lasers[116] suitable for what was called at that time high-density optical recording. A magneto-optical disk system using such a solid-state device was demonstrated in 1988 by a team of researchers at Nippon Electrical Company (NEC) Corp. of Japan[187].

The laser was emitting red light at a wavelength slightly higher than 650 nm.

Subsequent developments led to commercial semiconductor lasers that became available in the early '90s[108]. It is worth mentioning here several specific technical problems that had to be solved: the stability of the laser output, the quality of the readout spot, and the generation of sufficient light power. Red and even green laser light was used during many experiments with read-only optical discs, but worldwide research was also aiming at recordable and rewritable systems. In particular, the high-density media based on phase-change materials were considered very good candidates for use in future data storage systems[181] because no magnetic field was needed during recording and the required optical heads lacked the complexity of their magneto-optical counterparts. At about the same time, several companies pursued the miniaturization of optical and magneto-optical drives that would equally lead to small-size devices suitable for portable computers and to manufacturing expertise needed for mass-producing the more demanding components of a CD successor. The MPEG-2 standard for encoding digital audio and video data streams was also gaining increasing acceptance among experts and professional users, and thus paving the way toward the consumers' DVD-Video standard.

Encouraged by a very successful CD business, the European and Japanese companies spent tremendous resources to develop commercial high-density optical disc systems. Improving only the disc characteristics represented by itself a challenge apart. If phase-change media appeared sufficiently easy to be tuned for short-wavelength lasers, the mastering and replication of read-only discs raised complex problems to be solved. Nimbus Technology & Engineering of the U.K., a producer of both compact discs and CD manufacturing equipment, gave in January 1993 a public demonstration of 1.5-hour full-motion video stored on one CD. Several other companies announced afterwards that they were also in the possession of the technology needed to produce high-density optical recordings[115].

The pioneering work publicly shown by Nimbus was followed in December 1994 by the announcement of the MultiMedia Compact Disc (MMCD) developed by Philips Electronics of the Netherlands and Sony Corporation of Japan. The disc was able to hold 3.7 GB of computer data and it was claimed to be backward compatible with the already available CDs. This claim was mainly based on a cheap manufacturing process similar to that of read-only compact discs and on a similar readout technology that made use of one laser for both discs. Since the two MMCD advocates already enjoyed a history of good cooperation while establishing all CD specifications, an extension of their joint efforts toward a new standard appeared very motivating. In January 1995, a Japanese alliance formed by Hitachi Ltd., Matsushita Electric Industrial Co., Ltd.* Mitsubishi Electric Corp., Pioneer Corp., Toshiba Corp., and Victor

* The company changed its name in October 2008 to Panasonic Corporation.

Company of Japan, Ltd. (formerly known as Japan Victor Company or JVC), together with Thomson Multimedia of France countered the MMCD format and jointly proposed the Super Density Digital Video Disc (SD-DVD) capable of storing 5 billion bytes. Facing a stronger alliance that was also determined to create the successor of the compact disc, Philips and Sony created their own MMCD group by convincing several computer hardware manufacturers to join in. Among these companies were Acer Peripherals of Taiwan, Japan's Mitsumi, Ricoh Company, Ltd., and Teac, as well as Wearnes of U.S.A. A few consumer electronics vendors like Grundig of Germany, Bang & Olufson of Denmark, and Japan's Aiwa, Marantz and Alps also supported the MMCD proposal.

In the entertainment industry, the motion picture giants did not simply watch the new developments but insisted to play an essential, later proven historical role. In September 1994, a few months before the MMCD and SD-DVD were officially announced, several film studios formed the Hollywood Digital Video Disc Advisory Group and requested a set of features to become available in the next generation of optical disc systems. The Hollywood group was jointly established by Columbia Pictures (owned by Sony), Disney Enterprises, MCA/Universal (owned by Matsushita), Metro-Goldwyn-Mayer, Inc. (known as MGM), Paramount Pictures, Viacom International, Inc., and Warner Bross (with Toshiba as business partner of Time Warner, Inc.*). All these movie-maker giants were very much aware of the potentially huge market that could be created for home video entertainment and demanded a video disc with the following features[180]:

- picture quality superior to the existing consumer video systems, like VHS, Video CD, and even Laser Disc, and comparable to broadcast images;
- enough storage capacity on one side of the disc to accommodate a full length (120-135 minutes) film;
- capability to store multichannel surround sound compatible with the existing standards for high-fidelity audio equipment;
- support for minimum three dubbing languages synchronized with the video content;
- support for several screen aspect ratios like widescreen TV;
- support for various interactive features (title selection, scene search, etc.);
- copy protection capabilities, regional control, and parental lock.

The struggle between the MMCD and SD-DVD camps continued as the

*Time Warner, Inc. merged with the Internet provider America Online, Inc. in 2000 and became AOL Time Warner, but the new company dropped the abbreviation "AOL" from its name at the end of 2003.

two parties attracted other companies to support them. The competition became tougher when the former camp proposed a dual-layer MMCD capable to hold 7.4 GB of user data. The SD-DVD, on the other hand, was already proposed as a single-layer double-sided disc holding up to 5 GB only on one side and Matsushita raised the bar in May 1995 by announcing that a dual-layer double-side SD-DVD could also be successfully manufactured and read out. At this stage, many other companies started to worry about the reduced attention being paid to computer applications. Five American computer companies, namely Apple Computer, Inc., Compaq Computer Corp.*, Hewlett-Packard Company, IBM Corp., and Microsoft Corp. formulated their own objectives and persuaded the MMCD and SD-DVD camps to reach an agreement and share their best technologies. The following data storage requirements were elaborated:

- common physical format and file system for both computer and video applications;
- backward compatibility with the existing compact discs, especially with CD-DA and CD-ROM, and forward compatibility with recordable and rewritable systems to be developed in the future;
- low-cost manufacturing of both replicated media and data players, as it was already the case for read-only compact discs and CD-ROM drives;
- high-performance interactive features;
- provisions for future capacity enhancements while preserving the file system to be adopted;
- data reliability at least equal to that of CD-ROM, with no mandatory protective cartridge to encapsulate the disc.

The five computer companies, later also joined by Fujitsu Ltd. of Japan and Sun Microsystems, Inc. of U.S.A., recommended in August 1995 the adoption of the Universal Disk Format (UDF). This file system was already being developed by the Optical Storage Technology Association (OSTA), but was not yet finalized at that time and therefore not yet aiming at a new high-density optical medium. The discussions between the two camps intensified toward the end of 1995 and finally led to what is nowadays called the **Digital Versatile Disc**. In December 1995 a consortium of ten companies (Hitachi, JVC, Matsushita, Mitsubishi, Philips, Pioneer, Sony, Time Warner, Thomson, and Toshiba) was established to promote the new high-density optical disc formats. The DVD Consortium was also meant to manage the royalties that would be received in the future by the member companies for about 4000 patents related to the DVD system. Two years later, in May 1997, the DVD Consortium was replaced by the DVD Forum and became an alliance open

*Compaq Computer Corp. was acquired in 2001 by Hewlett-Packard Company.

to all companies, currently including all major DVD manufacturers as well as major DVD software developers and DVD media producers around the world.

It seems appropriate to address here the origin of the name digital versatile disc. Although it has never been endorsed formally by all companies that developed and proposed the DVD specifications, this name was suggested once by some of these companies to counteract the definition "digital video disc" independently invented by the world press. The acceptance of the term digital versatile disc spread out rapidly and to such an extent that no resistance was eventually encountered when journals and magazines started to use it. Furthermore, the versatility of the DVD with respect to various potential applications was soon proven to excel above that of any other storage medium developed by that time. This has ultimately made the acceptance of the name currently in use fairly deserved.

5.2. The DVD-ROM and DVD-Video

Held under pressure by the computer industry, the MMCD and SD-DVD champions agreed by the end of August 1995 to work out a common standard based on the best solutions proposed earlier by each party. Meanwhile, Philips and Sony proposed a new MMCD format capable to store 4.7 billion bytes in one layer of one side of the disc and potentially even employing two stacked layers on each side. The competing alliances were now supporting dual-layer, double-sided optical media with almost equal storage capacities.

During the last months of 1995, the MMCD and SD groups intensively discussed their proposals and were able to eventually reach a consensus. Before the end of the year, on December 12, the **Digital Versatile Disc Read-Only Memory (DVD-ROM)** was jointly announced. The adopted format, which has been used ever since in DVD drives for computer environments and in stand-alone players, specifies an optical medium that can store 4.7 billion bytes for various applications. Two layers containing relief structures are allowed on each side of the disc, which increases (but does not double) the disc capacity from 4.7 to 8.5 billion bytes on one side. The disc physical format[9] borrowed its parameters from both MMCD and SD-DVD. Note that the definitions of kilobyte, megabyte, and gigabyte used to quantify the storage capacity of compact discs, magneto-optical disks, and hard-disks received another interpretation at the inception of the DVD for reasons that have remained quite unclear until now. All DVD standards since then bear attributes like "4.7 GByte" to indicate a storage capacity equal to 4.7 billion bytes and not, as the technical community (with predilection the computer and data storage specialists) might expect, equal to 4.7×2^{30} bytes, where $2^{30} = 1{,}073{,}741{,}824$.

With the definitions adopted already in the previous chapter, a simple calculation will reveal that single-layer single-side DVD media can typically hold 4.376 GB, which is the correct numerical value that can be compared with the typical 650-MB storage capacity of compact discs.

As initially agreed by the MMCD and SD-DVD proponents, the read-only digital versatile disc has been standardized in three flavors: DVD-Video, DVD-ROM, and DVD-Audio. All three formats rely on the same physical structure of the media, which basically implies that the binary data is organized at low level similarly for all applications. In contrast with compact discs, the data stream on DVDs does not depend on any particular sampling frequency used to digitize the analog audio and/or video information. When the data spiral is scanned by the laser spot at nominal reference velocity, the user data throughput equals 11.08 Mbit/s. There is no specific time-related information equivalent to the CD subcode channel that is embedded in the data stream. Instead, a numerical identification label commonly designated as sector ID or sector address is associated with each two kilobytes (i.e., 2048 bytes) of user data. Accessing any information on disc can be performed by searching for a particular 2-kB sector after computing first its logical address and then its corresponding physical location on disc.

In brief, the information is much simpler organized on DVD media than on compact discs. Sixteen consecutive sectors form a so-called error correction code (ECC) data block whose bytes, originating from a continuous sequence on disc, can be mentally pictured as arranged in a rectangular matrix. A new error detection and correction mechanism different from CIRC and called Reed-Solomon product code (RSPC) operates upon each ECC block. If the laser spot remains in focus and on track, the RSPC has the theoretical capability to fully correct an amount of erroneous information equivalent to a 6.35-mm length of a defective track. Given the increase in storage density, a strong improvement of the error detection and correction capacity with respect to the CIRC performance was mandatory[176]. The conversion from user data to channel bits is achieved via the EFMPlus modulation scheme, which is again very different from the EFM code employed in compact disc and translates directly each byte into 16 channel bits. An ECC block containing 32 kB of user data and associated auxiliary information, like sector IDs, parity bytes, etc., will be converted to 619,008 channel bits that are scanned by the laser spot at a typical rate of 26.15625 Mbit/s corresponding to the overspeed factor 1X.* The channel bits form an uninterrupted sequence of pits and lands helically arranged just like on CD media from the central hub toward the outer edge of the disc. However, in order to meet the storage capacity requirements, the

*A definition similar to that used in CD systems applies here: the ratio between the linear velocity v at which the spiralled data track is scanned in practice and the reference velocity v_0 at which the optical disc is specified is called overspeed or X-factor.

pit/land structures on DVDs are 2.26 times shorter than their counterparts on compact discs whereas the track pitch is 2.16 times narrower.

A computer system or an audio/video application makes straightforward use of the 2-kB sectors grouped in 32-kB ECC blocks without imposing any other sort of restriction on the given physical arrangement of these bytes on disc. In fact, the common physical structure of all read-only DVD formats and the independence of this structure from the application requirements represent two of the strengths of the digital versatile discs. The data storage capacity of a 12-cm single-layer single-side DVD can be anywhere between 4.258 and 4.499 GB (with 1 GB = 2^{30} bytes), depending on the manufacturing tolerances of the track pitch, channel bit length, etc. The 8-cm single-layer single-side DVD media can hold between 1.316 and 1.411 GB. The playback time of a fully recorded DVD-ROM disc scanned at the overspeed factor 1X equals 57 minutes, with 11.08 megabits of user data emerging from disc every second. Video and audio discs, on the other hand, employ compression algorithms that will be discussed later throughout this section. The use of such algorithms extend the playback time beyond two hours in the case of average-compressed MPEG-2 data recorded in DVD-Video format.

From an application vantage point, while a DVD-ROM disc only contains raw computer data filling the 2-kB sectors, the binary information recorded on DVD-Video and DVD-Audio media within the same 2-kB sectors requires additional application-specific digital signal processing before being converted to the analog video and/or audio domain. Nevertheless, all read-only DVDs have equal raw storage capacities irrespective of the consumer application for which they are used. Recall from the previous chapter that the CD-ROM and the Video CD formats were built as extensions of the initial CD-DA data structure and provided thereby less raw storage capacity compared to the original audio disc. By contrast, the multipurpose 2-kB sectors on read-only DVDs excel by equally allowing them to be filled with raw computer data, digital audio, digital still and motion pictures, or multimedia files containing various combinations of digital information.

The DVD-Video represented from its inception a very attractive format for the consumer electronics market since it offered MPEG-2 encoded[129–137] picture quality comparable with the CCIR-601 television studio standard. Even at present, when large TV sets have become common in so many households, the native DVD-Video resolutions of 720 × 480 or 720 × 576 pixels at 30 NTSC or 25 PAL frames per second, respectively, can be displayed marvelously on 4:3 as well as 16:9 television screens. The image quality delivered by DVDs was meant to be superior to that of analog VHS tapes while the users were initially offered the same prerecorded content on both media for equal prices. Table 5.1 outlines the main specifications of the binary information recorded on DVD-Video discs. A maximum number of eight different languages can be

Multiplexed stream	Encoding system		MPEG-2	
	Maximum data rate		11.08 Mbit/s	
	Packet size		2,048 bytes	
Video stream	Number of streams		1	
	Encoding system		MPEG-1, MPEG-2	
	Television system		NTSC	PAL
	Frame rate [Hz]		29.97	25.00
	Aspect ratio		4:3 and/or 16:9	
	Resolution [pixels]	MPEG-1	352 × 240	352 × 288
		MPEG-2	352 × 240 352 × 480 480 × 480 544 × 480 704 × 480 720 × 480	352 × 288 352 × 576 480 × 576 544 × 576 704 × 576 720 × 576
	Bit rate	MPEG-1	fixed or variable (1.856 Mbit/s or less)	
		MPEG-2	fixed or variable (9.8 Mbit/s or less)	
Audio stream	Number of streams		1 or 2	
	Encoding system, bit rate, etc.		See Table 5.2	

Table 5.1. Characteristics of the data streams recorded on DVD-Video media.

simultaneously recorded and synchronized with the motion picture, which is equivalent to eight individual sound tracks. For people with impaired hearing each audio track may be accompanied by captions, but the format also supports up to 32 subtitles carrying still graphics with translations of the spoken language into other languages. The audiophiles have not been forgotten either, as a multitude of audio features is available, determined by the combinations listed in Table 5.2. The basic idea behind so many choices is to store on disc as much audio data as possible while increasing the overall performance of the reproduced sound. Note, however, that most of the storage capacity on disc is reserved for the video stream, which can top a data rate at playback in excess of 10 Mbit/s and leaves typically 384 or 448 kbit/s for compressed audio.

Feature	LPCM	MPEG-1	MPEG-2	Dolby Digital	DTS	SDDS
Sound mode	Mono, Stereo Dolby ProLogic		Mono, Stereo, Dolby ProLogic Dolby Surround, 5.1 Surround			
No. of audio channels	up to 8					
Coding system before compression	LPCM					
Quantization	16, 20, or 24 bits	20 or 24 bits		20 bits		16 bits
Sampling rate	48 or 96 kHz	48 kHz				
Compression system	—	MPEG-1 Layer II	MPEG-2 backward compatible	Dolby Digital (AC-3)	ADPCM	ATRAC
Typical compression ratio	—	6:1	10:1	10:1	4:1	5:1
Typical bit rate [kbit/s]	768 - 6144	64 - 384	64 - 912	64 - 448	192 - 1536	64 - 1280
Theoretical frequency response [1] [Hz]	20 - 24,000/48,000 (stereo) 100 - 7000 (surround)		20 - 24,000 (stereo and surround) 20 - 120 (subwoofer)			

[1] The theoretical frequency response extends up to half of the sampling frequency, which is either 24,000 or 48,000 Hz when sampling at 48 or 96 kHz, respectively. However, the frequency response of the human auditive system remains confined below these signal processing limits, with a spectral range of 20-20,000 Hz being generally accepted to suffice the most exigent requirements for recording and reproduction of high-quality sound.

Table 5.2. Characteristics of the audio information recorded on DVD-Video media.

Nevertheless, for the first time since the introduction of the compact disc, the music industry and the standards owners were offered the chance to distribute and provide, respectively, optical media holding digital audio of significantly better quality than available on CD*. This chance turned into a variety of new technical choices including the increase of the audio frequency range reproduced at playback, the addition of more audio channels than conveyed by the ubiquitous stereophonic system at that time, and the reduction of the amount of noise accompanying the recorded sound.

To begin with, a Dolby ProLogic option was added to the DVD-Video standard as a technology means for content providers to produce two more analog audio channels mixed with the existing stereo ones before generating the digital version of the content. A Dolby ProLogic module is required then also during reproduction to extract the additional information from the stereophonic signal retrieved from disc and drive thereby four loudspeakers. In the same category of the so-called matrix surround sound falls also the Dolby Surround technology, which mixes only one monaural channel derived from the existing stereophonic source with the stereo information. By contrast, true surround sound techniques make use of more than two completely separate audio paths. The MPEG-2 compression schemes, Dolby Digital, Digital Theater Systems (DTS), and the Sony Dynamic Digital Sound (SDDS) provide support for 5.1-channel surround sound. For all these techniques the audio information is recorded by six completely independent microphones and reproduced by an equal number of loudspeakers arranged as in Fig. 5.1 and designated as front left, center, front right, rear left (also called surround left), rear right (surround right) and subwoofer. The latter is also referred to as the low-frequency effects (LFE) channel and has a limited bandwidth, operating only for sounds below 120 Hz. The other five channels have full-bandwidth characteristics. Note that DVD-Video media can hold up to eight separate audio channels, but the 5.1 surround sound format effectively dominates the market.

The issue of increasing the range of audio frequencies that are recorded and reproduced has also been considered before setting the standard for DVD-Video applications. The mechanics of the human audititory system and the psychoacoustic processes that take place in the human brain confine the audible frequencies typically below 16 kHz[159, 160, 184]. In general, it is accepted that high-quality sound can be conveyed to listeners if both recording and reproduction equipment cover the frequency spectrum between 20 and 20,000 Hz. Basic signal processing rules imply that such audio signals must be digitized by sampling them at a rate at least equal to 40 kilohertz in order to restore them

*Previous attempts to improve the CD quality (for example, the HDCD format discussed in the previous chapter) had been limited by the technical specifications of compact discs. Other proposals, like DVD-Audio and SACD (to be addressed later throughout this chapter), would only be finalized a few years after the debut of the DVD-Video.

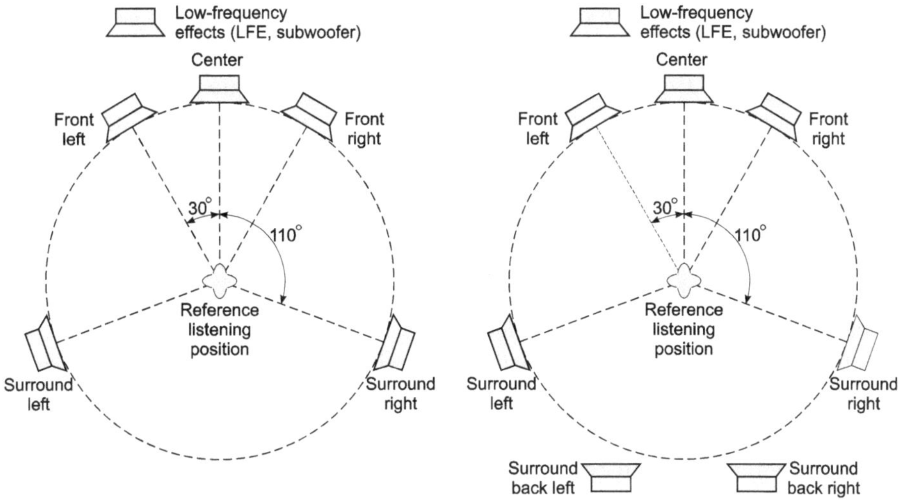

Fig. 5.1. Speaker arrangements in 5.1 and 7.1 surround sound configurations (left and right panes, respectively).

correctly during reproduction. By increasing the sampling rate beyond the theoretical minimum, a more accurate reconstruction of the original sound can be obtained at the expense of streaming more digital information and thereby wasting storage space on disc. The audio channels on DVD-Video media contain binary information streaming at either 48,000 or 96,000 samples per second (see also Table 5.2) and satisfy, hence, the most exigent audiophiles when compared to the sampling rate of 44.1 kHz used in compact disc.

A third important technical specification of the digital audio conveyed by the DVD-Video discs is related to the number of bits that represent a binary sample. Fewer quantization bits means more quantization noise due to rounding of each sampled analog level toward the most appropriate, usually the closest numerical code. This also leads to a smaller dynamic range since there are insufficient discrete codes to individually represent signal levels very close to each other. In order to increase the recording and reproduction quality a relatively high number of quantization bits per sample is needed, which in turn decreases the total storage capacity of the disc. A good trade-off may nevertheless be achieved as indicated in Table 5.2 by providing a fairly broad range of encoding and compression algorithms where the producer of digital content may choose from. Pulse code modulation (PCM) as used already for CD-DA plays a crucial role here as well. To distinguish between various types of PCM that have found application in electronics and telecommunications since the introduction of the compact disc, it is customary to designate the encoding technique of interest for CD and DVD as linear pulse code modulation (LPCM). The difference between any two consecutive LPCM numerical codes corresponds always to the same amount of amplitude variation measured at any place between the peak values of the sampled analog signal. Linear PCM has been chosen for DVD-Video applications as basic technique for converting the analog sound into 16-, 20-, or 24-bit samples. The resulting binary data stream may further be processed by one of the compression algorithms indicated in Table 5.2. Note once more that MPEG-2, Dolby Digital, DTS, and SDDS all provide support for multichannel surround sound. MPEG-2, in particular, must be used in the so-called backward compatibility mode that allows the MPEG-1 decoders to extract stereophonic sound from the 5.1-channel audio information recorded on the DVD-Video carriers.

In addition to all technical features described above, the DVD-Video discs are optionally recorded with a regional code corresponding to a particular geographical area on the globe. Eight such areas have been defined as indicated in Table 5.3. The use of regional codes was requested by the film industry to separately control the release of their movies in various parts of the world, which was believed to be needed for implementing particular market strategies and price policies. From another perspective, the region management was also regarded at that time as a means to prevent in some way the illegal spread of DVD-Video discs. It was anticipated that pirated media produced in some world regions would be rejected at playback if the assigned regional codes would not match those of the reproduction equipment used elsewhere, unless the original media was deliberately commercialized with region-free content.

No.	Region
1	Canada, U.S.A., Bermuda, Puerto Rico, the Virgin Islands, and U.S. territories in the Pacific Ocean
2	Japan, Europe, South-Africa, Turkey, Egypt, and the Middle East
3	Southeast and East Asia, including Hong-Kong, Indonesia, Macao, and South Korea
4	Australia, New Zealand, Pacific Islands, Mexico, Central and South America, and the Caribbean
5	Africa (excluding Egypt and South-Africa), Eastern Europe,Russia, former Soviet Union states, Indian subcontinent (Afghanistan, Bangladesh, India, Pakistan, etc.), North Korea, and Mongolia
6	China and Tibet
7	Reserved
8	Special international venues (airplanes, cruise ships, etc.)

Table 5.3. Geographical region numbers for DVD-Video discs.

The video information recorded on DVD-ROM for computer applications may also be protected by regional codes, but such standardization constraints do not apply to other forms of DVD-ROM software and neither to the DVD-Audio content released at a later stage after the introduction of DVD-Video.

The region management was never meant to offer a secure solution against counterfeiting, since it could not solve alone the complicated technical and legal issues related to copyrights. In practice, it turned out within only a few years after the DVD introduction that most players and computer drives could easily be modified to read out media produced all around the globe. Illegal copying had to be fought by advanced copy protection techniques that represented the topics of many discussions before the final DVD specifications were approved. Despite the agreements made in 1995 between the MMCD and SD-DVD camps to jointly propose the digital versatile disc format, no DVD-Video or DVD-ROM players were sold during the first three quarters of 1996. The entire film industry was holding out for a general consensus on copyright issues. Movie production houses were all worried about the eventual possibility of making analog or, even worse, perfect high-quality digital copies of the original DVD-Video discs. In order for the DVD license holders to manage their intellectual property rights, a new alliance was formed in December 1995 between Hitachi Ltd., Matsushita Electric Industrial Co., Ltd., Mitsubishi Electric Corp., Philips Electronics, Pioneer Corp., Sony Corp., Thomson Multimedia, Time Warner Inc., Toshiba Corp., and Victor Company of Japan (JVC). This alliance became the DVD Consortium. Various discussions concerning the copy protection technology to be adopted took place between hardware manufacturers affiliated with the DVD Consortium, on the one hand, and between these manufacturers and the film industry on the other hand. The DVD-Video format was still lacking at the beginning of 1996 a copy protection mechanism and the rush to implement an adequate technology started to dominate the already so many technical arguments.

The first consensus on the DVD copy protection was reached in October 1996 when the Consortium chose the implementation of a modified technology originally developed by Matsushita and Toshiba and finally designated as Content Scrambling System (CSS). The CSS has become since then part of the total package of DVD licenses and requires that some critical information on disc must be encrypted, with the encryption keys to be stored on the disc itself. A dedicated piece of hardware or software is then needed to reconstruct the video signal provided that the decryption algorithm has been licensed from its legal holders. The licensees receive themselves a revokable identification code that must match the keys stored on disc during playback. A failure of any licensee to comply with the agreed CSS technology would be immediately followed, apart from the legal procedures, by the revocation of the assigned code, leading to the inability of decrypting the video information from newly produced media.

In the meantime, several other details of the DVD specifications were also approaching the final status, with the first 1.0 version due for publication by the end of 1996 or the beginning of 1997. The first DVD-Video players were commercialized in November 1996 in Japan, followed one month later by the first four films released on digital versatile discs by Warner Home Video of U.S.A. During the first quarter of 1997 a few companies started to sell DVD-Video players on the U.S. market and, by the end of that quarter, about 40 movie titles were already available. The DVD-ROM players entered the world market only in April 1997, but they had not really convinced the early adopters until 1998 when more software vendors decided to distribute their products also in DVD format along with the ubiquitous CD-ROM release.

5.3. DVD Standards

The DVD standards had been initially divided into five categories pertaining to various applications and designated by capital letters from A to E. A set of documents bundled in a so-called book and bearing one of these capitals was associated with a particular digital versatile disc format as illustrated in Fig. 5.2. Each book was meant to be licensed separately as a collection of specifications that defined together the physical parameters of a given DVD format, the manner in which binary data was logically organized on disc, as well as how a particular application should make use of the recorded digital information or should record digital information. During the years that followed the introduction of the DVD-Video, the original letter-based identification practically lost its significance because new documents were added to the entire package while spin-off applications and even new DVD formats also started to be developed. The current classification of digital versatile discs, which is complementary to the original five books, is given in Fig. 5.3. The DVD-ROM and DVD-Video media fall under the category of read-only discs as illustrated in Fig. 5.3 and were first specified by Book A and Book B, respectively. Each of these books contained both the physical[9] and file system[10] specifications and were equivalent to each other up to this point. According to the current classification of the DVD Forum, however, each of the two documents[9, 10] and the corresponding updates[12, 22, 37, 48, 88] and[13, 23, 38, 65] form now **Part 1** and **Part 2** of the **DVD-ROM Book**, respectively. Supplemental information and optional specifications have also been added regularly to the DVD-ROM standard since its release. The original Book B also included a third document[11] that covered all video issues. This document formed later with its updates[16, 24, 39, 49, 74] and the corresponding supplemental information a separate set of documents called **Part 3: Video Specifications**, which is in line with the schematic tree

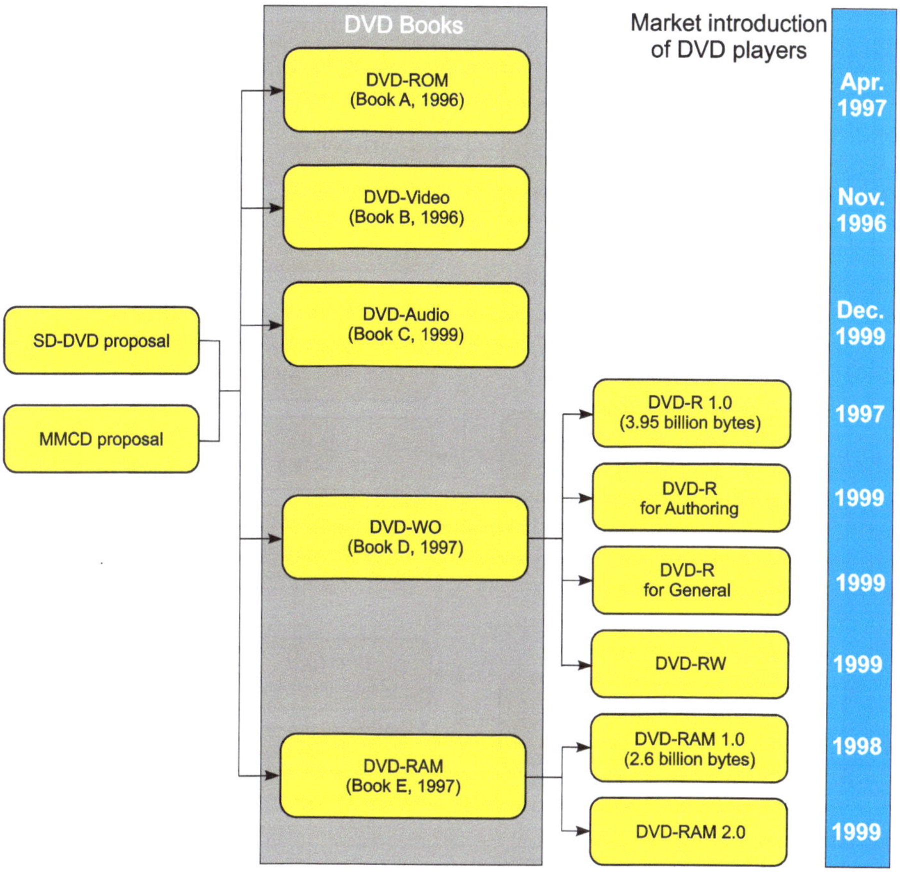

Fig. 5.2. Historical overview of the digital versatile disc formats as endorsed by the DVD Consortium before the end of the 1990s.

depicted in Fig. 5.3. The standards[9–11] were released in their first version simultaneously in 1996 and led to the introduction of the DVD-Video players and DVD-ROM drives for computer applications in 1996 and 1997, respectively. At logical level, the DVD-ROM data complies with the version 1.02 of the **Universal Disk Format (UDF)** specifications released in 1996 and obviously with all subsequent upgrades of the UDF[166]. It is also permitted but not mandatory to have both UDF and ISO 9660 file systems on one **UDF Bridge** disc, which represents a provision for backward compatibility with some old computers and their operating systems. Note that the DVD-Video media use only a restricted set of UDF features, which may be very well ignored by a video player but are sufficient for a DVD-ROM peripheral to retrieve the digital content in a computer environment. Nevertheless, either a separate MPEG-2 decoder board or MPEG-2 decoding software (running on a reasonably fast

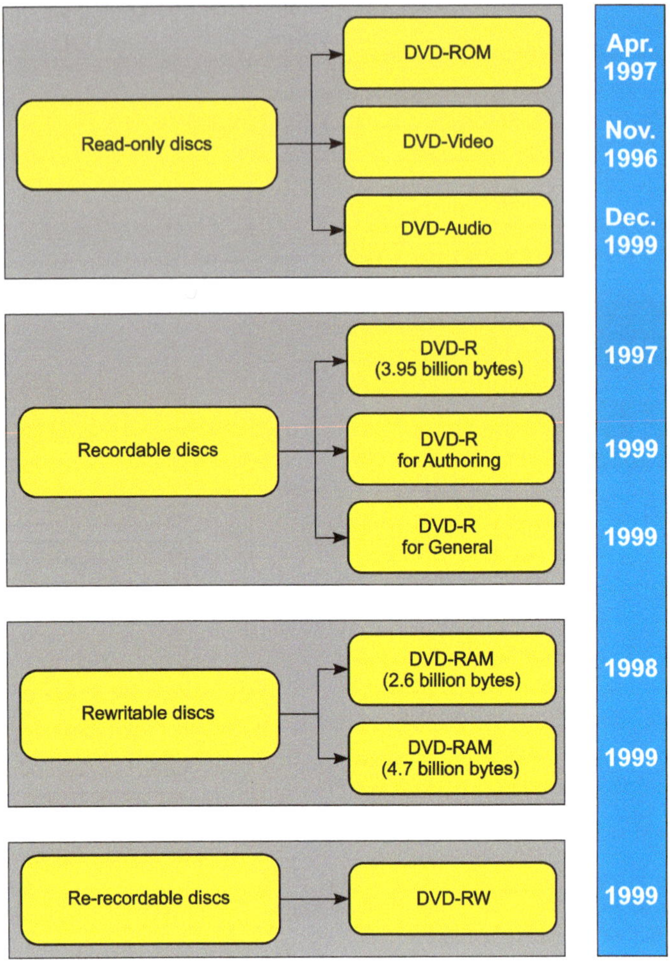

Fig. 5.3. Current classification of digital versatile disc formats endorsed by the DVD Forum.

computer) is additionally required to display the motion picture on the computer screen.

The third read-only DVD standard was finalized only in March 1999 when the version 1.0 of a dedicated format[26] for digital audio applications was released. The complete set of documents was previously known as Book C and included, besides the audio specifications published in 1999, the physical[9] and file system[10] descriptions mentioned already to be common for both DVD-Video and DVD-ROM. These three documents defined the **Digital Versatile Disc Audio (DVD-Audio)** format and led at the end of 1999 to the market introduction of dedicated optical media and players. The Working Group 4

(WG4) of the DVD Consortium, which was created to address and resolve all DVD-Audio issues, arrived at the final disc specifications after intensive discussions with representatives from the recording industry. Among the thirteen requirements to be met, the upcoming DVD media had to deliver an extremely high sound quality, to support multichannel surround sound, to hold at least 74 minutes of music, and to offer an efficient solution for protecting the copyrighted content. The release of the DVD-Audio format was delayed several times mainly due to copy protection issues, a situation that was very similar to postponing the release of the DVD-Video standard in 1996. Ultimately, the chosen technology for DVD-Audio copy protection was proposed by Verance Corporation of U.S.A., and represented a watermark in the form of a digital signature and the associated encryption key applied to the audio signal as imperceptible noise. It is the limitations of the human auditory system that make the watermark imperceptible. A second copy protection mechanism, called Content Protection for Prerecorded Media (CPPM) and developed by the 4C Entity formed by IBM Corp., Intel Corp., Matsushita Electric Industrial Co., Ltd., and Toshiba Corp., makes the DVD-Audio format suitable for distributing prerecorded copyrighted digital information. At present, the classification depicted in Fig. 5.3 contains the **Part 4: Audio Specifications** of the standard[26,27,40] for read-only DVD-Audio media along with all supplemental updates and optional specifications released since 1999.

The DVD-Audio format employs linear pulse code modulation (LPCM) as already used for CD-DA, but up to six channels at a much higher dynamic range (144 dB versus 96 dB in CD audio) can be reproduced. In other terms, the rate at which analog audio signals are sampled has been increased from 44.1 kHz in the CD system and a maximum of 96 kHz on DVD-Video discs to 192 kHz for the DVD-Audio format. The number of bits by which a digital audio sample is represented has also been raised from 16 used for CD-DA recordings to 24 in DVD-Audio. In order to store at least 74 minutes of 24-bit digital samples on a DVD-like physical layer while preserving the audio quality at reproduction, a suitable compression technique was needed. The DVD Consortium finally chose a technology known as Meridian Lossless Packing (MLP), which had been developed by Meridian Audio Limited of the U.K. together with several associated companies. The MLP makes use of very efficient encoding algorithms that are able to compact multichannel digital audio streams without loss of information. By contrast, the algorithms employed by other audio technologies, like Layer III (MP3) of the MPEG-2 standard, rely on perceptual coding and compress the digital information by throwing away those parts considered inaudible for the average listeners. The Meridian Loss-less Packing allows the record labels to store between 74 and 135 minutes of 5.1-channel surround music on a single DVD-Audio layer. Note, however, that using the maximum number of quantization bits while sampling simultaneously six independent channels at 192 kHz reduces the playback time considerably below 74 minutes. For this reason, most DVD-Audio audio streams are sampled only at 48 or 96 kHz and contain 20-bit samples. Other audio formats, like Digital Theater Surround (DTS) or Direct Stream Digital (DSD) are optionally supported. An overview of the DVD-Audio characteristics is presented in Table 5.4. By the end of 2004 the DVD Forum approved a revision of the DVD-Audio specifications by which five more compression technologies have become optional in addition to DTS and

Feature	DVD-Audio	DVD-Video	CD-DA
Disc size	8 cm, 12 cm		
Storage capacity	4.376 GB (single layer)		650 MB
Playback time [min]	74 - 160 [1)]	130 (on average)	74
Sampling frequency [kHz]	44.1, 88.2, 176.4 48, 96, 192	48 or 96	44.1
Quantization	16, 20, or 24 bits		16 bits
Audio coding	LPCM, MLP, DTS, DSD, MPEG-1 and -2, MPEG-4 AAC, ATRAC3plus, MP3, WMA	LPCM, MPEG-1, MPEG-2, Dolby Digital, DTS, SDDS	LPCM
No. of audio channels	up to 6	up to 8	2
Theoretical frequency response [2)] [Hz]	20 - 96,000	20 - 48,000	20 - 22,000
Bit rate [kbit/s]	1411.2 - 9600	64 - 1644	1411.2
Still and moving pictures	available		CD-Text CD-Graphics
Copy protection	CPPM watermarking	CSS regional code	none

1) A playback time of 74 minutes corresponds to six audio channels sampled at 96 kHz with a resolution of 24 bits and compressed using the MLP technology.

2) See the remark at the bottom of Table 5.2.

Table 5.4. Overview of several DVD-Audio, DVD-Video, and CD-DA characteristics.

DSD: MPEG-1/-2 Layer-II, MPEG-4 High-Efficiency (HE) Advanced Audio Coding (AAC), ATRAC3plus, MP3, and Windows Media Audio (WMA) Pro.

One of the notable features of the DVD-Audio format is the scalability of its six audio channels. It is possible to divide these channels into two groups and allocate to each a different sampling frequency and number of quantization bits, which increases the efficiency of using the total disc storage capacity. The DVD-Audio format also supports some graphical features derived from the core features of the DVD-Video specifications. These features allow for displaying still pictures, synchronized lyrics (i.e., karaoke), navigation menus on a television screen, automatic links to Web sites, and even video clips.

For several reasons the DVD-Audio discs could not be played back immediately after their inception in the DVD-Video players commercialized by that time. This incompatibility was primarily due to an information area on disc, called AUDIO TS directory, which had a different logical format than the early video players could decode. A second reason for the early playback incompatibility of DVD-Audio media with the installed base of video players was and is still given by the additional Meridian Lossless Packing circuitry that is mandatorily needed to process DVD-Audio signals. Even today, most DVD-Video players lack the built-in MLP electronics. Note also that neither a DVD-Audio nor a DVD-Video conventional player suffices for the reproduction of high-fidelity sound recorded on DVD-Audio discs. In both cases the user must feed the six analog audio channels output by the player to an amplifier featuring the same number of inputs and driving at least an equal number of independent loudspeakers. Such high-fidelity reproduction equipment, expensive and often designated as home theater systems, has only become more affordable for the average consumers after 2005. Further market limitations that are still in place today are due to the extremely low number of DVD-Audio titles released monthly, which has not led to more than 2500 titles available worldwide by the end of 2008.

Returning now to video applications, it was strongly felt in the early 2000s that the growing penetration of the high-speed computer networks would have an impact on how people will watch the video content in the near future. In response to the anticipated changes in home entertainment, several improvements to the digital versatile disc system were tested before the DVD Forum finally proposed in 2004 the so-called iDVD (Interactive DVD) specifications, sometimes also known as WebDVD. The official name became later **Part 5: Enhanced DVD Specifications**[66] and the content of this document allows designers to add HTML links and scripts to a DVD-Video disc or to enhance a Web site with audio/video information from a local DVD drive. The compliant products must mandatorily incorporate a minimum set of interactive functions among which the Web connectivity will allow both a player to request information from the Web and the Internet site to control the DVD player.

This section about read-only digital versatile discs cannot be concluded without having a look at the various storage capacity options offered by the physical specifications[9] common to DVD-ROM, DVD-Video, and DVD-Audio. Table 5.5 provides an overview of the possible combinations between the disc size, number of recorded sides, and number of information layers. The recordable and rewritable formats are also included in this table as a preamble to the specific issues that will be addressed in the next section. The DVD+R and DVD+RW, which are not endorsed by the DVD Forum but will be discussed in Sect. 5.8, are also included in Table 5.5. The reader should be aware of the fact that the playback durations are estimated based on the average bit rate required during the MPEG-2 encoding and decoding. This estimation corresponds to a particular compression ratio of the original video stream. By sacrificing video quality in favor of more MPEG-2 compression or by recording less audio data (for instance, only one dubbing language and two stereo channels), longer playing times become possible.

5.4. The DVD-R

In contrast with the DVD-Audio format which did not exist at all when the video-only players were launched in 1996, techniques for recording and erasing discs of storage capacities at least equal to that of the CD were already developed by that time. Matsushita, for example, had already accumulated experience with its phase-change dual (PD) rewritable disc since the beginning of the 1990s and was one of the driving forces behind the development of the new rewritable DVD media. Many other companies belonging to the DVD Consortium were already building up their CD-R/RW expertise based on the standards introduced by Philips and Sony. These companies extended then the accumulated CD-R/RW knowledge and arrived by themselves at feasible technical solutions suitable for eventually producing recordable and rewritable digital versatile discs.

Among the specific problems related to the development of recordable and rewritable DVDs, the compatibility with their read-only counterparts did not generally play a decisive role as it did for the CD family. More precisely, not all companies involved in establishing the specifications for the new discs shared the view that backward compatibility was essential. From this perspective it was thought that only write-once DVD media would have to be compatible with the DVD-Video and DVD-ROM discs, players, and computer drives. The developers of rewritable formats, on the other hand, adopted quite different standpoints and ignored in the beginning deliberately the compatibility with the read-only DVD systems. It will be seen later in this section that the DVD-RAM was meant to become the format of choice for computer applications

Disc specifications	Media format	Physical construction (12-cm discs, unless specified)	Nominal user storage capacity			
			GB	Billion bytes	Vs. CD	Video time
Read-only discs: DVD-ROM DVD-Video DVD-Audio	DVD-1	Single side, single layer (8-cm)	1.363	1.464	2.1×	41 min
	DVD-3	Single side, dual layer (8-cm)	2.601	2.793	4.1×	1 h 18 min
	DVD-5	Single side, single layer	4.376	4.699	6.9×	2 h 10 min
	DVD-9	Single side, dual layer	8.354	8.970	13.2×	4 h 09 min
	DVD-10	Double side, single layer	8.752	9.398	13.8×	4 h 20 min
	DVD-14	Double side, mixed layer	12.730	13.669	20.0×	6 h 19 min
	DVD-18	Double side, dual layer	16.708	17.940	26.3×	8 h 18 min
Recordable discs	DVD-R 1.0	Single side	3.679	3.950	5.8×	1 h 50 min
	DVD-R 2.0	Single side	4.376	4.699	6.9×	2 h 10 min
Rewritable discs	DVD-RAM 1.0	Single side	2.463	2.645	3.9×	1 h 13 min
	DVD-RAM 2.0	Single side	4.422	4.748	7.0×	2 h 12 min
Re-recordable discs	DVD-RW	Single side	4.376	4.699	6.9×	2 h 10 min
–	DVD+R	Single side	4.376	4.699	6.9×	2 h 10 min
	DVD+RW	Single side	4.376	4.699	6.9×	2 h 10 min

Notes: a) The comparison with CD takes into account a typical storage capacity of the latter equal to 650 MB, with 1 GB = 1024 MB.
b) The typical video playback times are based on an average MPEG-2 compression corresponding to 4.8 Mbit/s retrieved from disc.
c) Mixed-layer discs contain one layer on one side and two stacked layers on the other.
d) Dual-side formats are also specified for all recordable and rewritable discs, which effectively doubles their storage capacity.

Table 5.5. An overview of read-only, recordable, and rewritable DVD formats.

and it was for this reason optimized for data access and random read-write operations. By contrast, the DVD-Video was optimized for continuous readout of a recorded movie.

The write-once (WO) format is known as **Digital Versatile Disc Recordable (DVD-R)** and was originally defined by the so-called Book D. This document covered the disc physical[17] and file system[18] specifications and was made officially available in July 1997. The feasibility of a recordable DVD medium was proven for the first time, just like in the CD-R case, by Taiyo Yuden Co., Ltd. of Japan[107]. As indicated already in Table 5.5, version 1.0 of the DVD-R standard specifies an optical medium whose storage capacity does not match that of the read-only DVDs. Due to complex problems that had to be solved when recording data at high storage densities, the DVD-R was at its inception only capable to hold 3.95 billion bytes. The reduced storage capacity was due to a slightly larger track pitch and somewhat longer channel bits than standardized for read-only DVD media. In addition, the recording process had to use 635-nanometer laser beams and could not work with the 650-nm wavelengths already employed in DVD-Video and DVD-ROM units. The main reasons for this choice were given by the insufficiently small size of a 650-nm laser spot that was required to accurately inscribe the information on disc and partially by the difficulty of manufacturing suitable organic dyes sensitive at 650 nanometers. The optical readout, however, could take place with red laser light of either wavelength, which made the written DVD-R 1.0 media physically compatible with the installed base of read-only players.

As mentioned already at the beginning of this section, the DVD-R format was developed by promoting its compatibility with DVD-Video and DVD-ROM discs higher on the list of requirements. However, as it was also the case with the compatibility of the CD-Rs with their read-only counterparts, distinctive elements had to be introduced in the write-once DVD format to help the recorders find a particular location on the blank media. It is remarkable that, in contrast with the CD-Rs, the DVD-R discs feature not only a wobbled groove but also separate embossed patterns imprinted between grooves for synchronization and addressing purposes (see Fig. 5.4). The upgrade of the DVD-R specifications to version 2.0 took place in 2000 when the write-once format able to store 4.7 billion bytes was released. Compared to its predecessor, the new recordable disc featured a narrower track pitch and a shorter channel bit length, both equal to their counterparts on read-only discs. The new media came in two flavors: **DVD-R for Authoring**[33, 34] for professional use and **DVD-R for General** [35, 36, 46, 60]. Each set of standards comprises a **Part 1** and a **Part 2** describing the physical construction of the disc and the file system, respectively. While the format for general use was meant for consumer applications and had to be recorded with a 650-nm laser beam, the authoring format targeted the professional DVD customers and preserved the 635-nm specification

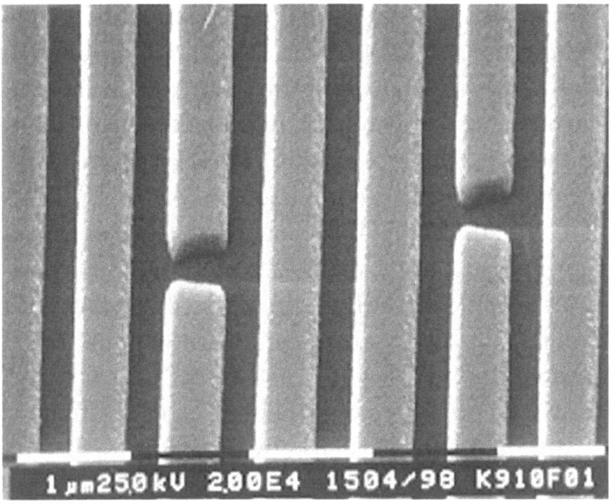

Fig. 5.4. Microscope image (20,000 magnification) of the DVD-R grooves (the dark bands) and land pre-pit imprints, viewed from the inside of the disc.

of the laser wavelength initially used in the DVD-R version 1.0. Another notable difference between the two types of media (and implicitly also between the related recording equipment) was that only professional, registered users and legitimate content providers were allowed to encrypt video data using the CSS algorithm. This restriction, changed nowadays to become less limiting and accommodate thereby more business scenarios, provided at that time protection against unauthorized replication of copyrighted content commercialized on DVD-Video discs. The advent of the legal distribution of audio and video via Internet, however, determined the DVD Forum at a later stage to create the possibility of recording a CSS-encrypted disc following the authorized downloading of the purchased movie. This technology is described in the **Part 1** and **Part 2** of the optional specifications of the **DVD-Download Disc for CSS Managed Recording**[84, 85].

The latest revisions[47, 55, 56, 61–63] of the DVD-R standard specify optionally video and data recording overspeeds up to 16X (although devices exist on the market to write the discs even at 20X). Note that, in order to use the DVD-Rs for video recording, an application layer was also needed to standardize the techniques of filling the DVD sectors with MPEG-2 data and prepare the recorded disc to be readable in legacy players and computer drives. Since the logical and video formats specified for DVD-Video did not easily allow common editing functions nor real-time recording, a third document entitled **DVD Video Recording for Rewritable and Recordable Discs**[31] was added in September 1999 to the set already containing the physical and file system specifications. Note that the current version of this document[87] does not only cover issues related to DVD-R media, but standardizes video recording techniques that are common to all recordable and rewritable DVD systems endorsed by the DVD Forum. Essentially, this standard describes the content layout on disc when audio and video streams are recorded in real time like, for example, when camcorders are used in conjunction with DVD-R drives. Unfortunately, the video recording process (often abbreviated as VR or DVD-VR) renders the disc incompatible with the DVD-Video standard[11] and consequently produces discs that remain unreadable by many conventional players. For this reason it is customary to record on DVD-R media in disc-atonce (DAO) mode or to make an incrementally-written disc compliant with the DVD-Video format before being ejected.

The first DVD-R recorders were introduced on the market by Pioneer Corp. in 1997 when no differentiation existed between media for general and authoring use. These devices were extremely expensive at that time and addressed the professional video recording market by using DVD-R media compliant with the version 1.0 of the specifications. The current professional applications still require such recorders, still highly priced, but relying at present on the DVD-R for Authoring specifications, version 2.0 (4.7 billion bytes). Another category of products is represented by the affordable video recorders commercialized for consumer purposes and using DVD-R for General discs. The same discs can also be written in computer data drives with DVD recording capabilities.

5.5. The Re-recordable DVD

A spin-off technology that emerged from the specifications initially known as Book D has led to the **Digital Versatile Disc Re-recordable**[28, 41, 50, 51, 58]. Despite its denomination, this format has been abbreviated since its inception as **DVD-RW** and is commonly pronounced as "DVD minus RW" or "DVD dash RW" (the latter pronunciation is recommended by the DVD Forum). Users may be confused by the mismatch between the attribute "re-recordable" and the already ubiquitous acronym DVD-RW. It was the judgement of several companies inside the DVD Forum that considered the DVD-RAM discs, introduced earlier than the DVD-RW, to deserve from a historical perspective being named rewritable. The same companies argued that DVD-RW, which emerged as a spin-off from the already-approved write-once specifications, must consequently be regarded as re-recordable. Looking back in time, the feasibility of a medium resembling many characteristics of the digital versatile disc and using the same recording material as CD-RW was proven already in 1996[177]. Since then, the real battle dealt with increasing the storage capacity on

*Since its establishment in 1961, ECMA has facilitated the timely creation of various standards in Information and Communications Technology and Consumer Electronics.

such media up to that of read-only DVDs. Some results were already achieved one year later[163], but the desired re-recordable DVD format was not yet in sight. An intermediate and quite reliable rewritable disc[158] with a storage capacity of 3.95 billion bytes was reported in 1997, although it was never standardized neither commercialized. The development efforts finally led to a DVD-RW format[185, 189] that could store 4.7 billion bytes (4.38 GB) and was compatible with the version 2.0 of the DVD-R. Phase-change materials were used to achieve about 1000 direct overwrite (DOW) cycles. Note also that, as a consequence of its spin-off from DVD-R, the new DVD-RW disc could be read out, written and erased with both 635- and 650-nm lasers. The recording speed was initially standardized to 1X, but an optional specification issued later under various revisions[52,59,67] allowed the media and drive manufacturers to write and erase up to 6X. The DVD-RW version 1.1 was also endorsed by European Computer Manufacturers Association (ECMA*) in an open standard[99] released in December 2002.

From a technical vantage point, the DVD-RW format was not directly compatible with the read-only DVDs when first commercialized. Being designed for appending new data and erasing or replacing the written information, the first DVD-RWs made use of dummy sectors that preceded and followed the written data blocks in order to guard them against being overlapped by new data during recording. These dummy sectors, which do not exist on read-only DVDs, are called "linking blocks" and were designed such that enough tolerance was allowed to replace an already written data sequence by a new one without affecting its neighbors, to safely erase a particular data sequence on disc, or to simply append new data. Although the linking blocks were optional, their absence required a high precision of linking new and old information on disc, which was hard to achieve in practice on DVD-RW media for quite some time. From this perspective, the DVD-RW format was often regarded in the past as being less attractive for computer applications since the latter usually write data randomly on disc. At present, however, most DVD-RW systems are capable to link data accurately and without loss areas (the so-called read-modify-write feature). For video applications only, the incompatibility between the DVD-VR video recording specifications[31] already discussed for DVD-R media and the equivalent specifications[11] for read-only discs remained also in place for DVD-RWs. It was, in fact, the "re-recordable" media that was meant to offer support for extended editing features. To accomplish this, the DVD-VR data stream was designed to circumvent those elements of the physical and logical structures of the DVD-Video format that were hampering the editing and real-time video recording. A so-called compatibility mode (also designated as video mode or VM) could be chosen during recording at the expense of using only fixed compression rates and of limiting the editing capabilities. During the years, as it was also the case for DVD-R media, most drives and quite some brands of consumer players have evolved and have incorporated capabilities to cope with re-recordable digital versatile discs written in DVD-VR mode.

For many reasons related to its specific physical and file system structure, the DVD-RW format was incompatible at its inception in 1999 with the installed base of DVD-Video players and DVD-ROM drives. The read-only systems already on the market by that time were not prepared to recognize and retrieve information from newer media, in general, which was a situation that also occurred when the CD-RW was introduced. However, the interest for both recordable and "re-recordable" discs began to grow in 2000 when Pioneer Corp. launched the DVD video recorder for consumer applications and Apple Computer, Inc. of U.S.A. decided to build DVD-R/RW drives supplied by Pioneer into their high-end Macintosh computers. The DVD-R/RW recorders remained extremely expensive for a while, but prices were soon heading toward affordable levels after two more rewritable DVD formats (to be discussed later) entered the challenge of providing on inexpensive optical discs, at home, more than two hours of digital video recording.

In order to improve the acceptance of the re-recordable DVD format among the potential users, several companies led by Pioneer Corp. established in May 2000 the RW Products Promotion Initiative (RWPPI). This organization, which originally counted 41 members and had several offices around the world, carried out events and campaigns and took care of surveys and investigations meant to promote the DVD-RW standard. Despite its relative success in the video recording markets, the DVD-RW format lacked from the beginning the flexibility required by data applications. This drawback, however, was not considered as such at the inception of the format since the DVD Forum used to promote the DVD-RAM for being used in computer environments, but it hampered the convergence between DVD video and data products.

5.6. The DVD-RAM

The **Digital Versatile Disc Random Access Memory (DVD-RAM)**, as mentioned already, was aiming from its early days to become the optical disc format of choice for computer applications. Initially specified in Book E, for which the version 1.0 was released in August 1997, the DVD-RAM standard emerged from a set of three proposals known at that time in the Work Group 5 (WG5) of the DVD Forum as Format A, B, and C. Format A was based on the PD media (already addressed in Sect. 4.4) which had been used in Japan for several years but was hardly available in the rest of the world. Format B was proposed by Philips Electronics and Sony Corp. and supported from the beginning by Hewlett-Packard, with all three companies aiming at a rewritable DVD system fully compatible with both DVD-ROM and DVD-Video media.

Format C, which eventually won by voting, was advanced by Matsushita[157] as a super-density RAM disc with a storage capacity of 2.6 billion bytes and was eventually amended to comply with many common specifications of the DVD family (for example, the initially proposed modulation code was abandoned in favor of EFMPlus). This format was also strongly supported by Toshiba Corporation and Hitachi Ltd.

Following the initial approach to categorize the DVD documents, Book E covered the physical[19] and file system[20, 21] specifications of the 2.6-billion-byte disc. Single- and double-side, single-layer DVD-RAM media that obeyed these specifications became available soon in plastic cartridges. The Japanese companies Hitachi, Matsushita, and Toshiba began shipping DVD-RAM drives for computers already in 1998, but the initial sales figures were not too promising. More storage capacity was needed as indicated already in Table 5.5, and this improvement took place in 1999 when the existing specifications were upgraded to version 2.0[29, 30]. Some computer vendors showed immediately a significant interest for the new discs, which were now capable to store 4.7 billion bytes while preserving both the excellent random access performance and the very large number of write-erase cycles introduced by the version 1.0 of the original Book E. At present, according to the classification depicted in Table 5.3, the DVD-RAM discs comply with **Part 1: Physical Specifications**[29, 42, 69] and **Part 2: File System Specifications**[30, 54]. The DVD-RAM format was also endorsed by the European Computer Manufacturers Association, which published two standards[92, 96] in 1999 and 2002, respectively. However, due to several specific format characteristics that will be explained below, the recorded DVD-RAM media was initially excepted from being played back in the vast majority of legacy DVD-ROM drives and DVD-Video players although the technology was already available before the end of the past century[151].

In the first place, the DVD-RAM discs use a technology called land-groove recording (see Fig. 5.5) that allows data to be written along grooves as well as in between. Since specialized optics and electronics are required to retrieve data from a land-groove physical structure, the vast majority of read-only DVD systems remained totally incapable in the early years to perform the readout operation. Secondly, the DVD-RAM discs feature embossed patterns that interrupt regularly the land-groove continuity to help the recorder locate any empty or written data block. These interruptions can only be handled by dedicated electronics, which was initially not incorporated in DVD-ROM drives nor in DVD-Video players. Another particularity of the DVD-RAM format is its zoned constant angular velocity (ZCAV) data structure. Unlike constant linear velocity for which all other optical disc systems have been designed, ZCAV media spin at a fixed rotational frequency and contain annular regions with a predermined number of data sectors within each region. Finally, DVD-RAM discs used to be supplied with protection cartridges[93, 97] that had to be

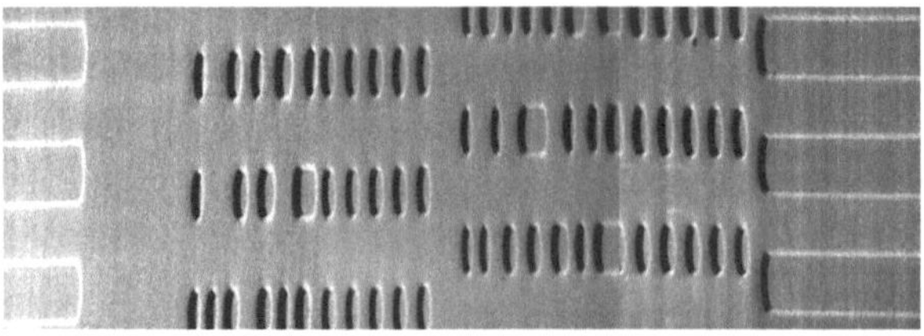

Fig. 5.5. SEM image of a DVD-RAM substrate viewed from the inside of the disc. The photograph has been electronically trimmed to reduce the length of the embossed headers, otherwise too long for one image.

accommodated as well by the mechanical construction of the conventional DVD-players. Although labeled as optional, the cartridge was very much desired in the early days because the land-groove DVD-RAM construction was known to deliver more erroneous data when playing back discs with surface defects. The cartridge did protect the optical medium against surface damages, but the downside of this practical requirement was that most manufacturers of consumer players and computer drives chose to exclude their support for DVD-RAM playback and avoid thereby the additional costs implied by the necessary mechanical modifications. At present, however, DVD-RAM media are also being sold without cartridges and they can be read out in many DVD-Video players and DVD-ROM drives.

As previously mentioned, the DVD-RAM physical and logical specifications have been designed deliberately for applications that read and write data very often and in a random manner, particularly in computing environments. Notwithstanding, DVD-RAM media have also been found suitable for real-time video applications. Stand-alone video recorders and even video cameras that use 8-cm DVD-RAM discs and are based on the DVD-VR standard[31, 43, 87] have entered successfully the consumer electronics market. When compared to DVD-RW, the DVD-RAM physical and logical formats are better prepared for random access real-time video recording and editing, but they suffer from their incompatibility with many DVD-Video players. As a sort of compensation, DVD-RAM media feature an impressive DOW performance since they can be practically recorded and erased successively more than 100,000 times compared to only 1000 overwrite cycles specified for the CD-RW and DVD-RW discs. As for the random access performance, the DVD-RAM format allows a host computer or video processor to read and write individual data sectors of 2048 bytes. This 2-kB operation mode matches the addressing capabilities of other data storage devices currently used in computer environments, such as the hard-disk drives. By comparison, recall that DVD-R and DVD-RW only offer sequential recording in data blocks of lengths equal to multiples of 32 kB in order for the written media to remain compatible with the read-only DVDs.

Yet another particularity of the DVD-RAM format is a well-designed defect management scheme that allows for checking off all bad sectors to avoid their potential usage at a later time. The DVD-RAM drive itself takes care of all defect management details and the bad sectors appear invisible for the host computer. The capabilities and the high reliability in operation of the DVD-RAM peripherals and media are comparable to those of the hard-disk and magneto-optical disk systems, which has basically been one of the main goals to be achieved from the beginning of the standardization process in the mid 1990s.

Returning now to the entertainment equipment, the DVD-VR technology in combination with real-time video recording on DVD-RAM have created several attractive features for consumers. Rich editing options are available even on camcorders to create menus and split, add, merge, or delete frames, titles, thumbnails, etc. On stand-alone devices it is possible for the user to legally make digital replicas of copyrighted content if permitted by the copy control information embedded inside the original material. This feature of DVD-RAM systems is based on the Content Protection for Recordable Media (CPRM) specifications. The technology was proposed to the DVD Forum by IBM Corp., Intel Corp., Matsushita Electric Industrial Co., Ltd., and Toshiba Corp. in the late 1990s and its implementation has become mandatory since 2000 for all manufacturers of DVD-RAM and DVD-RW equipment. The CPRM allows consumers to make only one copy (if permitted) of the audio/video content distributed through various digital communication channels like, for example, the digital TV broadcasts. A promotion group formed by several Japanese and Korean Companies and called RAMPRG was also formed in 2003 to raise the awareness of the DVD-RAM format among consumers and to educate the industry about its unique benefits.

5.7. Other DVD Forum's Specifications

In addition to the standards addressing individually the physical structure and the file system of only one type of recordable disc, the DVD Forum has also made available a few specifications that cover user applications for multiple sorts of media simultaneously. The **DVD-VR**, officially designated as the **DVD Specifications for DVD-RAM/DVD-RW/DVD-R for General Discs, Part 3: Video Recording**[31, 43, 87] has already been addressed in Sect. 5.4 and 5.5. Similarly, it is also possible to record only audio streams in real time, for which the appropriate technology has been described in the document entitled **DVD Specifications for DVD-RAM/DVD-RW/DVD-R for General Discs, Part 4: Audio Recording**[57, 64]. It becomes thereby possible to create DVD-Audio discs by employing either lossless or lossy coding algorithms (see

Table 5.4) operating upon multichannel surround sound inputs. The **DVD-AR** specifications, as they are also called, provide support for JPEG pictures to be displayed as static images on screens and allow menu-based navigation through up to 1000 songs. A complementary document is still under preparation[89] and addresses the professional audio recording devices (the application will become known as **DVD-PAR**). Last but not least, a technology called **DVD-SR** and covered by the **DVD Specifications for DVDRAM/DVD-RW/DVD-R for General Discs, Part 4: Stream Recording**[45] allows users to record in real time basically any digital signal. It becomes possible thereby to transfer audio/video digital information obtained from satellites, Internet, cable tuners, etc. to a simplified data stream that fits the DVD-Video structure as application data packets. However, the player will have to pass the stream during readout to an appropriate decoder.

Until now, only the single-layer recordable DVDs have been treated in this section. The consumers, however, have always indicated that more storage capacity is needed on optical discs and this also led, once the technology advanced sufficiently, to the introduction of the dual-layer (DL) recordable and rewritable DVDs. The **DVD-R for DL** is currently specified by two documents[70, 81] that naturally emerged from the version 2.0 of their Part 1 and Part 2 equivalents for single-layer media. Similarly, the **DVD-RW for DL** is also been specified by Part 1[82] and Part 2[86] covering the physical layer and the file system, respectively.

Last but not least, it must also be mentioned that all original specifications for recordable DVDs have been updated by the DVD Forum to allow the operation at higher overspeeds than 1X. It is not compulsory for a manufacturer to build drives complying with the highest recording speed for a particular DVD format, and for this reason the officially-called **Nx-speed** specifications are optional. Several successive revisions of these optional documents are available for each type of disc. As the technology progressed, the revisions defined the necessary system parameters and the technical requirements needed to increase the recording speeds from 1X to 16X for both the DVD-R for General[47,55,56,61–63] and the DVD-RAM[53,68,75–78] media, and from 1X to 6X for DVD-RWs[52, 59, 67]. The recording speed on dual-layer write-once discs also increased considerably[71–73, 79, 80] and reached 12X by the end of 2006. For DVD-RWs, although the standard only specifies over-speed factors up to 2X, many drive manufacturers have extended this limit in practice to 4X and even to 6X on media from selected suppliers only.

5.8. Miscellaneous DVDs and DVD-like Optical Media

The rich history of the compact disc has not only led to worldwide standards and the accompanying products, but has also produced several CD-like

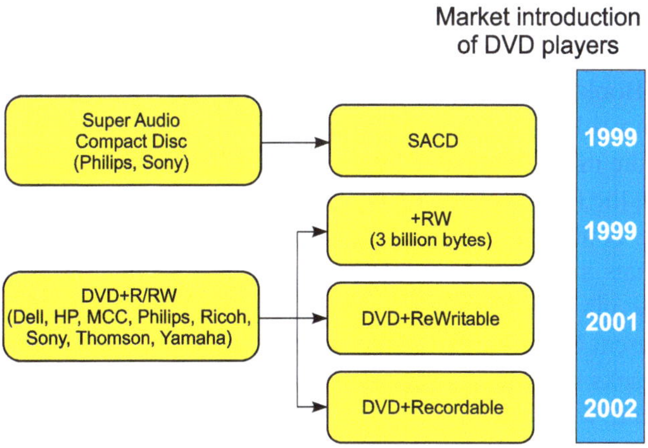

Fig. 5.6. Digital versatile disc standards endorsed by other organizations than the DVD Forum.

formats of which some succeeded as well on the market. From a similar perspective, the digital versatile disc does not represent an exception and has led to several DVD variations, with some of them being even supported strongly by important industry leaders as indicated schematically in Fig. 5.6.

To begin with, the market potential of a dual-layer CD/DVD medium was recognized by Philips Electronics and Sony Corporation already during their MMCD cooperation. The common view shared by these two companies had led them to the development of a new optical disc format called **Super Audio Compact Disc (SACD)** that was specified by documents released for the first time in 1999. Although not supported by the DVD Forum, the new format drew the attention of consumers in May 1999 when Sony introduced the first SACD player on the Japanese market. This example was followed soon by other manufacturers of audio equipment as the number of audio tiles started to increase as well. It is estimated that about 4500 SACD titles have become available since the format inception. The standardization aspects are managed by Philips and Sony through a set of documents describing the physical[172] and audio[173] specifications, and three separate specifications[171,174, 175] that cover the copy protection issues. At the heart of the copy protection technology lies the so-called Physical Disc Mark (PDM) that is embedded into the relief structure of the disc during the manufacturing process. Since the PDM is neither embedded in the data stream nor in the administration zones of the logical format, a bit-by-by copy of the recorded disc will be rendered unusable. Encryption of the data stored on SACD media is optional, but players must always be fitted with dedicated decryption electronics that will extract the PDM code from the readout high-frequency signal.

The hybrid SACD internal structure consists of a high-density (DVD type) semitransparent information layer that holds the SACD digital audio stream and a low-density information layer whose characteristics comply with the compact disc's Red Book. This particular construction allows the music producers to release audio titles for the owners of SACD reproduction equipment while preserving the media compatibility with the installed base of CD-DA players. In principle, the two information layers hold the same audio content (although bonus tracks could make them differ from each other) but encoded in different digital formats and providing different reproduction quality levels. A legacy compact disc player will recognize the CD-DA information stacked behind the semitransparent SACD layer inside the disc structure and will start decoding the audio tracks recorded as specified in the Red Book. The SACD player, on the other hand, will recognize both information layers but will prefer the signal reproduction from that layer closest to the disc surface. Note that the physical specifications[172] are not limited only to the hybrid dual-layer structure, but allow also the manufacturing of single- as well as dual-layer media containing only SACD content. In fact, only about half of the SACD albums released until mid-2003 carried information on both layers. One should also be aware that the SACD and DVD-Audio formats are not mutually compatible and can only be used each with its own dedicated reproduction equipment. Home entertainment systems that can handle DVD-Video media and one or both audio DVD formats are common at present.

A few relevant characteristics of the SACD format are summarized in Table 5.6. Historically, this format has been derived from the Direct Stream Digital (DSD) proposal of Sony Corporation, which incorporates analog-to-digital conversion (ADC) techniques very much studied and implemented by Philips in other products. The Sony proposal came in 1996 and at that time the DVD-Audio standard did not exist. The basic principle behind the DSD is sampling of the audio signal at very high frequencies, typically 15 times larger than in DVD-Audio or 64 times larger than in CD-DA. As a second characteristic, the DSD output data stream contains right after the the analog-to-digital conversion a sequence of single bits. Recall for comparison that a digital audio stream in CD-DA format contains words of 16 bits and the DVD-Audio format is based on words of 16, 20, or 24 bits, with a single data word being processed as an individual audio sample. Prior to being recorded on disc, the DSD informationis encoded in a lossless fashion using a technology called Direct Stream Transfer (DST) and the resulting data is arranged in sectors and data blocks that are similar to those used in all other DVD formats. Simply speaking, a SACD player retrieves the binary information just like any other DVD player but it performs dedicated DST decoding followed by the conversion of the bitwise DSD data into analog audio.

The SACD format claims that a high sampling rate combined with some technologies which shape the noise and push its frequency spectrum outside

Feature	SACD		CD-DA
	SACD layer	CD layer	
Disc size	8 cm, 12 cm		
Storage capacity	4.376 GB	650 MB	
Playback time [min]	80 [1)]	74	
Sampling frequency [kHz]	2822.4	44.1	
Quantization	1 bit	16 bits	
Audio coding	DSD	LPCM	
No. of audio channels	up to 6	2	
Theoretical frequency response [2)] [Hz]	20 - 100,000	20 - 22,000	
Bit rate [kbit/s]	2822.4	1411.2	
Still and moving pictures	available	CD-Text CD-Graphics	
Copy protection	PSP-PDM	none	

[1)] The playback time corresponds to an average compression ratio of 2.2:1
[2)] See the remark at the bottom of Table 5.2.

Table 5.6. Comparative overview of the SACD and CD-DA characteristics.

the audible range provide a superior sound quality. Another claim is that sharp transitions in the original audio waveforms can be reproduced more accurately from a bit-wise data stream clocked at very high sampling frequencies than from a low sampling rate, linear PCM data stream. The technical debate on sound quality and reproduction fidelity initially split the consumer electronics market into two camps, with DVD-Audio and SACD advocates trying to overrule the 20,000-Hz natural limit of the human auditive system. At present, however, the wide spread of several lossy coding technologies like MP3 have relaxed the race toward higher audio fidelity. A somewhat less disputable feature of both DVD-Audio and SACD systems is their capability to deliver 5.1-channel surround sound for which the users need relatively expensive 6-input/output amplifers (and speakers) during reproduction. Note, as a practical observation, that many consumers do not sit at one fixed spot when listening music and can therefore not fully enjoy the advantages of multichannel sound for which an optimal listener location is required (recall the speaker configurations from Fig. 5.1).

Despite their wide participation in the DVD Forum activities, Sony and Philips have never given up their own close and very old collaboration in optical storage. As indicated already in Fig. 5.6, the SACD standard for read-only media was followed by three other specifications for rewritable/recordable discs. The most spectacular and yet successful departure from the documents approved by the DVD Forum has emerged from Format B proposed once by Philips Electronics, Sony Corp., and Hewlett-Packard to become the rewritable DVD standard. Recall from Sect. 5.5 and 5.6 that DVD-RW was not considered evenly suitable for data and video applications, while DVD-RAM was not designed to be directly playback-compatible with the installed based of DVD-Video and DVD-ROM players. It was thought for quite a while during the second half on the 1990s that a cheap, rewritable and yet fully read-compatible DVD system would have little chance to be developed to serve all sorts of applications. Two reasons mainly accounted for this technological puzzle: (i) the difficulty of replacing directly a given data block on disc without affecting the physical continuity of the data stream along the track (recall the linking blocks used by the DVD-RW format); and (ii) the difficulty of replacing a given MPEG-2 video sequence by another one, while updating the logical information that gives the user selective access to all indexed written areas on disc.

The answers to the above questions did not come until 1999 when Royal Philips Electronics proposed a technology[182] that could effectively record, erase, and replace any block of MPEG-2 encoded information within the standardized DVD-Video data stream or, alternatively, any randomly chosen user data block within a DVD-ROM data sequence. This technology was first dubbed **plus Rewritable (+RW)** and did not use the DVD logo in order to avoid any conflict with the DVD Forum's standards equally endorsed by the +RW advocates. Three more Japanese companies, namely Mitsubishi Chemical Corp. (MCC), Ricoh Co., Ltd., and Yamaha Corp., strongly believed in the technical solution proposed by Philips and agreed to support the co-development initially started by Hewlett-Packard, Philips, and Sony. The short name +RW was soon replaced in journals and newspapers, thus not officially, by **Digital Versatile Disc plus ReWritable (DVD+RW)**. The six-company group published their rewritable DVD format specifications[94] in 1999 under the umbrella of the European Computer Manufacturers Association (ECMA). The +RW media could hold 3 billion user bytes, or 2.79 GB. Since these media could not store the total amount of data recorded on read-only DVDs, they became very soon obsolete, with Sony Corp. being the only manufacturer that released 3-billion-byte +RW discs and data drives during the year 2000 only.

The upgrade of the DVD+RW format took place within two years after the publication of the ECMA standard. The 6 companies afore-mentioned managed to increase the storage capacity of the disc up to that of a read-only DVD (that is, 4.7 billion bytes) and published the first version of the corresponding format specifications in 2001. The DVD Forum expressed no interest for the new format and discarded any proposal for including it among the DVD standards. Several ECMA representatives too regarded the new specifications as competing with the DVD-RW and DVD-RAM solutions and refused to support the DVD+RW proposal. The six DVD+RW format owners decided to keep this name on their internal documents but still avoid using the DVD logo that could only lead to legal confrontations. Aimed from an early development stage toward a two-way compatibility[3] with both DVD-Video and DVD-ROM applications, the DVD+RW turned very soon into the rewritable digital versatile disc of choice for many users. ECMA finally approved in 2002 the specifications for the 4.7-billion-byte version, at the same time with the approval of its DVD-RW counterpart. At present, the latest ECMA version of the DVD+RW standard[98, 103] allows recording speeds up to 8X on single-layer media and specifies also the dual-layer format rewritable at 2.4X. Hewlett-Packard, Mitsubishi, Philips, Ricoh, Sony, and Yamaha license together the DVD+RW format and supply all documents[110,112,113] to the licensees.

Several features of the DVD+RW system have been considered crucial for its present success. First, the format was designed to write a new data block on disc exactly from the point where the previous recording was halted, for which a positioning accuracy better than one channel bit was achieved as a significant accomplishment for that time[182]. Appending and/or erasing any number of recordings could consequently be performed without using linking sectors that would waste storage capacity. This so-called lossless linking technology formed the basis for obtaining, even after repeated erasures and replacements, an uninterrupted data stream similar to the one embossed on DVD-Video discs. Secondly, the DVD+RW system was also able to index on-the-fly the newly recorded information and rebuild the control areas on disc that allow a legacy DVD-Video player to access the data. A recorded disc could then be played back immediately in read-only systems without having to be finalized first, thereby avoiding a time-consuming operation for users. To assist the creation of discs fully compatible with the existing DVD-Video systems, the six DVD+RW format owners also published a document[168] that covered the real-time video recording (VR) aspects of the system. The DVD+VR, as it was often dubbed, provided a rich set of editing features and allowed for MPEG-2 compression at variable bit rate while preserving the playback compatibility of written media with the DVD-Video format. Users could insert, delete and append video titles and chapters, select video frames to build thumbnail-based menus, attach labels to thumbnails, etc., for which a remote control sufficed to operate a stand-alone consumer device. The defect management was yet another important DVD+RW feature that bridged the gap between consumer electronics and computer applications. The DVD-RAM media aside, which was developed specifically for computer data, no other optical disc system approved by the DVD Forum was designed to check off and bypass the damaged track areas. The defect management increased the reliability of reading and recording data and operated at the drive level, completely invisible for the host system. It was for this specific feature that computer vendors like Hewlett-Packard stood firmly behind this format. Last but not least, DVD+RW media was designed from the very beginning to cope with any recording overspeed between 1X and 2.4X in constant linear as well as constant angular velocity modes, thereby bridging the DVD video and data storage in an optimized and compatible manner.

For some time, the DVD+RW format was used without having implemented any particular copy protection technology, but the recorders had to be designed to inhibit the writing process when information flagged with copy control bits and/or CSS-protected information was input. In 2005, the **Video Content Protection System (VCPS)** jointly developed by Philips and Hewlett-Packard was added to the standardization package as an option needed only when recording copyrighted content from selected sources, such as digital video broadcasting.

The first DVD+RW video recorders were commercialized by Philips in September 2001 and proved that most DVD-Video players already on the market could play back written DVD+RW media. Ricoh Co., Ltd. followed soon with data drives that provided more practical evidence about the DVD+RW backward compatibility with the installed base of DVD-Video players and DVD-ROM peripherals. The DVD+RW strengths attracted immediately Thomson Multimedia of France and Dell Computer Corporation* of U.S.A., which joined in 2001 the cooperation established by that time only by the six format owners and formed together the 8C Group. Compaq Computer Corp. became in 2002 the third newcomer after its merge with Hewlett-Packard, and Microsoft Corp. became the ninth member of the DVD+RW steering group at the beginning of 2003. A few months later, however, Microsoft announced that it would support all recordable and rewritable DVD formats and gave up their privileged membership within the DVD+RW steering group. Backed by consumer electronics and computer industry leaders, the DVD+RW systems started to gain increasing attention from media manufacturers, from software developers, and obviously also from users. For some perfectionists it seems that the only disadvantage of the DVD+RW format until now has been its inability to gain support from the DVD Forum, with no book available from this international organization to cover a rewritable DVD medium fully compatible

*The company changed its name in 2003 to Dell Inc.

with the read-only counterparts and equally suitable for video applications and computer data. A notable characteristic of the DVD+RW format is also that it does not employ, apart from the wobbled groove, any sort of separate embossed patterns for synchronization and addressing purposes (recall that both DVD-RW and DVD-RAM formats make use of such relief structures). In fact, it was the goal of Philips and Sony to develop a rewritable DVD format based on cheap technologies already employed with success in the CD-RW systems.

At a later stage, a voluntary industry association known as the DVD+RW Alliance was formed between major manufacturers of DVD+RW hardware, software developers, as well as DVD+RW media producers around the world. Nevertheless, the set of "DVD plus" standards could not have been complete without including also specifications for a write-once medium. Two documents that cover the physical format[111] of the **Digital Versatile Disc plus Recordable (DVD+R)** and the related video recording issues[167], respectively, were released in their first version in 2002 by the eight-company group formed by Dell, Hewlett-Packard, MCC, Philips, Ricoh, Sony, Thomson Multimedia, and Yamaha. The DVD+R discs can be recorded in sessions, just like their CD-R predecessors, but have a DVD+RW-like physical format and make use of the lossless linking technology developed for their rewritable counterparts[178].

At present, the recording overspeed may range anywhere between 1X and 16X, but some drive manufacturers write selected media even up to 20X. Needless to say, the written DVD+R media are fully compatible with all legacy DVD-Video players, which has been achieved by employing the real-time DVD+VR video recording specifications[167]. A step further was made in October 2003 when the DVD+RW Alliance announced the dual-layer DVD+R media matching the storage capacity and the physical characteristics of the prerecorded dual-layer DVD-Video discs. The first version of the specifications[109] pertaining to the dual-layer DVD+Rs were released in December 2003.

Yet another addition to the initial DVD+RW technology proposed by Philips extended the Mt. Rainier concept already discussed at the end of Sect. 4.3. The new specifications[4] are referred to as **DVD+MRW** and provide drag-and-drop recording and erasing of computer data as well as defect management inside the drive. The latter feature should be associated, just like in the CD-RW case, with the inherent degradation of the phase-change media after successive rewrites. The drive detects the sectors worn out on disc and reallocates the space to compensate for the damage. The spare sectors use between 132 and 516 MB per disc, depending on the amount of written data. Since the defect management is implemented in the drive itself and not in the application software running on the host computer, it releases computer resources like microprocessor utilization time. Note that building Mt. Rainier capabilities only inside the drive does not suffice to support the DVD+MRW specifications.

The application software must also have knowledge about the new file format structure slightly different from the conventional DVD+RW. The Mt. Rainier requirements are at present fulfilled by most manufacturers of computers, computer devices, and relevant software.

Apart from the SACD and the DVD+RW Alliance's optical disc formats to which Royal Philips Electronics contributed fundamentally, several other DVD-like media have been proposed through the years. Some of these proposals were marketed with plenty of advertisement, but failed to become sustainable products. Considering first the read-only formats, a **DVDslim** disc was commercialized in Japan for a short period of time during 2002. The total thickness of the platter was practically equal to that of a 0.6-mm substrate, which reduced considerably the mechanical stiffness and even allowed the disc to be bent. The manufacturing costs could be reduced substantially by eliminating the dummy polycarbonate substrate that would have contributed otherwise to the conventional, total disc thickness of 1.2 mm. Further cost reductions were achieved also because the relatively expensive bonding process of the two substrates was not needed anymore. The DVD Forum reacted promptly by forbidding further commercialization of such media since they did not obey the DVD physical specifications.

The **hybrid DVD-Video/DVD-ROM** discs holding digital audio/video information mixed with raw computer data on one single information layer were also rejected by the DVD Forum. These discs were conceived by a few content providers by combining elements from two approved standards. When watching the movies by using a computer instead of a home entertainment system, the hybrid DVD-Video/DVD-ROM media were supposed to provide flexibility, functionality, and add new interactive features to the DVD-Video playback. The feedback from the various regional markets, however, was not very enthusiastic and led to practically no worldwide knowledge about these hybrid discs. The products disappeared very quickly from shelves, but the concept by itself could not be neglected.

The idea of mixing digital content for entertainment with computer-specific applications on the same disc was also studied by the DVD Forum itself. A **Combination DVD** approved in 2002 allowed the replication industry to create media with a read-only layer on one side and a recordable/rewritable layer on the other side. A very strict requirement was imposed, namely that each side remained compliant with the corresponding specifications already approved by the DVD Forum. For example, one side could contain a movie in DVD-Video format while the other side could be rewritable in DVD-RW format. It was obviously impossible to label a Combination DVD and maybe this was the main reason that hardly any media manufacturer released such products. From another perspective, however, the DVD Forum guaranteed in practice the compliance of such a hybrid disc with the endorsed standards.

Another attempt to combine the specifications of existing read-only optical media was made by the **DVD/CD Multi-Format** disc, probably better known as **DVDPlus**. This disc was believed to provide a smooth transition from CD to DVD especially in the Far East video markets by bonding the two distinct information layers in a dual-side stack. Such constructions were obviously substantially thicker than specified by any of the individual standards (1.2 and 0.6 mm for CD and DVD, respectively). The double-side DVDPlus was consequently heavier than conventional media and had the disadvantage of additionally loading the turntable motor and shortening thereby the lifetime of the drive. Slot loading mechanisms also had problems with heavier discs and often surprised the user unpleasantly by refusing to carry out an eject command. With some design efforts, a few replicators succeeded in producing Multi-Format discs of a total thickness just under the maximum limit of 1.5 mm allowed by the recognized international standards. This was achieved by lowering the total thickness of the CD part. Trusting the progress made in optical media manufacturing, the DVD Forum officially endorsed in 2004 a hybrid DVD/CD Multi-Format called **Single Thin Layer Disc** and approved a supplement to the DVD-ROM specifications. Somewhat independently, several replicators began to commercialize in 2004 the **DualDisc** featuring one side with prerecorded DVD-Audio content and another side with its CD-DA equivalent.

Inspired by the backward compatibility of the SACD with the ubiquitous compact disc players, the DVD Forum decided toward the end of 2002 to experiment with the **Hybrid CD/DVD disc** or simply **Hybrid DVD**. By contrast with its counterpart standardized by Philips and Sony, the prerecorded content on Hybrid DVDs was not necessarily limited to digital audio. Unfortunately, it was found that many tested DVD players could not decide what type of disc was inserted and either played from one of the layers chosen at random or did not play back the disc at all. In most cases the DVD equipment already on the market decided to reproduce the CD information, which was obviously not desired since the disc featured another layer with high-quality audio/video content. It became clear that most manufacturers of DVD equipment designed their products to comply with the existing standards but without anticipating any future format extensions. To avoid a commercial disaster that would be caused by the incompatibility of a DVD-like disc with the installed base of DVD-Video and -Audio players, the Forum decided not to pursue the Hybrid DVD option. Other concerns were related to the intellectual property legacy related to the SACD, which did not belong to the DVD Forum.

An 1.2-mm dual-side optical medium with two wavelength-dependent semi-transparent information layers was also considered as an alternative to both Single Thin Layer Disc and Hybrid DVD. When read out with an infrared laser from one side, the DVD semireflective film at half depth inside the disc would let the laser light pass through to be focused behind on the CD information layer (situated at 1.2 mm from the incident surface). Conversely, the CD's semireflective information layer close to the other disc surface hardly disturbed the incident red light and allowed the laser beam to read the information stored at 0.6 millimeters below the surface. When viewed from the CD-readable part, such an optical medium displayed the 1.2-mm polycarbonate substrate (divided in two halves by the DVD's reflective film) whereas only a thin but hard, transparent cover layer protected the CD's semireflective film at the DVD readable surface. No particular name was associated with this dual-layer hybrid DVD. The concept was abandoned soon because of the increased manufacturing costs compared to replicating conventional discs.

Remaining in the area of read-only DVD-like media, a well-represented category is formed by the optical discs used to distribute software for electronic games. The new generation of game consoles started to rely in the early 2000s on high-density optical media instead of re-using the CD format, and employed proprietary copy protection technologies meant to prevent the illegal spread of the officially-released software. For reasons easy to understand, any detailed description of such technologies is safely kept away from the large community of users and technical people. Among the vendors of game platforms, Sony introduced a modified version of the DVD-5 format for its Playstation 2 (PS2) machine, Nintendo Co. Ltd. of Japan made use of 8-cm DVD-like media for its GameCube console, while Microsoft Corp. introduced a version of the DVD-9 for their the Xbox units. In all cases the physical and logical specifications[9, 10] of the read-only digital versatile discs approved by the DVD Forum were adapted to an extent that could provide means for protecting the prerecorded information. The hardware platforms used for games, most of them still on the market at present, make use of a copy protection system but are also capable of playing back unprotected media, like legacy DVD-Video discs, and display their content on the TV screen used otherwise for games. Last but not least, Sony's PlayStation Portable uses yet other derivative of the DVD format: the 60-mm **Universal Media Discs (UMD)** holding 1.8 billion bytes and introduced in 2003.

One of the most controversial read-only DVD-like formats is the **Digital Video Express (DIVX)**, pronounced "divix." In its original concept, the DIVX as information carrier held MPEG-2 video content prerecorded and organized logically in a manner very similar to the video information replicated on DVDs, but allowed only a limited set of DVD features and was protected by a powerful encryption algorithm. The users were attracted to participate in a pay-per-view entertainment system organized by video rental companies and supported by leading movie studios and hardware vendors. During their several years of glory at the end of the 1990s, DIVX media could be purchased for a few dollars per piece (US $4.50 when the system was launched in June 1998), which represented a cheap deal compared to the sales price of a DVD-Video. However, for this fee, the buyers could only watch the film during a fixed period of time, usually during the next 48 hours from the acquisition. Once this period had elapsed, additional viewing periods could be purchased using the built-in modem also featured by any DIVX player. Note that a DIVX disc did not have to be returned and its viewing period only started when the disc was first played back. The DIVX system was initially backed by Circuit City, one of the largest consumer electronics retailers in the U.S., and by film studios like Disney, Paramount, and Universal. Facing the success of the DVD-Video systems, the DIVX business model was officially ended at the end of 1999. A reincarnation of the old name, now written as **DivX**, took place in 2001 when some technically-oriented consumers discovered the appeal of a new and very powerful video encoding technology called MPEG-4[138–143]. Within a matter of weeks, many people started to offer and download MPEG-4 video streams through computer networks, convert these streams to MPEG-1, and record the files in Video CD format onto CD-R or CD-RW media. The MPEG-4 encoding eventually became one of the fundamental technologies behind the DVD's successor: Blu-ray Disc.

The DIVX rental system did not represent a bad business model in itself, but the DVD-Video demand surged unexpectedly fast and triggered a significant price erosion of the prerecorded media and playback hardware. It was this price erosion that displaced DIVX from rental stores. The "nostalgia" for allowing consumers to watch a movie only a limited time period from the initial purchase returned in 2003 when self-destructing DVDs were introduced by Disney's home video division, Buena Vista Home Entertainment. Known as EZ-D DVDs, these media turned black after 48 hours from being removed from their vacuum-sealed packages. **EZ-D DVDs** did not promote a new video format, but were used just as DIVX to support a specific business model. Although primarily intended for consumers who were bothered by the inconvenience of returning a rental disc, the EZ-D technology also served the advertising markets. Some voices argued that such business models were environmentally unfriendly since the discs were thrown away after being rendered unplayable. Also unfavorable to this business were several surveys conducted just after the launch of the self-destructible DVDs and indicating that between 50 and 70% of the inquired customers would not be interested to rent products that "magically" render themselves unusable. Nevertheless, it became interesting also for the DVD Forum to create an optional specification[83] for those companies interested to release time-limited DVDs fully compliant with the installed base of video players and computer drives.

Perhaps the most controversial DVD-like formats until now have been those promoted by various academic institutions, governmental organizations, and industry representatives from China and Taiwan. Following the production and export in the early 2000s of optical disc drives, players, and media without fulfilling their obligations as licensees, several manufacturers in the Far East were pulled into legal disputes with the Japanese, American, and European DVD license holders. At that time all assorted royalties for a DVD player added up to $15-20 payable to the 6C group (the DVD 6C Licensing Agency formed by Hitachi, Matsushita, Mitsubishi, Toshiba, JVC and AOL Time Warner, later also joined by IBM), the 3C group (consisting of Philips, Pioneer, and Sony), and the MPEG licensing authority representing several companies. Royalty payments were also expected by Dolby Laboratories and by the licensers of various copy protection technologies. Many companies in the Far East thought about developing their own successor of the Super Video CD already discussed in Sect. 4.4. Several Chinese makers and researchers joined their efforts and submitted the **Advanced Versatile Disc (AVD)** physical format to their government. Defined as a read-only optical medium, the AVD was proposed in two flavors: single-sided single-layer discs with a storage capacity of 6 GB and single-sided dual-layer counterparts capable to hold 11 GB (judging after the DVD definitions, one should probably replace the gigabyte by billion bytes). A higher storage density than for DVD was achievable by reducing both the channel bit length and the track pitch, but there was insufficient information disclosed about other possible improvements, like the channel modulation code or the error correction. The AVD initiative obtained further support in April 2002 from the Taiwan Advanced Optical Storage Research Alliance (TAOSRA) and from the Industrial Technology Research Institute (ITRI) of Taiwan. **Enhanced Video Disc (EVD)** became the new appellative for the optical disc format and system to be developed through joint Chinese and Taiwanese efforts. Seeking to evade the royalty payments also for video processing while the prices of DVD-Video players started to drop, the EVD advocates acquired in 2003 a set of video compression algorithms from the American company On2 Technologies. These algorithms were known as being very efficient and were offered for significantly lower license fees than MPEG-2. A proprietary surround sound technology dubbed Enhanced Audio Compression (EAC) was developed to circumvent the Dolby licenses, while features providing Karaoke entertainment and supporting computer applications, games, Internet connectivity, etc. were also taken into account. It was estimated that royalty fees of only US $4 per player would have to be paid to international patent holders, but the lack of EVD software producers and the potential incompatibility with the DVD-Video discs (for which playback licenses were still required) curbed the enthusiasm of many manufacturers in the Far East already in 2004. Only two companies introduced EVD players in the beginning of that particular year, but the number of supporters grew to about 20 by 2006. Numerous promotions at international fairs gave hope that the DVD-Video format would be replaced by EVD in China ultimately in 2008.

The Chinese government also announced plans to build 20,000 movie theaters to stimulate the production of Hollywood films in EVD format. The issue, however, was that the introductory prices of US $250 in 2004 and the subsequent price reductions followed closely the prices of the DVD-Video players also produced massively in the Far East. Even if the EVD could offer high-definition video with resolutions up to 1920×1080 pixels, the compatibility with DVD-Video was absolutely required because media in the latter format was becoming widely available. By continuing to make use of the necessary licenses, no price reductions could be achieved as initially desired and the interest for EVD dropped at both the consumers and manufacturers side. In addition, the lack of an appropriate copy protection system discouraged further investments planned by some content providers and the number of EVD titles did not exceed several hundreds by the end of 2008.

The Chinese and Taiwanese ambitions to develop and commercialize a proprietary optical disc format did not stop with the announcement of the EVD. More interesting, even with HD-DVD preliminarily approved by the DVD Forum, ITRI submitted to this association in 2004 a separate proposal known as **Finalized Versatile Disc (FVD)**. Based on a red laser and meant to store 6 and 11 GB on single- and dual-layer media, respectively, FVD adopted the Microsoft's Windows Media Video Series 9 (WMV9) compression technology to handle HDTV resolutions up to 1280×720 pixels. The technical appraisal conducted by the DVD Forum, however, did not lead to a recommendation that could further promote FVD as a worldwide standard.

Yet another Chinese attempt to play a role in optical storage came from a company called Kaicheng HD Electronics Co., Ltd. in Beijing, which proposed in January 2004 a format called **High Definition Movie Player** and peculiarly abbreviated **HDV**. It was apparently the disputes inside the EVD alliance that had led to this new format. HDV was promised to consumers with plenty of prerecorded content, at lower prices than EVD and allegedly made use of a modified MPEG-2 technology to achieve lower bit rates at high-definition video quality. No players, though, followed the promises made early in 2004.

The Chinese announcements of new optical media formats continued to surprise the optical storage communities. In February 2005, New Medium Enterprises Inc. (NME) announced "a truly evolutionary technology" that would enable consumers, according to the company's long-term vision, to watch high-definition video at prices equivalent to those of DVD systems. The new disc was christened **Versatile Multilayer Disc (VMD)** and was claimed to be capable of storing in excess of 100 GB of data. The underlying technology would allow the information to reside in up to 20 layers on a single disc with no quality loss in the content stored. The first VMD players would hit the market by Christmas 2005 for less than US $150, but would use discs of only 20 GB in the beginning. These ambitions were trimmed two years later when other optical disc formats for high-definition video became commercially available: the HD-DVD endorsed by the DVD Forum and the Blu-ray Disc endorsed by a group of companies led again by Royal Philips Electronics and Sony.

Another high-density rewritable optical disc was developed independently from the DVD Forum by NEC Corporation of Japan. Initially known as **Multi-Media Video File (MMVF)**, this format became soon obsolete but competed once directly with DVD-RW and DVD-RAM in dedicated video recorders. A 12-cm MMVF disc could hold 5.2 billion bytes of data on one side, could withstand at most 1000 direct overwrite (DOW) cycles, and used a land-groove recording technology similar to the one employed in DVD-RAM media. A particular characteristic of the MMVF system was that it performed the readout and recording/erase operations with a 640-nanometer laser beam, which broke further away the compatibility with any type of DVD equipment. Before reaching the consumer electronics market, the NEC's format was renamed **Multimedia Video Disc (MVDisc)** and was solely intended for video applications, namely for video editing. The first MVDisc recorders reached the consumers at the end of 1999. Nevertheless, their incompatibility with any other sort of optical media became a serious drawback since the users could not play back DVD-Video nor CD discs on their MVDisc machines.

Finally, a write-once optical medium that relied on DVD-like technologies but did not comply with any of the specifications discussed until now is known as **DataPlay**. The development of the physical format and the realization of the entire system were initiated by DataPlay of U.S.A. but were completed in cooperation with several other companies. Among the latter, semiconductor vendors like ST Microelectronics of France and Intel Corporation contributed to the IC design, Eastman Kodak Co. developed the recordable media, and Samsung Electronics Co., Ltd. of Korea participated together with Toshiba Corp. in the design of the total drive. DataPlay discs, which could be used with legacy DVD equipment, only had an outer diameter of 32 mm but could hold 500 MB of information on both sides or 1 GB in dual-layer dual-size configurations. The miniature coin-size dimensions allowed these media to be used in many portable applications, such as handheld computers, cellular phones, car audio systems, digital still cameras, digital camcorders, etc. The disc was protected by a cartridge that resembled the mechanical constructions used for floppy and magneto-optical disks. Although based on red-laser DVD technology, the DataPlay systems employed a proprietary error detection and correction scheme that was claimed to enhance the reliability of the recorded and retrieved information. The recording material was very sensitive and required much less laser power during writing (about 2 mW) than even at present needed in DVD-R and DVD+R drives. The DataPlay system also featured a proprietary copy protection mechanism. At market launch in 2002, the miniaturized DataPlay drives made use of recordable media but could not prove themselves sufficiently competitive from an application point of view against the recordable and rewritable CD and DVD formats. An attempt made by a few content providers to supply music and video on DataPlay discs did not help the concept either and the advent of solid-state memories and of the miniaturized hard-disks finally brought the 3-cm DataPlay endeavor to an end.

References

Some documents listed below are not publicly available and can only be obtained from the licensers of the standard in question.

[1] Braat, J. and Bouwhuis, G. "Position Sensing in Video Disk Readout". *Applied Optics*, Vol. 17, No. 13, pp. 2013-2021, 1978.

[2] Braat, J.J.M. and Bouwhuis, G. "Optical Video Disks with Undulating Tracks". *Applied Optics*, Vol. 17, No. 13, pp. 2022-2028, 1978.

[3] Buma, C., Brondijk, R., and Stan, S.G. "DVD+RW: 2-Way Compatibility for Video and Data Applications". *IEEE International Conference on Consumer Electronics. Digest of Technical Papers*, June 13-15, 2000, Los Angeles, U.S.A., pp. 88-89.

[4] Compaq Computer Corp., Microsoft Corp., Royal Philips Electronics, and Sony Corp. *CD-MRW. Defect Management & Physical Formatting. Version 1.2*, 2004.

[5] Dobashi, H., Tanabe, T., and Yamamoto, M. "Crosstalk-Suppressed Readout System Using Shading Band". *Japanese Journal of Applied Physics*, Vol. 36, Part 1, No. 1B, pp. 450-455, 1997.

[6] DVD Format Logo License Entity. *DVD-RAM Disc (2.6 Gbytes). Test Specification. Version 1.0,* 1999.

[7] DVD Format Logo License Entity. *DVD-R Disc (3.95 Gbytes). Test Specification. Version 1.0*, 1999.

[8] DVD Format Logo License Entity. *DVD-Video Disc. Test Specification. Version 1.0*, 1999.

[9] DVD Forum. *DVD Specifications for Read-Only Disc. Part 1: Physical Specifications. Version 1.0*, 1996.

[10] DVD Forum. *DVD Specifications for Read-Only Disc. Part 2: File System Specifications. Version 1.0*, 1996.

[11] DVD Forum. *DVD Specifications for Read-Only Disc. Part 3: Video Specifications. Version 1.0*, 1996.

[12] DVD Forum. *DVD Specifications for Read-Only Disc. Part 1: Physical Specifications. Version 1.01*, 1997.

[13] DVD Forum. *DVD Specifications for Read-Only Disc. Part 2: File System Specifications. Version 1.01*, 1997.

[14] DVD Forum. *DVD Specifications for Read-Only Disc. Part 3: Video Specifications. IEC958 to Convey non-PCM Encoded Audio Bitstreams. Version 1.01*, 1997.

[15] DVD Forum. *DVD Specifications for Read-Only Disc. Part 3: Video Specifications. Jacket Picture Format. Version 1.01*, 1997.

[16] DVD Forum. *DVD Specifications for Read-Only Disc. Part 3: Video Specifications. Version-up Information (1.0 to 1.1)*, 1997.

[17] DVD Forum. *DVD Specifications for Recordable Disc. Part 1: DVD-R (3.95 Gbyte). Physical Specifications. Version 1.0*, 1997.

[18] DVD Forum. *DVD Specifications for Recordable Disc. Part 2: DVD-R (3.95 Gbyte). File System Specifications. Version 1.0*, 1997.

[19] DVD Forum. *DVD Specifications for Rewritable Disc. DVD-RAM (2.6 Gbyte). Part 1: Physical Specifications. Version 1.0*, 1997.

[20] DVD Forum. *DVD Specifications for Rewritable Disc. DVD-RAM (2.6 Gbyte). Part 2: File System Specifications. Version 1.0*, 1997.

[21] DVD Forum. *DVD Specifications for Rewritable Disc. DVD-RAM (2.6 Gbyte). Part 2: File System Specifications. Version-up Information (1.0 to 1.1)*, 1998.

[22] DVD Forum. *DVD Specifications for Read-Only Disc. Part 1: Physical Specifications. Version 1.02*, 1999.

[23] DVD Forum. *DVD Specifications for Read-Only Disc. Part 2: File System Specifications. Version 1.02*, 1999.

[24] DVD Forum. *DVD Specifications for Read-Only Disc. Part 3: Video Specifications. Version 1.11*, 1999.

[25] DVD Forum. *DVD Specifications for Read-Only Disc. Part 4: Audio Specifications. Packet PCM: MLP Reference Information. Version 1.0*, 1999.

[26] DVD Forum. *DVD Specifications for Read-Only Disc. Part 4: Audio Specifications. Version 1.0*, 1999.

[26] DVD Forum. *DVD Specifications for Read-Only Disc. Part 4: Audio Specifications. Version-up Information (1.0 to 1.1)*, 1999.

[27] DVD Forum. *DVD Specifications for Re-Recordable Disc. DVD-RW (4.7 Gbyte). Part 1: Physical Specifications. Version 1.0*, 1999.

[28] DVD Forum. *DVD Specifications for Rewritable Disc. DVD-RAM (4.7 Gbyte). Part 1: Physical Specifications. Version 2.0*, 1999.

[29] DVD Forum. *DVD Specifications for Rewritable Disc. DVD-RAM (4.7 Gbyte). Part 2: File System Specifications. Version 2.0*, 1999.

[30] DVD Forum. *DVD Specifications for Rewritable Disc. DVD-RAM (4.7 Gbyte). Part 2: File System Specifications. Version 2.0*, 1999.

[31] DVD Forum. *DVD Video Recording for Rewritable and Recordable Discs. Part 3: Video Recording. Version 1.0*, 1999.

[32] DVD Forum. *GuideBook for DVD-Video Text Data Usage. Version 1.0*, 1999.

[33] DVD Forum. *DVD-R for Authoring. Part 1: Physical Specifications. Version 2.0*, 2000.

[34] DVD Forum. *DVD-R for Authoring. Part 2: File System Specifications. Version 2.0*, 2000.

[35] DVD Forum. *DVD-R for General. Part 1: Physical Specifications. Version 2.0*, 2000.

[36] DVD Forum. *DVD-R for General. Part 2: File System Specifications. Version 2.0*, 2000.

[37] DVD Forum. *DVD Specifications for Read-Only Disc. Part 1: Physical Specifications. Version 1.03*, 2000.

[38] DVD Forum. *DVD Specifications for Read-Only Disc. Part 2: File System Specifications. Version 1.03*, 2000.

[39] DVD Forum. *DVD Specifications for Read-Only Disc. Part 3: Video Specifications. Version 1.12)*, 2000.

[40] DVD Forum. *DVD Specifications for Read-Only Disc. Part 4: Audio Specifications. Version-up Information (1.1 to 1.2)*, 2000.

[41] DVD Forum. *DVD Specifications for Re-Recordable Disc. DVD-RW (4.7 Gbyte). Part 2: File System Specifications. Version 1.0*, 2000.

[42] DVD Forum. *DVD Specifications for Rewritable Disc. DVD-RAM (4.7 Gbyte). Part 1: Physical Specifications. Version-up Information (2.0 to 2.1)*, 2000.

[43] DVD Forum. *DVD Video Recording for Rewritable and Recordable Discs. Part 3: Video Recording. Version-up Information 1.1*, 2000.

[44] DVD Forum. *Recommendations for DVD-Video Karaoke. Version 1.0*, 2000.

[45] DVD Forum. *DVD Specifications for DVD-RAM/DVD-RW/DVD-R for General Discs. Part 5: Stream Recording. Version 1.0*, 2001.

[46] DVD Forum. *DVD-R for General. Part 2: File System Specifications. Version-up Information (2.0 to 2.1)*, 2002.

[47] DVD Forum. *DVD-R for General. Part 2: Optional Specifications. 4X Speed DVD-R. Revision 1.0*, 2002.

[48] DVD Forum. *DVD Specifications for Read-Only Disc. Part 1: Physical Specifications. Version 1.04*, 2002.

[49] DVD Forum. *DVD Specifications for Read-Only Disc. Part 3: Video Specifications. Version 1.13)*, 2002.

[50] DVD Forum. *DVD Specifications for Re-Recordable Disc. DVD-RW (4.7 Gbyte). Part 1: Physical Specifications. Version-up Information (1.0 to 1.1)*, 2002.

[51] DVD Forum. *DVD Specifications for Re-Recordable Disc. DVD-RW (4.7 Gbyte). Part 2: File System Specifications. Version 1.01*, 2002.

[52] DVD Forum. *DVD Specifications for Re-Recordable Disc (DVD-RW). Optional Specifications. 2X Speed DVD-RW. Revision 1.0*, 2002.

[53] DVD Forum. *DVD Specifications for Rewritable Disc. DVD-RAM (4.7 Gbyte). Optional Specifications. 3X Speed DVD-RW. Revision 1.0*, 2002.

[54] DVD Forum. *DVD Specifications for Rewritable Disc. DVD-RAM (4.7 Gbyte). Part 2: File System Specifications. Version-up Information (2.0 to 2.01)*, 2002.

[55] DVD Forum. *DVD-R for General. Part 2: Optional Specifications. 6X Speed DVD-R. Revision 2.0*, 2003.

[56] DVD Forum. *DVD-R for General. Part 2: Optional Specifications. 8X Speed DVD-R. Revision 3.0*, 2003.

[57] DVD Forum. *DVD Specifications for DVD-RAM/DVD-RW/DVD-R for General Discs. Part 4: Audio Recording. Version 1.0*, 2003.

[58] DVD Forum. *DVD Specifications for Re-Recordable Disc. DVD-RW (4.7 Gbyte). Part 1: Physical Specifications. Version-up Information (1.1 to 1.2)*, 2003.

[59] DVD Forum. *DVD Specifications for Re-Recordable Disc (DVD-RW). Optional Specifications. 4X Speed DVD-RW. Revision 2.0*, 2003.

[60] DVD Forum. *DVD-R for General. Part 1: Physical Specifications. Version-up Information (2.0 to 2.1)*, 2004.

[61] DVD Forum. *DVD-R for General. Part 2: Optional Specifications. 12X Speed DVD-R. Revision 5.0*, 2004.

[62] DVD Forum. *DVD-R for General. Part 2: Optional Specifications. 16X Speed DVD-R. Revision 6.0*, 2004.

[63] DVD Forum. *DVD-R for General. Part 2: Optional Specifications. 2X Speed DVD-R. Revision 4.0*, 2004.

[64] DVD Forum. *DVD Specifications for DVD-RAM/DVD-RW/DVD-R for General Discs. Part 4: Audio Recording. Version-up Information 1.1*, 2004.

[65] DVD Forum. *DVD Specifications for Read-Only Disc. Part 2: File System Specifications. Version 1.05*, 2004.

[66] DVD Forum. *DVD Specifications for Read-Only Disc. Part 5: Enhanced DVD Specifications. Version 0.9*, 2004.

[67] DVD Forum. *DVD Specifications for Re-Recordable Disc (DVD-RW). Optional Specifications. 6X Speed DVD-RW. Revision 3.0*, 2004.

[68] DVD Forum. *DVD Specifications for Rewritable Disc. DVD-RAM (4.7 Gbyte). Optional Specifications. 5X Speed DVD-RW. Revision 2.0*, 2004.

[69] DVD Forum. *DVD Specifications for Rewritable Disc. DVD-RAM (4.7 Gbyte). Part 1: Physical Specifications. Version-up Information (2.1 to 2.2)*, 2004.

[70] DVD Forum. *DSpecifications for Recordable Disc for Dual Layer (DVD-R for DL). Part 1: Physical Specifications. Version 3.0*, 2005.

[71] DVD Forum. *DVD-R for DL. Part 1: Optional Specifications. 4X-Speed DVD-R for DL. Revision 1.0*, 2005.

[72] DVD Forum. *DVD-R for DL. Part 1: Optional Specifications. 4X-Speed DVD-R for DL. Revision 2.0*, 2005.

[73] DVD Forum. *DVD-R for DL. Part 1: Optional Specifications. 4X-Speed DVD-R for DL. Revision 3.0*, 2005.

[74] DVD Forum. *DVD Specifications for Read-Only Disc. Part 3: Video Specifications. Version 1.14)*, 2005.

[75] DVD Forum. *DVD Specifications for Rewritable Disc. DVD-RAM (4.7 Gbyte). Optional Specifications. 16X Speed DVD-RW. Revision 6.0*, 2005.

[76] DVD Forum. *DVD Specifications for Rewritable Disc. DVD-RAM (4.7 Gbyte). Optional Specifications. 6X Speed DVD-RW. Revision 3.0*, 2005.

[77] DVD Forum. *DVD Specifications for Rewritable Disc. DVD-RAM (4.7 Gbyte). Optional Specifications. 8X Speed DVD-RW. Revision 4.0*, 2005.

[78] DVD Forum. *DVD Specifications for Rewritable Disc. DVD-RAM (4.7 Gbyte). Optional Specifications. 8X Speed DVD-RW. Revision 4.0*, 2005.

[79] DVD Forum. *DVD-R for DL. Part 1: Optional Specifications. 4X-Speed DVD-R for DL. Revision 4.0*, 2006.

[80] DVD Forum. *DVD-R for DL. Part 1: Optional Specifications. 4X-Speed DVD-R for DL. Revision 5.0*, 2006.

[81] DVD Forum. *DVD-R Specifications for Recordable Disc for Dual Layer (DVD-R for DL). Part 2: File System Specifications. Version 3.0*, 2006.

[82] DVD Forum. *DVD-RW for DL. Part 1: Physical Specifications. Version 2.0*, 2006.

[83] DVD Forum. *DVD Specifications for Read-Only Disc. Part 1: Optional Specifications. Tiem Limited DVD-ROM. Revision 1.0*, 2006.

[84] DVD Forum. *DVD-R for General. Part 1/Optional Specifications: DVD Download Disc for CSS Managed Recording*, 2007.

[85] DVD Forum. *DVD-R for General. Part 2/Optional Specifications: DVD Download Disc for CSS Managed Recording*, 2007.

[86] DVD Forum. *DVD-RW for DL. Part 2: File System Specifications. Version 2.0*, 2007.

[87] DVD Forum. *DVD Specifications for DVD-RAM/DVD-RW/DVD-R for General Discs. Part 3: Video Recording. Version 1.2*, 2007.

[88] DVD Forum. *DVD Specifications for Read-Only Disc. Part 1: Physical Specifications. Version 1.05*, 2007.

[89] DVD Forum. *DVD Specifications for DVD-RAM/DVD-RW/DVD-R for General Discs. Part 6: Professional Audio Recording. Version 0.9*, 2008.

[90] ECMA-267. *120 mm DVD - Read-Only Disk. Third Edition*, 2001.

[91] ECMA-268. *80 mm DVD - Read-Only Disk. Third Edition*, 2001.

[92] ECMA-272. *120 mm DVD Rewritable Disk (DVD-RAM). Second Edition*, 1999.

[93] ECMA-273. *Case for 120 mm DVD-RAM Disks*, 1998.

[94] ECMA-274. *Data Interchange on 120 mm Optical Disk Using +RW Format - Capacity: 3,0 GB and 6,0 GB. Second Edition*, 1999.

[95] ECMA-279. *80 mm (1,23 Gbytes per side) and 120 mm (3,95 Gbytes per side) DVD-Recordable Disk (DVD-R)*, 1998.

[96] ECMA-330. *120 mm (4,7 Gbytes per side) and 80 mm (1,46 Gbytes per side) DVD Rewritable Disk (DVD-RAM). 3rd Edition*, 2005.

[97] ECMA-331. *Cases for 120 mm and 80 mm DVD-RAM Disks. Second Edition*, 2004.

[98] ECMA-337. *Data Interchange on 120 mm and 80 mm - Optical Disk using +RW Format - Capacity: 4,7 and 1,46 Gbytes per Side. Recording Speed up to 4X. 4th Edition*, 2008.

[99] ECMA-338. *80 mm (1,46 Gbytes per side) and 120 mm (4,70 Gbytes per side) DVD Re-recordable Disk (DVD-RW)*, 2002.

[100] ECMA-349. *Data Interchange on 120 mm and 80 mm Optical Disk using +R Format - Capacity: 4,7 and 1,46 Gbytes per Side (Recording speed up to 16X). 4th Edition*, 2008.

[101] ECMA-359. *80 mm (1,46 Gbytes per side) and 120 mm (4,70 Gbytes per side) DVD Recordable Disk (DVD-R)*, 2004.

[102] ECMA-364. *Data interchange on 120 mm and 80 mm Optical Disk using +R DL Format - Capacity: 8,55 and 2,66 Gbytes per Side (Recording speed up to 16X). 3rd Edition*, 2007.

[103] ECMA-371. *Data Interchange on 120 mm and 80 mm Optical Disk using +RW HS Format - Capacity: 4,7 and 1,46 Gbytes per Side Recording speed 8X). 2nd Edition*, 2008.

[104] ECMA-374. *Data Interchange on 120 mm and 80 mm Optical Disk using +RW DL Format Ü Capacity: 8,55 and 2,66 Gbytes per Side (Recording speed 2,4X). 2nd Edition*, 2008.

[105] ECMA-382. *120 mm (8,54 Gbytes per side) and 80 mm (2,66 Gbytes per side) DVD Recordable Disk for Dual Layer (DVD-R for DL)*, 2008.

[106] Eguchi, N. and Akiyama, Y. "High Density Optical Disk System Using Variable Five Modulation and Second Harmonic Generation Green Laser". *Japanese Journal of Applied Physics*, Vol. 32, Part 1, No. 11B, pp. 5307-5311, 1993.

[107] Hamada, E., Fujii, T., Tomizawa, Y., and Iimura, S. "High Density Optical Recording on Dye Material Discs: An Approach for Achieving 4.7 GB Density". *Japanese Journal of Applied Physics*, Vol. 36, Part 1, No. 1B, pp. 593-594, 1997.

[108] Hatakoshi, G., Nitta, K., Itaya, K., Nishikawa, Y., Ishikaw, M., and Okajima, M. "High-Power InGaAlP Laser Diodes for High-Density Optical Recording". *Japanese Journal of Applied Physics*, Vol. 31, Part 1, No. 2B , pp. 501-507, 1992.

[109] Hewlett-Packard Co., Mitsubishi Chemical Corp., Royal Philips Electronics, Ricoh, Co., Ltd., Sony Corp., Yamaha, Corp. *DVD+RW 4.7 Gbytes. Basic Format Specifications. System Description. Part 1: Dual Layer, Volume 1: 2.4X to 16X. Version 1.21*, 2007.

[110] Hewlett-Packard Co., Mitsubishi Chemical Corp., Royal Philips Electronics, Ricoh, Co., Ltd., Sony Corp., Yamaha, Corp. *DVD+RW 4.7 Gbytes. Basic Format Specifications. System Description. Part 1: Dual Layer, Volume 1: 2.4X. Version 1.01*, 2007.

[111] Hewlett-Packard Co., Mitsubishi Chemical Corp., Royal Philips Electronics, Ricoh, Co., Ltd., Sony Corp., Yamaha, Corp. *DVD+RW 4.7 Gbytes. Basic Format Specifications. System Description. Part 1: Single Layer, Volume 1: 1X to 16X. Version 1.31*, 2007.

[112] Hewlett-Packard Co., Mitsubishi Chemical Corp., Royal Philips Electronics, Ricoh, Co., Ltd., Sony Corp., Yamaha, Corp. *DVD+RW 4.7 Gbytes. Basic Format Specifications. System Description. Part 1: Single Layer, Volume 1: 2.4X & 4X. Version 1.31*, 2007.

[113] Hewlett-Packard Co., Mitsubishi Chemical Corp., Royal Philips Electronics, Ricoh, Co., Ltd., Sony Corp., Yamaha, Corp. *DVD+RW 4.7 Gbytes. Basic Format Specifications. System Description. Part 1: Single Layer, Volume 1: 8X. Version 1.01*, 2007.

[114] IEC 62107. *Super video compact disc–Disc-interchange systemspecification*, 2000.

[115] Iida, T., Higuchi, T., Sasaki, H., Ogoshi, K., and Yokozeki, S. "High Recording Density Optical Disc Mastering". *Japanese Journal of Applied Physics*, Vol. 32, Part 1, No. 11B, pp. 5447-5448, 1993.

[116] Ikeda, M., Nakano, K., Toda, A., Mori, Y., and Kojima, C. "AlGaInP Visible Semiconductor Lasers". *Japanese Journal of Applied Physics*, Vol. 26, Supl. 26-4, pp. 101-105, 1987.

[117] Immink, K.A.S. "EFMPlus: The Coding Format of the MultiMedia Compact Disc". *IEEE Transactions on Consumer Electronics*, Vol. 41, No. 3, pp. 491-497, 1995.

[118] ISO/IEC 10918-1. *Information technology–Digital compression and coding of continuous-tone still images: Requirements and guidelines*, 1994.

[119] ISO/IEC 10918-2. *Information technology–Digital compression and coding of continuous-tone still images: Compliance testing*, 1995.

[120] ISO/IEC 10918-3. *Information technology–Digital compression and coding of continuous-tone still images: Extensions*, 1997.

[121] ISO/IEC 10918-4. *Information technology–Digital compression and coding of continuous-tone still images: Registration of JPEG profiles, SPIFF profiles, SPIFF tags, SPIFF colour spaces, APPn markers, SPIFF compression types and Registration Authorities (REGAUT)*, 1999.

[122] ISO/IEC 13346-1. *Information technology–Volume and file structure of write-once and rewritable media using non-sequential recording for information interchange. Part 1: General*, 1995.

[123] ISO/IEC 13346-2. *Information technology–Volume and file structure of write-once and rewritable media using non-sequential recording for information interchange–Part 2: Volume and boot block*, 1999.

[124] ISO/IEC 13346-3. *Information technology–Volume and file structure of write-once and rewritable media using non-sequential recording for information interchange–Part 3: Volume structure*, 1999.

[125] ISO/IEC 13346-4. *Information technology–Volume and file structure of write-once and rewritable media using non-sequential recording for information interchange – Part 4: File structure*, 1999.

[126] ISO/IEC 13346-5. *Information technology–Volume and file structure of write-once and rewritable media using non-sequential recording for information interchange–Part 5: Record structure*, 1995.

[127] ISO/IEC 13490-1. *Information technology–Volume and file structure of read-only and write-once compact disk media for information interchange–Part 1: General*, 1995.

[128] ISO/IEC 13490-2. *Information technology–Volume and file structure of read-only and write-once compact disk media for information interchange–Part 2: Volume and file structure*, 1995.

[129] ISO/IEC 13818-1. *Information technology–Generic coding of moving pictures and associated audio information: Systems*, 2000.

[130] ISO/IEC 13818-10. *Information technology–Generic coding of moving pictures and associated audio information–Part 10: Conformance extensions for Digital Storage Media Command and Control (DSM-CC)*, 1999.

[131] ISO/IEC 13818-2. *Information technology–Generic coding of moving pictures and associated audio information: Video*, 2000.

[132] ISO/IEC 13818-3. *Information technology–Generic coding of moving pictures and associated audio information–Part 3: Audio*, 1998.

[133] ISO/IEC 13818-4. *Information technology–Generic coding of moving pictures and associated audio information–Part 4: Conformance testing*, 1998.

[134] ISO/IEC 13818-5. *Information technology–Generic coding of moving pictures and associated audio information – Part 5: Software simulation*, 1997.

[135] ISO/IEC 13818-6. *Information technology–Generic coding of moving pictures and associated audio information – Part 6: Extensions for DSM-CC*, 1998.

[136] ISO/IEC 13818-7. *Information technology–Generic coding of moving pictures and associated audio information–Part 7: Advanced Audio Coding (AAC)*, 1997.

[137] ISO/IEC 13818-9. *Information technology–Generic coding of moving pictures and associated audio information–Part 9: Extension for real time interface for systems decoders*, 1996.

[138] ISO/IEC 14496-1. *Information technology–Coding of audio-visual objects–Part 1: Systems*, 2001.

[139] ISO/IEC 14496-2. *Information technology–Coding of audio-visual objects–Part 2: Visual*, 2001.

[140] ISO/IEC 14496-3. *Information technology–Coding of audio-visual objects–Part 3: Audio*, 2001.

[141] ISO/IEC 14496-4. *Information technology–Coding of audio-visual objects–Part 4: Conformance testing*, 2000.

[142] ISO/IEC 14496-5. *Information technology–Coding of audio-visual objects–Part 5: Reference software*, 2001.

[143] ISO/IEC 14496-6. *Information technology–Coding of audio-visual objects–Part 6: Delivery Multimedia Integration Framework (DMIF)*, 2000.

[144] ISO/IEC 16448. *Information technology – 120 mm DVD – Read-only disk. Second Edition*, 2002

[145] ISO/IEC 16449. *Information technology – 80 mm DVD – Read-only disk, Second Edition*, 2002

[146] ISO/IEC 16824. *Information technology–120 mm DVD rewritable disk (DVD-RAM)*, 1999.

[147] ISO/IEC 16825. *Information technology–Case for 120 mm DVD-RAM disks*, 1999.

[148] ISO/IEC 16969. *Information technology–Data interchange on 120 mm optical disk cartridges using +RW format– Capacity: 3,0 Gbytes and 6,0 Gbytes*, 1999.

[149] ISO/IEC 20563. *Information technology–80 mm (1,23 Gbytes per side) and 120 mm (3,95 Gbytes per side) DVD-recordable disk (DVD-R)*, 2001.

[150] ITU-R BS.775-1. *Multichannel Stereophonic Sound System With and Without Accompanying Picture*, 1994.

[151] Izumi, K., Ohnishi, K., Inoue, M. "DVD-RAM Disc Reproducing Techniques for DVD-ROM Drives". *IEEE International Conference on Consumer Electronics. Digest of Technical Papers*, June 22-24, 1999, Los Angeles, U.S.A., pp. 266-267.

[152] Kayanuma, K., Iwanaga, T., Inada, H., Okanoue, K., Katayama, R., Yoshihara, K., Yamanaka, Y., Tsunekane, M., and Okada, O. "High Track Density Magneto-Optical Recording Using a Crosstalk Canceller". *Optical Data Storage. Proceedings SPIE*, Vol. 1316, pp. 35-39, 1990.

[153] Kobayashi, S.; Horigome, T.; Kock, J.P. de; Yamatsu, H.; Ooki, H. "Single Carrier Independent Pit Edge Recording". *Proceedings SPIE*, Vol. 2514, pp. 73-81, 1995.

[154] Kock, J.P. de, Kobayashi, S., Ishimoto, T., Yamatsu, H., and Ooki, H. "Sampled Servo Read Only Memory System Using Single Carrier Independent Pit Edge Recording". *Japanese Journal of Applied Physics*, Vol. 35, Part 1, No. 1B, pp. 437-442, 1996.

[155] Kozlowsky, W.J., Lenth, W., and Risk, W.P. "Compact Blue Lasers for Optical Recording Applications". *Optical Data Storage. Proceedings SPIE*, Vol. 1316, pp. 194-198, 1990.

[156] Milster, T.D., Wang, M.S., Li, W., and Walker, E. "Optical Filtering in the Collection Path of a Data Storage Device". *Japanese Journal of Applied Physics*, Vol. 32, Part 1, No. 11B, pp. 5397-5401, 1993.

[157] Miyagawa, N., Ohno, E., Nishiuchi, K, and Akahira, N. "Phase Change Optical Disk Using Land and Groove Method Applicable to Proposed Super Density Rewritable Disc Specifications". *Japanese Journal of Applied Physics*, Vol. 35, Part 1, No. 1B, pp. 502-503, 1996.

[158] Muramatsu, E., Yamaguchi, A., Horikawa, K., Kato, M., Taniguchi, S., Jinno, S., Yamaguchi, M., Kudo, H., and Inoue, A. "The New Rewritable Disc system for Digital Versatile Disc". *Japanese Journal of Applied Physics*, Vol. 37, Part 1, No. 4B, pp. 2257-2258, 1998.

[159] Muraoka, T., Iwahara, M., and Yamada, Y. "Examination of Audio Bandwidth Requirements for Optimum Sound Signal Transmission". *Journal of the Audio Engineering Society*, Vol. 29, No. 1/2, pp. 2-9, 1981.

[160] Muraoka, T., Yamada, Y., and Yamazaki, M. "Sampling Frequency Considerations in Digital Audio". *Journal of the Audio Engineering Society*, Vol. 26, pp. 252-256, 1978.

[161] Nagashima, M. "High density optical disk with V-shaped grooves". *Applied Physics Letters*, Vol. 43, No. 2, pp. 144-146, 1983.

[162] Nagashima, M., Takenaga, M., Yamashita, T., Yoshikane, T., Kagata, S., Okino, Y, and Kamio, K. "Experimental estimation of an optical disk system with V-shaped grooves". *Applied Physics Letters*, Vol. 43, No. 1, pp. 4-6, 1983.

[163] Nagata, K., Saimi, T., Furukawa, S., Nishiuchi, K., Yamada, N., and Akahira, N. "4.7 GB Phase-Change Optical Disk for an Authoring System of Digital Versatile Disc". *Japanese Journal of Applied Physics*, Vol. 37, Part 1, No. 4B, pp. 2236-2240, 1998.

[164] Oka, M., Kashiwagi, T., and Kubota, S. "A Triple-Density Optical Disk System Using a Frequency-Doubled Nd;YAG Laser". *Optical Storage and Scanning Technology. Proceedings SPIE*, Vol. 1139, pp. 149-154, 1989.

[165] OSTA. *Read Compatibility Specifications for DVD-ROM and 2.6 GB and DVD-RAM Devices. Revision 1.0*, 1999.

[166] OSTA. *Universal Disk Format Specification. Revision 2.60*, 2005.

[167] Royal Philips Electronics. *DVD+R Video Format Specifications. System Description. Version 3.0*, 2005.

[168] Royal Philips Electronics. *DVD+RW Video Format Specifications. System Description. Version 3.0*, 2005.

[169] Royal Philips Electronics and Sony Corp. *Super Audio CD System Description. Part 1. Physical Specification. Version 1.2*, 2001.

[170] Royal Philips Electronics and Sony Corp. *Super Audio CD System Description. Part 2. Audio Specification. Version 1.2*, 2001.

[171] Royal Philips Electronics and Sony Corp. *Super Audio CD System Description. Part 3-General. Copy Protection Specification. Version 1.2*, 2001.

[172] Royal Philips Electronics and Sony Corp. *Super Audio CD System Description. Part 1. Physical Specification. Version 2.0*, 2004.

[173] Royal Philips Electronics and Sony Corp. *Super Audio CD System Description. Part 2. Audio Specification. Version 2.0*, 2004.

[174] Royal Philips Electronics and Sony Corp. *Super Audio CD System Description. Part 3-General. Copy Protection Specification. Version 2.0*, 2004.

[175] Royal Philips Electronics and Sony Corp. *Super Audio CD System Description. Part 3-General. Copy Protection Specification. Version 2.1*, 2007.

[176] Schouhamer Immink, K.A. "The Digital Versatile Disc (DVD): System Requirements and Channel Coding". *SMPTE Journal*, pp. 483-489, August 1996.

[177] Shinotsuka, M., Shibaguchi, T., Abe, M., and Ide, Y. "Potentiality of the Ag-In-Sb-Te Phase Change Recording Material for High Density Erasable Optical Discs". *Japanese Journal of Applied Physics*, Vol. 36, Part 1, No. 1B, pp. 536-538, 1997.

[178] Stan, S.G. and Spruit, H. "DVD+R: A Write-Once Optical Recording System for Video and Data Applications". *IEEE International Conference on Consumer Electronics. Digest of Technical Papers*, June 18-20, 2002, Los Angeles, U.S.A., pp. 256-257.

[179] Tatsuno, K., Takahashi, M., Muraoka, K., Sugiyama, H., Nakamura, J., Andou, T., and Miyai, T. "High Storage Density Optical Recording with a Stable Micro Green Second Harmonic Generation Laser Consisting of Nd:YVO4 and KTP". *Japanese Journal of Applied Physics*, Vol. 31, Part 1, No. 2B , pp. 601-604, 1992.

[180] Taylor, J. *DVD Demystified. Second Edition*. McGraw-Hill., 2001.

[181] Uchino, K., Takada, K., Ohno, T., Yoshida, H., and Kobayashi, Y. "High-Density Pulse Width Modulation Recording and Rewritable Capability in GeSbTe Phase-Change System Using Visible Laser Beam at Low Linear Velocity". *Japanese Journal of Applied Physics*, Vol. 32, Part 1, No. 11B, pp. 5354-5360, 1993.

[182] Vlerken, J.J.L.M. van and Kimpe, W. de. "Format Detection for DVD+ReWritable 4.7 GB". *IEEE International Conference on Consumer Electronics. Digest of Technical Papers*, June 13-15, 2000, Los Angeles, U.S.A., pp. 162-163.

[183] Wang, M.S. and Milster, T.D. "Crosstalk Cancellation Using Differential Wax-Wane Focus Servo Technique". *Japanese Journal of Applied Physics*, Vol. 32, Part 1, No. 11B, pp. 5277-5283, 1993.

[184] Watkinson, J. *The Art of Digital Audio. Third Edition*. Focal Press, 2001.

[185] Yamaguchi, M., Togashi, T., Jinno, S., Kudo, H., Muramatsu, E., Taniguchi, S., and Inoue, A. "4.7 GB Phase Change Optical Disc with In-Groove Recording". *Japanese Journal of Applied Physics*, Vol. 38, Part 1, No. 3B, pp. 1806-1810, 1999.

[186] Yamanaka, Y., Hirose, Y., and Kubota, K. "High Density Optical Recording by Superresolution". *Japanese Journal of Applied Physics*, Vol. 28, Supl. 28-3, pp. 197-200, 1989.

[187] Yamanaka, Y., Kubota, K., Fujii, H., Kobayashi, K., Suzuki, T., and Gokan, H. "High Density Magneto-Optical Recording Using 0.67 μm Band High Power Laser Diode". *IEEE Transactions on Magnetics*, Vol. 24, No. 6, pp. 2300-2304, 1988.

[188] Yasuda, K., Ono, M., Aratani, K., Fukumoto, A., and Kaneko, M. "Premastered Optical Disk by Superresolution". *Japanese Journal of Applied Physics*, Vol. 32, Part 1, No. 11B, pp. 5210-5213, 1993.

[189] Yoshida, M., Shimoda, Y., Ishii, H., and Inoue, A. "4.7 Gbyte Re-Writable Disc System Based on DVD-R System". *IEEE Transactions on Consumer Electronics*, Vol. 45, No. 4, pp. 1270-1276, 1999.

Chapter 6

BLU-RAY DISC

6.1 Video after 2000: Blu-ray Disc as the ultimate high-definition format

J.A.M.M. van Haaren[1] and M. Kuijper[2]

[1] *Philips Research Laboratories Eindhoven*
[2] *Philips & Lite-On Digital Solutions, Eindhoven*

In the two decades that followed the initial CD-press conference in 1979, the optical storage industry had grown up. A complete family of CD formats had been defined and developed. There were read-only, write once and rewritable discs on the market. The price of these discs had gradually come down to very affordable levels. Many suppliers offered popular optical disc drives. The CD family was well know, all over the world, and very successful

A second generation optical disc format with a higher storage capacity and with a new compelling application had been successfully launched in Japan (1996), USA (1997) and Europe (1998). This Digital Versatile Disc (DVD) was rapidly replacing the video cassette recorder for playback of standard definition video. In fact the market transition from video cassettes to optical discs (six years, from 1997 to 2003) had been faster than that of other previous consumer electronics technology transitions, like for instance the replacement of the vinyl records by compact discs (from 1982 to 1991). The acceptance in the market had been fueled by distinct advantages of optical disc media over video tape cassettes. The discs were smaller, more robust, had lower cost and offered the compelling feature of random access.

This rapid adaptation of optical discs for standard definition video had not only been a technical achievement. It was also a sign that different parties in the market could negotiate, establish and launch a standard that was enthusiastically accepted in a global market.

The creation of this new format had, however, not been without difficulties. There had been technical challenges and political changes. A lot of technical work needed to be done due to the use of a shorter laser wavelength (CD: 780 nm, DVD: 650 nm), and the development and manufacturing of new lasers, lenses, actuators and (de)coders. The DVD discs had the same physical dimensions as the CD. There was a market expectation that the discs would play in the same drive, and eventually also for the same price. A novelty in DVD compared to CD was the use of double-layer discs to increase the storage capacity per disc. The disc mastering and replication industry had delivered on the challenge to create these higher capacity discs at small extra cost.

There had been political changes in the industry as well. The balance of powers had changed. Japan had established itself as a powerhouse for innovative consumer electronics, and next to Sony, a number of Japanese companies had acquired strong positions in this field. As an illustration of this: the DVD format had been developed by a consortium of ten companies: Hitachi, JVC, Matsushita, Mitsubishi, Philips, Pioneer, Sony, Thomson, Time Warner, and Toshiba. This consortium also owns the standard and the DVD-logo.

The content industry ("Hollywood") and the computer companies (mainly IBM) had taken a much stronger role compared to the situation in the early eighties when the standard was set for CD. Among other effects this led to an increased interest in copy-protection technology.

For the rewritable DVD format, several approaches had been developed, and three different consortia for three different standards would launch products. Next to the DVD-RW and DVD-RAM standards, Philips was promoting the so-called DVD+RW system, in a small but powerful alliance.

At the end of the nineties, there were good reasons to believe that at some moment a market opportunity for higher performance optical discs would emerge. This higher performance would consist of a higher capacity, a higher data rate, a better digital rights management system, and more advanced interactive features. Various trends in the market and in technology pointed towards this. Data capacities in computer hard disk drives grew rapidly, and in the mid-nineties popular hard disk drives in the 3.5 inch (desktop PC size) and 2.5 inch (laptop) format surpassed optical discs in capacity; CD in 1993 and DVD in 1997.

More bonus material became available with releases of movies and television programs for home play. These bonuses - trailers, extra movie clips, unused scenes, background documentaries and interviews, games - turned out to be compelling arguments for consumers to buy movies on discs. People started to build DVD-collections of movies, just like they had built libraries of books, and collections of vinyl records or CDs. Video displays also got better. They held a promise of growing in size via projection displays and thin, low-weight flat-panel (LCD) displays. High definition screens were introduced and attention was given to elegant, attractive designs of televisions. These high-quality displays became more affordable. And, last but not least, powerful video compression technology became mainstream: MPEG-2 was used in DVD, while better and more efficient compression was developed and standardized in MPEG4/part10 (also known as H.264 and as AVC).

In several industrial research groups, scientists and engineers were working on their proposals for a third generation optical disc format. Many of them had an almost intuitive drive to push technology to higher capacities, in the belief that "there would always be more to store". These research groups presented their progress at the two yearly international conferences in this field: ISOM[1] and ODS[2]. ISOM is a Japan/Asia based annual conference, in autumn. The ODS is an annual USA-based conference in the spring. However, once every three years (1993, 1996, ...), the ODS and ISOM conference merge into a single conference, organized in the summer at Hawaii. In the history of Blu-ray Disc, the joint ISOM/ODS conferences in 1996, 1999 and 2002 were especially important. There may have been a tendency under scientists and engineers to present their best papers at these joint conferences, perhaps because of the attractive geographical location, but certainly also because in the years in which they were held, they were the only major optical storage conference.

In the development of Blu-ray Disc, a public announcement triggered a two-party collaboration, just like in the development of CD a quarter of a century earlier[3]. And again the two collaborating and competing companies were Sony and Philips. However, now the public announcement was made at Sony. In early 1997, they announced a high numerical aperture (NA), two-element objective lens that could be used for higher density optical recording [4]. The paper appeared in January 1997, but had already been presented at the joint ISOM/ODS conference in 1996. Sony organized a press event showing a prototype in October 1997, at their home base Shinagawa (Tokyo).

Many things were still open at that time: the wavelength of the laser, the numerical aperture of the lens, the physical principle for optical recording, the coding, the mastering and replication technology. At Philips, the first attempt to go beyond the just proposed DVD rewritable, with only 3 GB storage capacity[5], resulted into a DVD-like disc that was read out by a lens with a significantly increased numerical aperture. As the NA=0.85 lens reduced the tilt margins, the system required an advanced tilt servo system.

These optics and servo results raised the interest from Sony, who had just presented their thin cover layer disc proposal, and both companies agreed to evaluate possibilities to collaborate. In a series of meetings, alternately in Tokyo and Eindhoven, the two teams explored the opportunities and limitations of their embodiments of third generation optical recording. The first contours of the VDR (Video Disc Recorder), soon renamed to DVR (Digital Video Recorder), emerged: a NA=0.85 objective lens, red laser, thin substrate and land-groove format. The exchange of components, substrates and discs between the two companies not only generated a sense of common interest, but also created an atmosphere of strong, but healthy, competition between the 'conculleagues' in Eindhoven and Shinagawa.

The Sony-Philips meetings started in 1997 and they ran in a typical pace of one meeting every two to three months, until the creation of the Blu-ray Disc Founders group (later renamed Blu-ray Disc Association) in 2002. Though contacts with other companies (Nichia, Pioneer, Thomson, Matsushita) were sought right from the start as well, the bilateral technical meetings of Sony and Philips were the heart of the Blu-ray Disc development in the period 1997-2001. Results were presented at these meetings. Unsolved issues were discussed and put as challenges for research at the home-base labs after the meetings.

In order to stand a chance in the market, a possible new format needed to be distinctly better than DVD. Two options were considered: an extension of the CD/DVD-paradigm and magneto-optical recording. The technology that prevailed stayed close to CD and DVD. A choice for a technology close to CD and DVD included the perspective to a complete family of disc formats: read-only, write-once, and rewritable, the former two being difficult to realize using magneto-optical technology. Also from an economic perspective an extension of the CD/DVD paradigm was the preferred option, as it was more likely to allow future use of recent investments by industry, both in capital and in expertise. A system close to CD and DVD also offered a realistic opportunity to work on backwards compatible drives. In this way consumers could play earlier format discs in new drives. Such a system was preferred, but in 1997 it was not certain whether it would actually be possible to combine legacy design choices with a sufficiently large performance step.

An important design-choice was the laser wavelength. The size of the optical spot probing the disc scales proportional to the wavelength of the light. So, shorter wavelengths would be better, and the target was a blue laser. The question to be answered concerned the feasibility of blue solid state lasers. There had been blue-laser research projects at several companies, including Philips. The actual solution was found by Shuji Nakamura and his team of the Japanese company Nichia Corporation. In 1997 Nakamura and co-workers developed a gallium nitride based solid-state blue-violet laser with a wavelength of 405 nm[6]. After this, the question was whether this research prototype could be turned into a mass-market product, with a long life time, sufficient power, and an acceptable price. The teams at Sony and at Philips developed the confidence that this would be possible, and decided to base their research on this belief. Meanwhile, many of the initial experiments on the third generation optical storage technology were however done either with a red-light laser, after which an appropriate arithmetic scaling could be done, or with a large table-top gas-laser emitting blue-violet light.

The combination of the higher numerical aperture with the shorter wavelength allowed a higher optical resolution, and therefore an increase of the storage capacity. The scaling of the spot diameter is proportional to the wavelength, and inversely proportional to the numerical aperture. The combination of NA = 0.85 and a 405 nm wavelength would lead to a factor of 4 capacity increase with respect to the DVD system.

A higher numerical aperture also leads to a need to rethink the design of the disc layout and the optics. System tolerances like the disc-tilt margins, the tolerance for disc cover layer thickness variations and the depth of focus scale proportional to the wavelength and inversely proportional to higher orders of the numerical aperture. The aggressive reduction of some of these margins is more detrimental than the benefits of the stronger lens and the shorter wavelength. A system architecture that takes this into account is complicated but crucial. It is important to have a common understanding and alignment on the trade-offs and the conclusions of such an architectural analysis in an early stage. At the same time, it is important to leave sufficient opportunities and encouragement to the engineers to make progress in their fields and to see their progress reflected in a better performance of the end-product, improved manufacturability via better technical margins, and finally in a more competitive product at lower costs for the consumers. Sect. 6.2 reproduces a paper that gives an insight in this balancing act. The paper deals with the problem which substrate thickness should be chosen for the optical disc, and whether the objective should be in an actuator with active tilt control. The conclusion is that for a disc with 0.1 mm substrate thickness, such an active tilt control is not needed. And this is indeed the design choice that was made for Blu-ray Disc.

The Sony-Philips competitive meetings rapidly spread over many disciplines: the physical and logical layout of the disc, the physical format of the wobble for timing-recovery, channel coding of the user data, error correction and drive design. As a result, each meeting was a demonstration of the highest capacity reached, the fastest recording done, the most robust detection of the wobble, the most efficient coding. The collaborative effort was showcased to the world with the joint paper "Optical Disc System for Digital Video Recording", claiming a 9.2 GB rewritable disc using red laser on a Land-Groove disc (Sect. 6.3). It describes the architectural choices that had been made by mid 1999, and was presented at the ISOM/ODS conference in Hawaii in July 1999. This paper reports the jointly developed 17PP channel modulation scheme as well as the Picket code for error correction. The paper also contains some of the dilemmas and unsolved or unproven points. The experiments in the paper were done with a red laser. Consequently the maximum disc capacity was only 9.2 GB. The paper also contains extrapolations on the performance of the system when it would have been based on blue lasers. The authors express confidence that they will be able to obtain capacities of 22 GB or higher with a blue laser, but they did not show that experimentally in this paper.

Because high-power blue diode lasers were not readily available at the time, a blue recorder was built with an impressive Krypton gas-laser and equipped with one of the first blue NA 0.85 lenses of Sony. It measured over $4 \times 1 \times 1$ meters and was mounted on an optical bench (a 'zerk' (tombstone) in the laboratory jargon) to reduce the effect of vibrations. Initial blue DVR recordings immediately showed a roadmap to 22 GB and perhaps higher[7]. Piles of fresh discs that were delivered weekly by the Philips Optical Disc Technology Centre, were measured on this machine on parameters like maximum speed, overwrite capability, achievable storage capacity, and disc life time. Each experiment involved a range of phase change material compositions, stack designs, cover layer properties, but also mastering settings like variations in depth and width of the tracks. Once the first Nichia blue laser diodes became available, all other DVR testers were converted from red to blue.

In mid 2000 the first deep-UV mastering machine became operational at the Philips Optical Disc Technology Centre. This paved the way for mastering grooves that were less than 600 nm separated. Now, a groove-only format with track pitch of around 320 nm was possible. Within a few weeks the less preferred land-groove format was abandoned and groove-only became the credo for DVR. Again, all disciplines had to redevelop their format or disc properties and benchmark the new settings with Sony. During the Optical Data Storage (ODS) Topical Meeting in Santa Fe (New Mexico) in April 2001, both Sony and Philips presented their papers on the 23 GB groove-only DVR format. The Philips-paper is reproduced in Sect. 6.4. Though it is a short paper, it represents a huge amount of experimental teamwork at Philips Research, Philips Optical Storage and the Philips Optical Disc Technology Centre. The presentation at this conference was a major milestone for the Philips team.

To the surprise of many, rival Matsushita[8] also presented a blue laser based, NA = 0.85, 0.1 mm cover layer thickness, groove-only format at the same conference. The three companies met in the following weeks and decided to team up. Blu-ray Disc (BD) was born.

Sect. 6.5 contains again a conference submission, now at ISOM/ODS 2002. This paper has authors from three major companies in this field: Sony, Panasonic and Philips. As such it was an important signal that a standard for third generation optical storage had been developed (see Fig. 1). At that moment in time a number of companies had already met to support this standard. Nine companies presented themselves as BD founders: Sony, Panasonic, Pioneer, Thomson, LG Electronics, Hitachi, Sharp, Samsung and Philips. They gave a press conference on 19 February 2002 in Tokyo, where they announced the

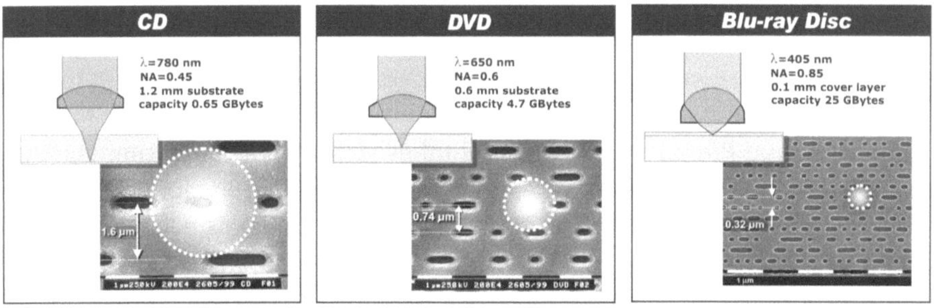

Fig. 1. Schematic diagrams for the three optical storage generations: CD, DVD and BD, with wavelength λ, Numerical Aperture NA and disc substrate thickness. The circles on the scanning electron microscope pictures of the read-only discs indicate the spot size.

Blu-ray Disc standard and its support with future products. About one year later, the first products were released on the Japanese market.

The Sony-Panasonic-Philips paper in Sect. 6.5 presents the address format. Here an innovative solution was presented that contained elements of rivaling formats in DVD recorders. The rewritable Blu-ray Disc format has a predetermined, small amplitude (10 nm) wobble superimposed on each track. This wobble is used for write-clock generation and for retrieving accurate timing and address information. This is a key attribute of the new Blu-ray Disc format, and it is important to note that this is a three-company result. The paper describes a wobble format based on a combination of minimum-shift keying and sawtooth modulation that is very robust against various distortions.

The sub-micron structures defining the physical disc format were transferred to replicated discs via so called master discs. The production of these masters involves the careful, high-resolution writing of the information on a non-structured photo-resist layer. This is followed by development, galvanic coating and then reproduction of masters from the original, via a so-called family-process. This is a highly specialized, multidisciplinary activity and is a crucial technology for any optical disc format. A format can only become popular and successful, if this mastering can eventually be done at sites all over the world. Such a global access to mastering technology is especially needed for the read-only discs. These mastering machines are high-accuracy machines bought from professional equipment manufacturers, like at that time ODME/Toolex and later Singulus. In Philips, research and development work on mastering was done at Philips Research and at the Philips Optical Disc Technology Centre, in close collaboration with ODME and its business successors. The main motivation for this work was in establishing technology that allowed standardization for read-only discs.

For the previous formats (CD and DVD) it had been possible to build mastering machines with significantly shorter wavelengths than the wavelengths used in the optical drives. In the case of Blu-ray Disc, just wavelength reduction would not be sufficient. For some time it was even thought that optical techniques would not be able to deliver the required resolution for BD-masters. Especially the masters for the read-only discs were difficult to make. Companies like Pioneer (Japan) and Nimbus (UK) were promoting electron-beam mastering for this purpose. There were also small, exploratory projects at Sony and Philips on this. If electron-beam mastering was needed, than that would be a paradigm change and a significant hurdle in the acceptation of the BD-format by the industry.

An optical solution was highly desirable to avoid such a drastic change in mastering technology. In an effort to develop such a solution, a start was made from the deep-UV (257 nm) gas lasers and optics that had already been introduced in the Optical Disc Technology Centre. The use of a powerful deep-UV light source required special skills and care in the construction of an experimental mastering system.

Additional technology was needed to arrive at the resolutions to write sufficiently accurate BD-masters. The Philips team followed its own approach for this. Sect. 6.6 reports that the Philips-team focused a deep UV-laser beam through a thin film of water in between the photoresist and the objective lens. This water film enhanced the numerical aperture of the mastering lens, in this way reducing the optical spot size. The film needed to be stable between the objective and the rotating disc. The water was dispensed and removed via a small device close to the objective lens. This liquid immersion mastering technology was developed at Philips Research, and it was extensively used for the standardization of BD-ROM. The paper in Sect. 6.6 indicates that even capacities above 30 GB would be possible.

It turned out to be difficult to develop a supplier for the liquid immersion lens heads needed for this liquid immersion mastering. The lens heads needed at Philips were hand-made special modification of purchased UV-mastering lenses. This technology could only become successful at other places as well, if there was an independent lens supplier. Attempts to achieve this, failed. In the end another approach, the phase-transition mastering technology, proved to be more practical, as it did not need such special lens heads. Phase transition mastering became the technology of choice for the Blu-ray Disc replication industry[10].

The first Blu-ray Disc format was introduced in 2002[11]. First products appeared in 2003. At that time, the laboratory set-up recorders had been re-engineered to mass-manufacturable products. As an example, the NA=0.85 doublet lenses that had been used in the initial experiments had been replaced by singlets. Philips had even been successful in combining the lens requirements for CD, DVD and BD writers in a single all-in-one lens and a single detector, the so-called triple writer. This triple writer used a specially designed diffractive element that accurately compensates the spherical aberration for the three relevant wavelengths. The prototype was first publicly demonstrated at the 2005 Consumer Electronics Show in Las Vegas (USA).

The initial Blu-ray Disc format was a recording format primarily aimed at video-recording with a set-top box directly connected to the TV. The disc was enveloped in a cartridge to provide protection against dust, fingerprints and scratches, to which the fine structure in the BD disc appeared to be more sensitive.

Subsequent updates of the BD-format between 2002 and 2006 contained some significant enhancements, both at the physical, hardware level as well at the systems level. The most visible physical enhancement was in the area of disc robustness and data safeguarding. TDK developed a protective hard-coat to be deposited on top of the thin plastic cover layer. Consequently, the discs could be used without the cartridge that had initially been used to protect the data. This was a sign that several of the practical problems that were foreseen from the initial work on high NA doublet lenses and thin-cover layer systems (Sect. 6.2), were solved in a convenient and cost-effective way at the time of mass-market introduction. It allowed the return to a disc with appearance similar to the compact disc. This had been a strong wish of the consumer. Other updates concerned higher layers in the architecture, like the file system. This file system was adapted so that it would be suitable not only for video-recording in a consumer electronics set-top box environment, but also for data applications and use in a PC environment.

Next to the rewritable format (BD-RE), a write-once (BD-R) and a pressed format (BD-ROM) were introduced. For BD-ROM, a new video application format was developed to be able to offer the best consumer experience in this next generation video publishing format. After several years of experience with the DVD video, it was felt that there were many opportunities to offer better and more sophisticated possibilities in a Blu-ray Disc video publishing format. To define requirements for that format, a series of meetings was held between engineers of the BD Founders group (at that moment exclusively coming from the hardware industry) with the major Hollywood studios. These studios later also joined the Blu-ray Disc Association.

Main conclusions from those meetings were the following. Apart from copy protection, the best possible video quality would have to be a basic element for new formats. The new format should include advanced interactivity and also web connectivity, to allow new consumer experiences and to enable new business models. And the new format should be flexible for new, creative applications. In a way, BD should be able to deal with ideas for bonus material, interactivity and internet access that were beyond imagination at the time the standard was set.

It was concluded that these requirements could only be met by creating a programmable environment, as opposed to the deterministic, fixed-function based command set as used in DVD. After long consideration, the BD companies agreed to proceed with the development of a Java-based programmable platform, using the same base technology that was already used for interactive broadcast (GEM)[12]. The creation of such a platform was a major multi-company effort.

It was also clear that this meant a huge step for the authors of DVD titles, who were not trained to be programmers. For this reason, and also to offer synergies in simultaneous authoring for BD and DVD, it was decided to develop a two-tier format, the base being deterministic (like DVD, but with far more functionality) extended by the BD-Java format. To avoid incompatibilities, both were made mandatory for players, whereas a choice could be made for the discs. Several codecs were added to the format to offer more choices to the publishers, such as VC-1 for video (in addition to MPEG-2 and AVC) and Dolby Digital and DTS for audio. The BD-J format was first published in 2004 and the first BD-J discs came to the market in March 2007.

After the press conference at which BD was launched in February 2002, a rivaling format was announced as well. HD DVD was mainly supported by Toshiba and Microsoft. Unfortunately a format war emerged. In early 2008, the HD DVD format withdrew. This made Blu-ray Disc the winning format for high-definition video on physical media, and for 25-50 GB capacity removable media in PCs.

Today, more than two hundred fifty companies are members of the Blu-ray Disc Association. The support for Blu-ray Disc covers all industry: equipment makers, manufacturers of blank media and of players and their components, a range of content providers from movies and concerts to games, and many representatives of the personal computer industry. Almost all major companies in each of these domains are part of the BDA, and the same holds for many small companies.

Blu-ray Disc is the preferred format in the market for next generation physical media for content distribution. Its sales figures increase rapidly. BD is popular because it offers new exciting experiences to consumers. This is not only because of a six times higher resolution than DVD. It also offers superior sound quality (7.1 channels surround sound). BD also offers new navigation interfaces, games that are integrated in movies, enhanced interactivity and options to enrich content already bought with new live events and to get access to additional bonus material via the internet.

Blu-ray Disc started in 1997 from the ambition of small group of researchers and engineers in Tokyo and Eindhoven to break records in optical disc storage capacity and data rate. Eleven years later and after the work of many multi-disciplinary teams all over the world, BD is established as the format of choice for physical distribution of high-definition video material and for removable storage for at least the coming decade. And it is likely to become the ultimate format for optical discs.

References

[1] ISOM stands for International Symposium on Optical Memory. See www.isom.jp.

[2] ODS refers to the Optical Data Storage Topical Meeting. This is an annual conference organized by the Optical Society of America, most often in the United States.

[3] See Chapter 2 of this book.

[4] K. Yamamoto, K. Osato, I. Ichimura, F. Maeda and T. Wanabe, Jpn. J. Appl. Phys. **36** (1997) 456-459.

[5] B.A.J. Jacobs and J.P.W.B. Duchateau, Jpn. J. Appl. Phys. **36** (1997) 491-494.

[6] S. Nakamura, M. Senoh, S. Nagahama, N. Iwasa, T. Yamada, T. Matsushita, H. Kiyoku, Y. Sugimoto, T. Kozaki, H. Umemoto, M. Sano and K. Chocho, Jpn. J. Appl. Phys. **36** (1997) L1568-L1571.

[7] B. Tieke, M. Dekker, N. Pfeffer, R. van Woudenberg, G.-F. Zhou and I. P. D. Ubbens, Jpn. J. Appl. Phys. **39** (2000) 762-765.

[8] Matsushita is also known under one of its brand names: Panasonic. In fact per 1 October 2008, the Matsushita company renamed itself to Panasonic. In the rest of this section we will use the name Panasonic.

[9] E. R. Meinders, R. Rastogi, M. van der Veer, P. Peeters, H. E. Majdoubi, H. Bulle, A. Millet and D. Bruls, Jpn. J. Appl. Phys. **46** (2007) 3987-3992.

[10] This liquid immersion mastering technology has been further developed for wafer steppers for ASML. The technology was successfully introduced into the market in 2004, under the trade name HydroLith, and is currently the state-of-the-art technology for accessing the 38 nm node in semiconductors industry.

[11] See www.blu-raydisc.com.

[12] GEM stands for Globally Executable MHP, and MHP stands for Multimedia Home Platform. GEM and MHP are open standard for interactive digital TV. See www.mhp.org.

6.2 High numerical aperture optical recording: active tilt correction or thin cover layer?

Yourii V. Martynov, Benno H.W. Hendriks, Ferry Zijp, Jan Aarts, Jan-Peter Baartman, Gerard van Rosmalen, Jean J.H.B. Schleipen and Henk van Houten

Philips Research Laboratories, Prof. Holstlaan 4, 5656 AA Eindhoven, The Netherlands

Abstract
Playback of a 12 cm diameter replicated ROM disc with a 0.6 mm substrate thickness and a storage capacity of 10 GB has been achieved using a light path with a dual-lens objective with NA=0.85 and active tilt control. The disc tilt margin exceeds ±0.7 degree. Also backward compatibility with digital versatile disc (DVD) has been demonstrated. Active tilt correction is not required for read out of a disc with a 0.1 mm transparent cover layer. This technique has also been studied experimentally. The merits and disadvantages of the two approaches are discussed.

6.2.1 Introduction

To achieve a storage capacity of 9 to 10 GB on a 12 cm optical disc there are two basic options, each giving about a factor of two increase in data density with respect to the digital versatile disc (DVD). The first is to replace the red laser of DVD (650 nm) by a blue laser. A breakthrough allowing such an innovation is the blue-violet (410 nm) diode laser based on GaN. According to announcements by Nichia,[1] it is expected that a laser with sufficient power, beam quality, and lifetime will soon come on the market. However, it could still take several years before mass produced optical recording systems based on blue lasers would be feasible.

The second possibility is to use an objective lens with a higher numerical aperture, replacing the NA=0.60 of DVD by NA=0.85. As in the transition from CD to DVD, the price to pay for a higher NA is a collapse of the disc tilt margin, a tightening of the disc thickness tolerance, and a shorter depth of focus. The same holds true for a shorter wavelength, but as the system tolerances scale with high powers of NA, and only linearly with wavelength, the increase in NA would seem to be more difficult to realise technically. First of all, a manufacturable NA=0.85 lens with sufficient free working distance has to be designed as a doublet, i.e. by combining two lenses, with at least one aspherical surface. Secondly, the tilt margin needs to be widened.

We have proposed to compensate for disc skew by using an actuator actively tilting the second lens of the high-NA objective with respect to the optical axis[2, 3]. In this paper, we discuss an experimental study of this active tilt correction (ATC) method. To this end we have developed an actuator with three degrees of freedom (separation between the elements of the dual lens and two angles determining the orientation of the second lens). The read out of a 10 GB ROM disc with low jitter and ample tilt margin will be demonstrated. In addition, we demonstrate that the ATC approach allows excellent backward compatibility with DVD.

An alternative solution has been reported by Yamamoto *et al.*[4] They have proposed to address the information layer through a 0.1 mm thin cover layer on a 1.1 mm thick plastic substrate. This cover layer can be manufactured by a spin coating process, or by bonding of a thin plastic sheet on the disc. A tilt tolerance better than the one achieved with DVD can thus be obtained for read out with an NA=0.85 objective, without active tilt correction. Given an appropriate optical design, thickness variations of the thin cover layer can be actively compensated using an actuator in which the separation between the two lenses can be adjusted. This approach has also been studied experimentally.

In this paper, the two approaches to high-NA optical recording mentioned above are assessed and compared, based on a theoretical analysis and an experimental investigation of the optical system tolerances.

6.2.2 Active tilt control

6.2.2.1 Optical tolerances and lens design for ATC

In this paper we study dual-lens objectives of the type shown in Fig. 1. The objective consists of a first lens with refractive index n_1 and thickness d_1, followed by a second lens with refractive index n_3 and thickness d_3. The surface of the second lens facing the disc is flat. The air gap d_4 (free working distance) between the second lens and the disc is large compared to the wavelength. To ensure compatibility with DVD in the case of ATC, a disc with d_5=0.6 mm made of polycarbonate (refractive index n_5=1.58) was considered. Sufficient disc tilt tolerance at NA=0.85 can be obtained by actively tilting the second lens depending on the disc tilt. When the disc is tilted by an angle β, the laser beam picks up comatic aberrations when entering the disc. Due to the fact that the focused beam enters almost perpendicularly to the hemispherical surface of the second lens, almost no aberration is introduced by this surface when the second lens is tilted by an angle α (in the same direction as the disc). The planar surface of the tilted lens, however, gives rise to comatic aberration which is proportional to the angle α but opposite in sign to the comatic aberration introduced at the air-disc interface. Consequently, by proper adjustment of the tilt angle α of the second lens with respect to the tilt angle β of the disc, both contributions can be made to cancel almost perfectly. In particular, in this optimal tilt correcting case the third-order coma wave front aberration contribution vanishes. The wave front aberration introduced by a combination of the tilted disc and tilted second lens is now determined by higher-order coma contributions only, leading to a significant increase in disc tilt tolerance. It has been calculated[3] that for small angles β (less than a few degrees) the optimal relation between α and β is given by $\alpha=\mu\beta$, where the parameter μ is independent of α and β, and is a function of the refractive indices of the second lens (n_3) and of the disc (n_5), and of the thickness of the air gap (d_4) and of the disc (d_5). The optimal value for μ is found to decrease with increasing value of the disc thickness and increasing value of the air gap. This is to be expected because for increasing air gap or with increasing disc thickness the relative contribution to the coma introduced by the air-disc interface decreases with respect to that introduced by tilting the second lens.

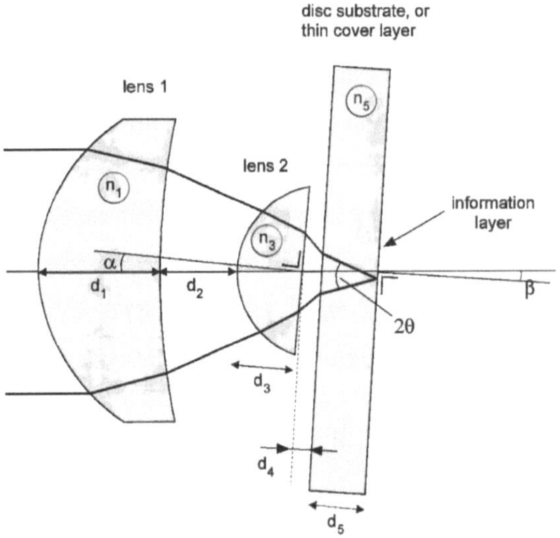

Fig. 1. Schematic drawing of a dual-lens objective; the numerical aperture is given by NA=n_5 sin(θ).

To study the ATC concept experimentally, we have chosen to design a lens set with NA=0.85 and free working distance d_4=0.05 mm. Although larger free working distances are, of course, preferred in order to be less sensitive for head/disc crashes, it is large enough for dust and fingerprints to cause no serious problems. The small free working distance allows for a relatively simple design of the NA=0.85 dual-lens objective, consisting of a plano-aspherical first lens followed by a plano-spherical second lens. We will return to the topic of free working distance in Sect. 6.2.3.1 and Sect. 6.2.4. The entrance pupil diameter value was set at 3.2 mm. The lens system is designed to read discs with substrate thickness of 0.6 mm (hence the same as for DVD). Apart from the standard radial and focusing actuator an additional actuator with three degrees of freedom controls both the distance d_4 and the tilt angle α (in two directions) of the second lens (see Sect. 6.2.2.2). By proper adjustment of the distance between the two lenses spherical aberration caused by disc thickness variations can be compensated in a way similar to the one described by Yamamoto *et al.*[4]

The plano-aspherical lens has been made by the glassphotopolymer replication process.[5] In this process the aspherical surface is made by pressing a mould with the prescribed aspherical surface against the spherical surface of a plano-convex glass lens on which a drop of resin has been applied. The resin, which acquires the shape of the aspherical mould, is then hardened by UV light. The second lens is a simple truncated glass sphere. In Table I the main tolerances of this lens system are given (jointly with those for a typical DVD lens and for the lens used in the thin-coverlayer approach, discussed in Sect. 6.2.3.1). The root mean square of the optical path difference (OPD_{rms}) of the total ATC lens system was less than 30 mλ (as verified experimentally using Twyman-Green interferometry).

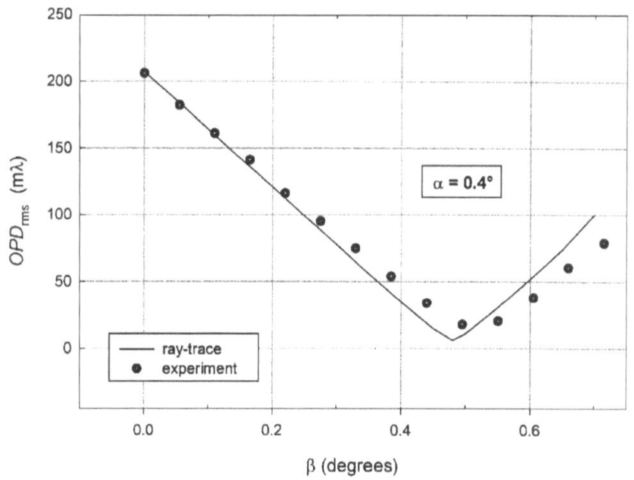

Fig. 2. Calculated OPD_{rms} as a function of the tilt angle β of the disc when the second lens is tilted by an angle α=0.4° for the ATC lens system used in the experiments. Also shown are the results from experiments (only the coma wave front aberration is taken into account here).

Figure 2 presents a simple experimental illustration of coma compensation by tilting the second lens of the ATC dual-lens objective. In this experiment the second lens was deliberately tilted by a fixed angle of 0.4° and the OPD_{rms} of this system was measured in a Twyman-Green interferometer using a spherical mirror, with a transparent glass plate, optically equivalent to a 0.6 mm polycarbonate disc, mounted in front of the mirror. In such a configuration comatic aberrations can be measured and monitored. Calculated and measured values of wavefront aberration are plotted as a function of the tilt angle β of the disc. The figure shows the averaged result of two interferometric measurements, corresponding to two orientations of the mirror (rotated over an angle of 180°). In this way, the comatic wavefront aberration due to the mirror imperfections are cancelled. The small remaining differences between theory and experiment are due to residual lens aberrations and imperfect alignment of the two lenses. One can clearly see that optimal coma cancellation is obtained for $\beta \approx 0{:}5°$.

Parameter	DVD	ATC	thin cover layer
NA	0.6	0.85	0.85
Entrance pupil diameter	3.3 mm	3.2 mm	3.3 mm
Free working distance	1.35 mm	0.05 mm	0.3 mm
Number of aspherical lens surfaces	1	1	3
Substrate/cover layer thickness variation (no correction)	15 μm	2.5 μm	2.5 μm
Substrate/cover layer thickness variation (correction by adjusting dist. between two lenses)	-	38 μm	85 μm
Field of view	0.6°	1.2°	0.9°
Decentering of the dual lens objective	-	25 μm	26 μm
Tilt objective with respect to the second lens	-	0.17°	0.032°
Disc tilt (no tilt correction)	0.15°	0.04°	0.22° [a)]
Disc tilt ($\alpha = \beta$)	-	0.18° [a)]	0.03°
Disc tilt ($\alpha = \mu\beta$)	-	1.09°	1.95°
		μ=0.83	μ=0.13

a) Experimentally tested in this paper.

Table I. The 15 mλ OPD_{rms} tolerances for the dual lens objective for the case of ATC (0.6 mm substrate thickness) and the thin cover layer approach (thickness 0.1 mm). For comparison, the relevant parameters for the DVD are given as well.

Table I shows that the ATC dual-lens objective is tolerant for disc thickness variations, de-centering of the second lens, and has a large field of view in accordance with the results of Ref. 3. Furthermore, the table reveals that the tolerance for disc tilt can indeed be significantly improved by actively tilting the second lens. *Optimal* tilting the second lens with respect to the tilt of the disc improves the disc tilt tolerance by a factor of 27 compared to the tilt-rigid system, whereas keeping the second lens parallel to the disc improves the disc tilt tolerance by a factor of 4.5. Even in the latter case the tolerance for disc tilt is larger than in the case of DVD read out with an NA=0.6 objective and no tilt correction. The value of proportionality constant μ is found to be 0.83 consistent with experimental results presented in Fig. 2. The ATC objective has been designed in such a way that for correction of spherical aberration due to disc thickness variations it is sufficient to keep the air gap between its second lens and the disc constant.

6.2.2.2 ATC actuator

An optimal tilt-correcting system for read out of a disc through a 0.6mm substrate is not easy to implement. One option is to perform a direct measurement of the value of coma introduced by the lens system and the disc. This is difficult because the coma aberration cancels when the reflected beam from the disc returns through the lens. Another option is to measure the *absolute* value of the disc tilt and then turn the lens by an angle proportional to that value with the desired proportionality coefficient μ. This would require exact knowledge of all the gain and transfer characteristics of the actuator determining the lens tilt. Besides, in both cases one would have to perform a separate measurement of spherical aberration to compensate for disc thickness variations.

Although the above described techniques are currently under investigation, for our research prototype we have chosen to implement the sub-optimal correction strategy, where the lens facing the disc is kept at a fixed distance from the disc surface and parallel to it at all times (i.e. $\alpha=\beta$). Both coma and spherical aberration are then corrected automatically to a degree sufficient to compensate for disc manufacturing tolerances. This condition is fulfilled if the distance between the second lens and the disc surface is actively controlled at three points in a plane perpendicular to the axis of symmetry of this lens. To measure the distances we employed three miniature auxiliary lightpaths focussing their beams on the (upper) surface of the disc and generating focal error signals. Three small lenses were mounted in a single sub-frame together with the second element of the dual-lens objective and driven by three independent motors. The advantage of this control system is that it, in fact, comprises three standard focussing servo loops and is easy to implement.

Besides the tilt-correcting part the usual focussing and tracking movements must be possible. To accomplish this, we used a commercial two-dimensional (2D) CD-ROM focusing and tracking actuator (model CDM12 from Philips). The tilt-correcting actuator containing the dual-lens objective is placed in this 2D actuator instead of the conventional objective lens.

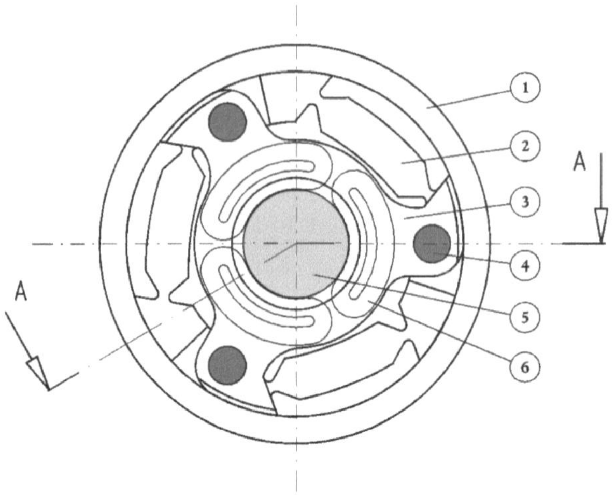

Fig. 3. Bottom view of the two-stage actuator design for the two-lens ATC objective. Dash-dotted lines with the two arrows denoted A-A define an actuator cross section depicted in Fig. 4.

The tilt-correcting actuator is drawn in Fig. 3, viewed from the disc towards the lens. A cross section is shown in Fig. 4.

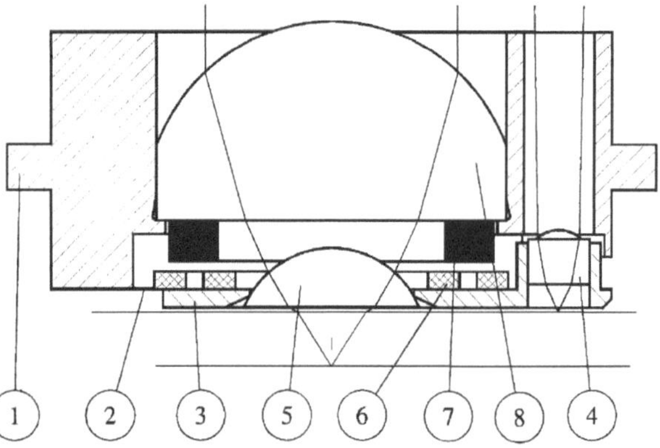

Fig. 4. Cross section of two-stage actuator, as indicated in Fig. 3. Main readout beam and one of the three auxiliary beams are shown.

The first lens (8) is mounted in a frame (1). This frame is connected to the CDM12. A permanent magnet ring (7) is mounted on the frame. The second lens (5) is mounted on a sub-frame (3), together with three coils (6) and three auxiliary lenses (4). The two frames are connected to each other with three leaf springs (2). The tilt-correcting actuator must keep the second lens at a constant distance of 50 μm from the disc surface within 1.1 μm, and parallel to the disc surface within 0.035°.

The sub-frame has 3 degrees of freedom, because the tilt actuator does not permit motion parallel to the disc surface, nor does it allow rotation along the optical axis. The movements in three remaining degrees of freedom are allowed because of flexible suspension in three metal springs, and are actuated by means of the three coils (see Fig. 3). The center region of the chosen configuration can tilt and move in the focal direction by bending (and torsion) of the springs. The stiffness for vertical displacement and tilt is low. The springs have a widened middle section to increase in-plane stiffness. The 10 μm-thick springs were manufactured using standard wet-etching technology.

The three-degree-of-freedom motor consists of a vertically magnetised annular magnet fixed to the frame and three banana-shaped coils placed on the sub-frame, directly underneath the magnet. When an electrical current flows through one of these coils, the resultant force is directed vertically. Together,the three coils can therefore apply the required force in the direction of the optical axis, and the torque's in any tilt direction. The three objective lenses of the three auxiliary lightpaths are also mounted onto the sub-frame. The polyamide lenses are plano-aspheric, 0.6 mm thick, and 0.7 mm in diameter. Because the accuracy required from the auxiliary focusing systems is comparable to that of a CD light path focussing, the NA of the auxiliary lenses was set close to that of a CD objective lens (NA=0.4). The lenses for our research prototype were manufactured by direct turning from bulk polyamide. Although the numerically-controlled lathe delivers highly reproducible products, all the lenses were independently characterised before mounting them into the actuators. The alignment accuracy in focal direction, relative to the second lens of the objective, is 2 μm. This requires special alignment and connection techniques. For the current prototype, the assembly is done manually with specialized equipment in a clean working environment. As light sources for the three auxiliary lightpaths we used commercially available laser-detector-grating units (LDGUs) that incorporate a 780 nm semiconductor laser, Foucault focus detector and a holographic grating, coupling the returning light to the detector. A system of mirrors and three collimator lenses was used to couple the light from the three LDGU's into the focusing lenses. It consisted of a pyramid-like member with three reflecting surfaces and a circular hole in the middle for transmitting the main readout beam and three separate folding mirrors. A photograph of the assembled ATC actuator is shown in Fig. 5(a).

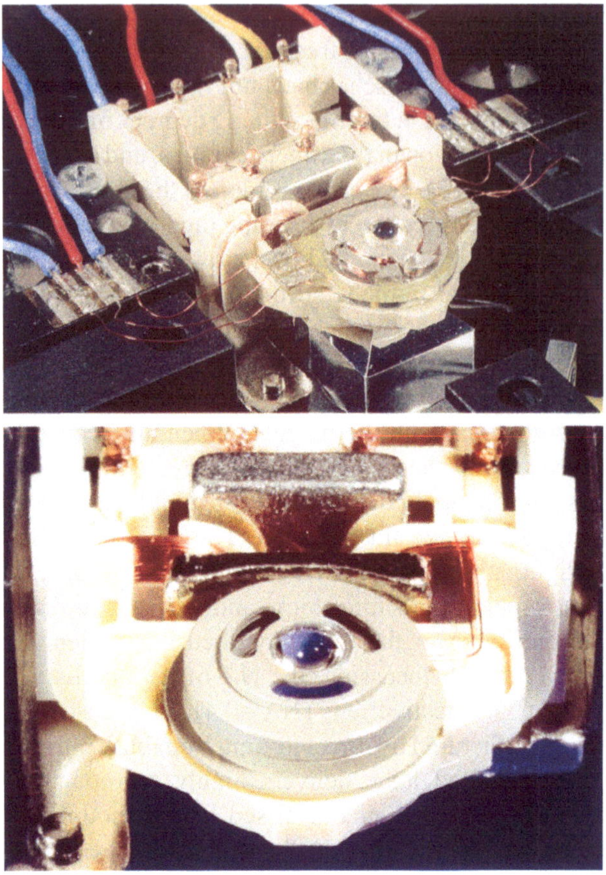

Fig. 5. (a) Photograph of the ATC actuator mounted in a Philips CDM12 CD-ROM actuator. A sub-frame with the second element of the dual-lens objective and three auxiliary lenses suspended in three leaf springs is clearly visible. Underneath the lens holder a pyramid-shaped mirror coupling the beams into auxiliary lenses can also be seen. (b) Photograph of the actuator for the thin cover layer approach mounted in a CDM12 CD-ROM actuator. Through the banana-shaped holes in the top cap the second lens (lens1, Fig. 1) can be seen. Two wires come out of the assembly for driving the linear motor actuating this second lens.

6.2.2.3 10 GB ROM disc and DVD ROM disc read out

The objective lens in the actuator together with three auxiliary lightpaths have been incorporated into a test player equipped with a 640 nm laser and means for focussing, tracking and data detection. To test the system performance we used a 12 cm diameter glass-photo-polymer ROM disc, containing an EFM+ data pattern (the modulation code of DVD). The track pitch of 0.5μm and minimum pit length of 0.278 μm correspond to a user capacity of 10 GB.

The glass thickness was optically measured to be 0.62 mm, the substrate was flat within 0.1°. During readout the position of the front lens of the dual-lens objective with respect to the disc surface was dynamically controlled in a manner described in the previous subsection. Using this 10 GB ROM disc, we have obtained a good readout signal quality, with a data-to-clock timing jitter of the equalised signal below 7%.

To check play back compatibility of the ATC system, we verified that a DVD ROM disc could also be read out by the same set-up.[6] Because the numerical aperture of 0.85 was much higher than the NA=0.6 of the DVD system, an excellent read-out signal could be obtained without any equalisation: the data-to-clock jitter was about 5.5%. Digital eye patterns from the 10 GB ROM disc and from the DVD ROM disc are presented in Fig. 6.

To test the tilt-correcting capability of our actuator we measured the dependence of jitter on disc tilt for both the 10 GB ROM disc and the DVD ROM disc. The tilt windows were sufficiently broad (see Fig. 7) and were mainly limited by the mechanical stroke of the tilt-correcting actuator.

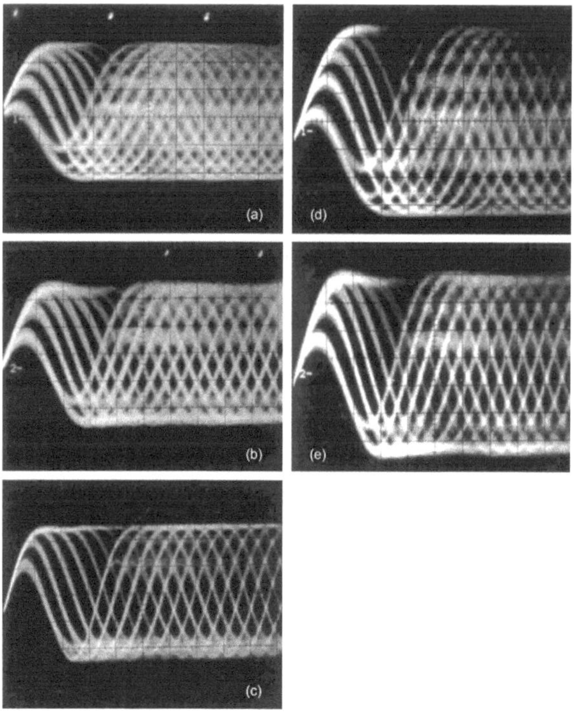

Fig. 6. Digital eye patterns of various discs. 10 GB ROM disc with 0.6 mm cover layer read with ATC system before (a) and after (b) equalisation; 4.7 GB DVD ROM disc (c) without equalisation, no equalisation is necessary in this case; 10 GB ROM disc with 0.1 mm cover layer read with 3-asphere lens combination system before (d) and after (e) equalisation.

6.2.3 Thin cover layer

6.2.3.1 Optical analysis

An alternative to the ATC approach is to increase the disc tilt tolerance by decreasing d_5, as in the thin-cover-layer approach (d_5 now corresponds to the cover layer thickness rather than the substrate thickness). Since the comatic wave front aberration is linearly proportional to the disc thickness, the reduction of disc tilt tolerance due to the increase of the NA can be compensated for by reducing the cover layer thickness by an appropriate factor. For a system with NA=0.85 we find from Table I that the cover layer thickness should be smaller than 160 μm to obtain a disc tilt tolerance larger than the tolerance obtained for DVD.

Since we have chosen to keep the second lens parallel to the disc in the ATC concept, the free working distance (FWD) was limited to approximately 50 μm, in order to keep coma caused by disc tilt within acceptable bounds. In the case of optimal tilt correction as well as in the thin-cover-layer approach, this limitation to the FWD is no longer present. A lens design with a FWD larger than 50 μm is attractive because it makes head/disc crashes less likely and it allows better protection of the lens facing the disc. A larger FWD has some disadvantages with respect to design tolerances, however. With larger FWD the tolerances to de-centering of the two lenses, as well as the field of view, become smaller. These tolerances can be controlled to some extent by using a larger number of aspherical surfaces in the two-lens system. However, the tolerance to tilt misalignment between the two lenses also decreases with increasing FWD and it is independent of the number of aspherical surfaces. In the case of the tilt-rigid thin cover layer system this tolerance puts a severe constraint on the assembling process of the actuator for relatively large FWD. Taking 0.5 mrad as an acceptable tolerance for the tilt error between the two lenses (with a 15 mλ OPD_{rms} as a criterion) a maximum FWD of approximately 300 μm can be obtained.

For the thin-cover-layer concept we have chosen to use a lens system with NA=0.85, a FWD of 300 μm and a cover layer thickness of 0.1 mm. Apart from the lens surface facing the disc, which is flat, all surfaces of the lens system are aspherical. The entrance pupil diameter has been set at 3.3 mm. In Table I the main tolerances of this lens system are given. The table shows that, despite the (relatively) large FWD, the tolerances for disc thickness, field of view and decentering of the two lenses remain more or less comparable to the 50 μm FWD design of the ATC dual lens objective, while the tolerance for disc thickness variation is large. Due to the 6 times thinner cover layer than in the ATC concept, the disc tilt tolerance (for the case $\alpha=0$, i.e. a tilt rigid dual lens objective) is improved by about the same factor as in the case of ATC ($\alpha=\beta$)

in combination with a 0.6 mm substrate. For completeness, we have also listed in Table I the parameters for the thin cover layer approach *in combination with* active tilt correction. It is interesting to note that keeping the second lens parallel to the disc in this case would result in a *smaller* disc tilt tolerance than when the system is kept tilt-rigid. This is a result of the low value of μ for this system ($\mu = 0.13$) Applying the optimal tilt correction can still improve the disc tilt tolerance by an additional factor of 9.

The lenses were again made using the glass-photo-polymer replication process. The OPD_{rms} of the resulting lens system used in the experiments was less than 35 mλ. Realization of a similar lens set (NA=0.85, 0.1 mm thin cover layer) has been reported earlier by Osato *et al.*[7] Their objective lens, manufactured by glass moulding, had a larger entrance pupil diameter of 4.5 mm.

6.2.3.2 *Actuator for the thin-cover-layer approach*

As in the ATC prototype, the two-lens system for read out of a disc with a thin cover layer was mounted in a commercial Philips CDM-12 2D actuator, used for focusing and tracking. The resulting prototype is shown in Fig. 5(b). The main objective lens (lens 1; see Fig. 1) is additionally actuated in focal direction. This may be used to correct for cover layer thickness variations in the order of $\pm 15\,\mu$m. Allowing for an additional adjustment for compensation of errors in the lens set, the required stroke of this actuator is $\pm 50\,\mu$m. Optical tolerances require the tilt between the two lenses to be less than 0.5 mrad, and the decentering of both lenses to be below $\pm 20\,\mu$m. The realised linear guide inside the two lens actuator has a tilt accuracy better than 0.1 mrad, and remains centered within 1 μm. This leaves sufficient decentering tolerance for a relatively simple assembly process. Each lens, including the bi-asphere (lens 1, Fig. 1), contains a flat reference surface. During assembly, the tilt between these two surfaces can be adjusted within 0.1 mrad using a Fizeau interferometer. The actuator uses a voice coil motor that fits inside the dual lens housing. The complete assembly has a diameter of less than 7 mm, is only 3 mm high and weighs 230 milligrams.

6.2.3.3 *Experiment on ROM disc*

The thin-cover-layer actuator has been tested in the same test player described in Sect. 6.2.2. A glass-photo-polymer ROM disc was used with a thin laminated plastic sheet of 100 μm thickness as the transparent cover layer. Track pitch and minimum pit length of the data embossed on the disc are 0.5 μm and 0.278 μm, respectively, again corresponding to a user capacity of 10 GB. The thickness variation of the cover layer was $\pm 2\,\mu$m.

Since this thickness variation is so small, it turned out that the actuator governing the separation of the two lenses in the dual-lens objective only needed to be adjusted for static compensation of the cover layer thickness and for static compensation of manufacturing errors in the lens set. This was accomplished by applying DC current to the actuator coils while monitoring the data-to-clock timing jitter. In this way, any residual spherical wavefront aberration of the system could be minimised. (The actuator would also allow dynamic compensation of spherical aberration using a spherical aberration sensitive servo signal, but this turned out to be unnecessary in the present experiment). Fig. 6 shows the digital eye patterns obtained, before and after equalisation. Due to residual wavefront aberrations in the lens system of our prototype (the OPD_{rms} was somewhat less than 35 mλ), the measured data-to-clock timing jitter had a minimum of 7.3% after equalisation. In Fig. 7 the variation of jitter with disc tilt is presented for the thin-cover-layer approach. From this figure we may conclude that sufficient tilt window can be obtained using a large FWD, high numerical aperture (NA=0.85), tilt-rigid actuator, using the thin cover layer approach. With some further improvement of the dual lens objective it should be possible to obtain bottom jitter values better than 7%.[7]

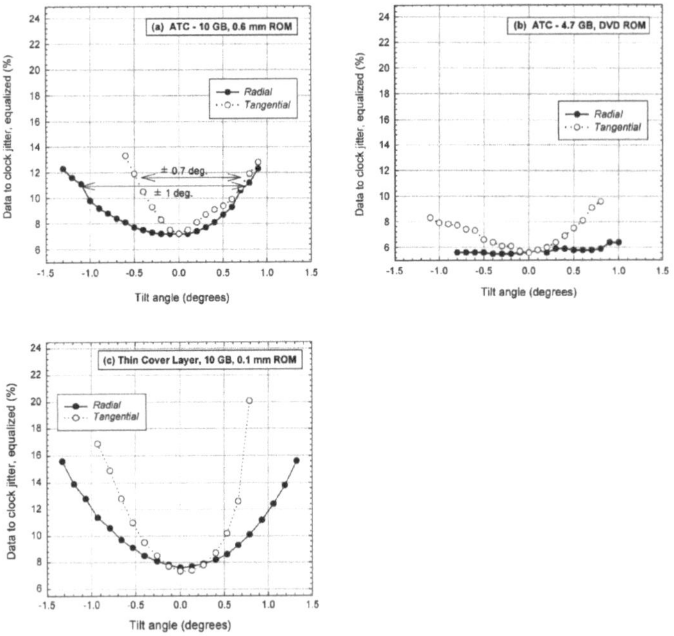

Fig. 7. Data to clock jitter versus disc tilt for a 10 GB ROM disc with 0.6mm cover layer read with the ATC system (a); a 4.7 GB DVD-ROM disc read out with the same ATC system (b); and a 10 GB ROM disc with a 0.1mm cover layer read out with a dual lens objective with static spherical aberration correction (c).

6.2.4 Conclusions

In this paper we have shown that both the active tilt correction (ATC) and the thin-cover-layer approach can be used with ample margins for the play back of a 12 cm optical ROM disc with user capacity of 10 GB. The advantage of ATC is that it allows easy backwards compatibility with DVD, as demonstrated. On the other hand, the ATC actuator is rather complicated from a system point of view because it has three additional degrees of freedom (two angles and the separation between the two lenses) that must be actively controlled besides the two of a conventional optical pick up (radial tracking and focussing). In addition, the method of servo signal generation employing three auxiliary light beams used in our research prototype is, of course, not acceptable in a product for the mass market. A more practical solution for the generation of the error signals still has to be demonstrated for the ATC approach.

In the case of the thin cover layer approach, a simple compatibility solution with DVD has not yet been identified. On the other hand, it has the considerable advantage that an actuator with only one additional degree of freedom can be used. This actuator does not need high-bandwidth control: adjustment once per disc insertion is likely to be more than adequate. Indeed, the thickness variations of the cover layer can be so small (within a few microns), that even a rigid dual lens objective may prove to be feasible. Given the low cost, "mass-manufacturing" character of the optical disc drive industry, this is a very compelling advantage. At first sight, the disc technology seems more complicated in the case of the thin-cover-layer approach. In practice, however, there are various convenient ways to make the thin cover layer reliably, and with adequate tolerances. In addition, the thicker substrate (1.1 mm versus 0.6 mm in the case of ATC) is easier to replicate by injection moulding, and it has adequate mechanical strength without the need for bonding two substrates back to back. A price to pay for the thin cover layer approach is an increased sensitivity to dust and fingerprints, because of the small diameter of the optical spot at the cover layer surface. A cartridge will therefore be necessary. Taking all considerations into account, a system based on NA=0.85 and read out through a thin cover layer seems to be the most attractive technology option for a next generation high-capacity optical recording system. ATC, however, might still be of interest for magneto-optical recording systems, in particular in the case of direct overwrite methods based on magnetic field modulation, in which case read out through a thick substrate can not easily be avoided because the information layer has to be close to the coil used to generate the high-frequency modulated magnetic field.

Acknowledgements

The authors wish to thank Messrs. C.L.Adema, F.C. van Dorst, F.M.A. van Gaal, C.A.N. van der Vleuten, P.J.C.H. Smulders and P.G.J.M. Nuyens for their contribution to the construction and characterisation of the lenses and actuators.

References

[1] S. Nakamura, M. Senoh, S. Nagahama, N. Iwasa, T. Yamada, T. Matsushita, H. Kiyoku, Y. Sugimoto, T. Kozaki, H. Umemoto, M. Sano and K. Chocho: Jpn. J. Appl. Phys. **36** (1997) 1578.

[2] B. H. W. Hendriks: Proc. SPIE **3482** (1998) 90.

[3] B. H. W. Hendriks: Appl. Opt. **37** (1998) 8195.

[4] K. Yamamoto, K. Osato, I. Ichimura, F. Maeda and T. Watanabe: Jpn. J. Appl. Phys. **36** (1997) 456.

[5] J. Braat: 'Proc. of Huygens' Principle 1690-1990 Symp. The Hague/ Scheveningen, 1990

[6] Because the cover glass refraction index of the 10 GB disc was lower than that of polycarbonate for which the lens system has been designed (1.52 instead of 1.58) the optimal distances between the front lens and the disc surface for this disc and for DVD were 35 μm and 50 μm, respectively. An additional adjustment of auxiliary focussing lightpaths was therefore required.

[7] K. Osato, K. Yamamoto, I. Ichimura, F. Maeda and Y. Kasami: Conf. Proc. Optical Data Storage, WA2-1/131, 1998.

6.3 Optical Disc System for Digital Video Recording

Tatsuya Narahara[1], Shoei Kobayashi[1], Masayuki Hattori[2], Yoshihide Shimpuku[2], Gijs J. van den Enden[3], Joost A. H. M. Kahlman[3], Marten van Dijk[3] and Roel van Woudenberg[3]

[1] *Giga Byte Laboratories, Sony Corporation, 6-7-35 Kitashinagawa, Tokyo 141-0001, Japan*
[2] *Media Processing Laboratories, Sony Corporation, 6-7-35 Kitashinagawa, Tokyo 141-0001, Japan*
[3] *Philips Research Laboratories, Prof. Holstlaan 4, 5656 AA Eindhoven, The Netherlands*

Abstract

We have developed a new error correction method (Picket: a combination of a long distance code (LDC) and a burst indicator subcode (BIS)), a new channel modulation scheme (17PP, or (1, 7) RLL parity preserve (PP)-prohibit repeated minimum transition runlength (RMTR) in full), and a new address format (zoned constant angular velocity (ZCAV) with headers and wobble, and practically constant linear density) for a digital video recording system (DVR) using a phase change disc with 9.2 GB capacity with the use of a red ($\lambda=650$ nm) laser and an objective lens with a numerical aperture *(NA)* of 0.85 in combination with a thin cover layer. Despite its high density, this new format is highly reliable and efficient. When extended for use with blue-violet ($\lambda \approx 405$ nm) diode lasers, the format is well suited to be the basis of a third-generation optical recording system with over 22 GB capacity on a single layer of a 12-cm-diameter disc.

6.3.1. Introduction

Two major technological breakthroughs have been achieved in the last two years, which together allow a large increase in optical disc capacity. First, high numerical aperture *(NA)* objective lenses have become feasible by using two-element lenses: $NA=0.85$ lenses can be applied with sufficient system margin when readout is performed through a thin transparent cover layer of 0.1 mm thickness,[1–3] instead of reading out through a 0.6-mm-thick substrate as is done for the digital versatile disc (DVD). Second, tremendous progress in the field of blue-violet diode lasers has been made over the last two years:[4] diode laser samples with a wavelength around, $\lambda=400$–410 nm and sufficient lifetime and output power have recently been realized, and will soon be commercially available. These combined breakthroughs allow a reduction of the focussed spot size (which is proportional to (λ/NA)),[2] by a factor of about 5 when compared with DVD, thus allowing a capacity of 22 GB on one layer of a 12-cm-diameter single-sided disc. This opens the way for (real-time) recording of bit-hungry high-quality video streams.

In this paper, we present the system design criteria (Sect. 6.3.2), the disc structure (Sect. 6.3.3), and a full format description and verification for a rewritable optical disc for the digital video recording system (DVR) with recording and playback through a thin cover layer with a red laser and an *NA* of 0.85, yielding a disc capacity of 9.2 GB. More specifically, we present a new error correction method (Sect. 6.3.4), a new channel modulation code (Sect. 6.3.5), and a new address format (Sect. 6.3.6), together with their experimental integration and evaluation (Sect. 6.3.7). This full format has an increased efficiency compared to conventional optical disc formats and is highly reliable, despite its high density.

More details on the various enabling technologies for the DVR optical disc system are described in a number of associated papers, presented together with our paper at the Joint International Symposium on Optical Memory and Optical Data Storage 1999.[5-9] Specifically, these papers discuss the cover layer technology,[5] various options for phase-change media with 9.2 GB disc capacity with the use of a red laser[6,7] and with 22 GB capacity with a blue-violet diode laser,[7,8] and the feasibility of dual-layer recording[9] in the high-*NA* thin cover layer approach. The red format development described in detail in this paper allows extension to capacities of 22 GB and more when blue-violet diode lasers (λ around 405 nm) are implemented in our system.

6.3.2 Requirements for digital video recording

The format presented in this paper is intended for use in a optical disc based digital video recorder. Although a detailed discussion of the digital video recording application is beyond the scope of this paper we will give a few general comments to emphasize the importance of a suitable disc format for this application.

Key issues for recording of high-definition (HD) video and to enable advanced features are a high (user) data rate and a high (user) capacity. In the HD application, a high data rate and a high capacity are imperative to be able to deal with the HD-rate and realize sufficient playing time. For special features such as dual stream operation, *e.g.* simultaneous recording and play-back from one disc using a single optical pickup, the user data rate of both streams must be sustained without interruption. Accesses have to be performed because the data may be scattered over the disc and, as a consequence, the net user rate will be lowered as a result of read and write actions with seeks and accesses in between. Various parameters such as seek time, fragmentation, read data rate and write data rate have to be taken into account to estimate the resulting user rate. However, it is clear that the time lost in seek operations has to be compensated by a higher disc rate. Therefore, a high disc rate and a disc format which allows fast access are very important.

We made some rough estimates of the required performance for a few application areas:

1) Recording of 4 h of DVD-quality video requires a 9 GB disc capacity.
2) Dual-stream operation of two DVD-quality video streams requires data rates of 30-35 Mbps and a disc capacity of 9 GB (for 2 h of each stream) or more
3) Recording of 2 h of HDTV video (*e.g.* according to the Japanese BS4B standard at 24 Mbps) requires a 22 GB disc capacity and a data rate of 35 Mbps.
4) Editing of digital video camcorder recording (e.g. DV at 28 Mbps raw data rate) requires a 30–50 Mbps data rate (depending on the editing options) combined with fast random access.

These application areas are summarized in Table I, together with the key parameters (required disc capacity and data rate) and the system options for realizing them. The application, capacity and data rate requirements imply that the disc format needs to be highly efficient and should support fast access.

	Source video rate data	Disc capacity	Required data rate to/from disc	System
4 h of DVD video	10 Mbps max.; 4.5 Mbps average	9 GB	10–15 Mbps	650 nm, *NA* = 0.85
2 streams of DVD-quality video (2 h each)	2–10 Mbps	9 GB	33 Mbps	650 nm, *NA* = 0.85
2 h of HD video	24 Mbps (BS4B)	22 GB	24–35 Mbps	405 nm, *NA* = 0.85
Video editing (Digital Video, DV)	28 Mbps	22 GB	30–50 Mbps	405 nm, *NA* = 0.85
2 streams of HD video	2–24 Mbps	40 GB	80 Mbps	405 nm, *NA* = 0.85

Table I. Video application areas, user requirements and system options. Note DVD allows variable bit rate (VBR) video, thus two source data rates are given: the maximum and (typical) average. The other examples (BS4B and DV) are constant bit rate (CBR) video streams

Additionally, the optical disc system needs to be highly robust and reliable.

Therefore, for full featured digital home video recording without the loss of picture quality, rewritable optical discs with capacities of 9 GB and data rates of 33 Mbps are required at least. In the near-future, 22 GB and 50 Mbps will be required for a video recorder with sufficient recording time for HDTV recording and new user features.

6.3.3 Disc structure

A rigid 1.1-mm-thick poly-carbonate substrate is covered with a phase-change stack, deposited in reversed order compared to the standard CD-RW or DVD+RW phase-change stacks (Fig. 1). On top of this stack a 0.1-mm-thick cover layer is applied by spin coating or foil lamination.[5] This thickness of 0.1 mm allows for sufficient tilt margin at $NA = 0.85$: when using a blue-violet laser, the tilt margin is approximately equal to that of DVD (650 nm, $NA = 0.60$, 0.6 mm substrate). The cover layer can be made with a thickness variation well within $\pm 3\,\mu$m. With this thickness uniformity, there is no need for dynamical spherical-aberration correction, so a rigid dual-lens objective can be used. The substrate serves as a stiff and rigid carrier, containing the mastered information (embossed data and grooves). We use standard astigmatic or Foucault wedge focussing methods and the radial push-pull method for tracking.

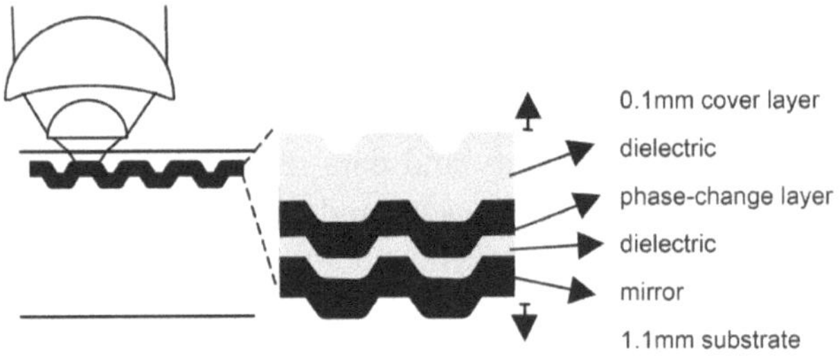

Fig. 1. Disc structure

6.3.4 Error detection and correction

Compared to DVD ($NA = 0.60$, 0.6 mm substrate thickness), the $NA = 0.85$, 0.1 mm cover layer disc system has one drawback: the spot size on the entrance surface of the disc is reduced from approximately 0.50 mm diameter (0.20 mm^2) to 0.14 mm diameter (0.015 mm^2). This results in increased sensitivity to dust and scratches on the disc surface, which may cause burst errors, on top of the usual random errors during readout of the recording layer. Our so-called picket code is a new error detection and correction method that uses two correction mechanisms to handle these errors effectively: a longdistance code (LDC) combined with a burst indicator subcode (BIS).

6.3.4.1 Long-distance code (LDC)

The LDC has 304 [248, 216, 33] Reed-Solomon (RS) code words. Each 9.5 RS code word contains the user data bytes of one logical 2K information block (with 4 additional bytes used for extra error detection). The LDC has sufficient parity symbols and interleaving length for correcting random errors, multiple long bursts and short bursts of errors. The burst error correction capability is strongly enhanced by using erasure correction on the erroneous symbols flagged by the BIS code described below.

6.3.4.2 Burst indicator subcode (BIS)

The LDC is multiplexed with the synchronisation patterns and the BIS. The BIS has 24 [62, 30, 33] RS code words (Fig. 2). The latter carries address and control information strongly protected by these BIS-RS code words. In fact, the BIS code can be properly decoded (i.e. all its errors can be corrected) with extremely high probability. The location of its corrected bytes and erroneous synchronization patterns serve as "pickets" indicating the likely position of long burst errors in the LDC data between these pickets: when subsequent pickets have "fallen", it is highly likely that all the data located physically in between these pickets was also detected erroneously. The LDC can use this information to perform erasure correction (see above).

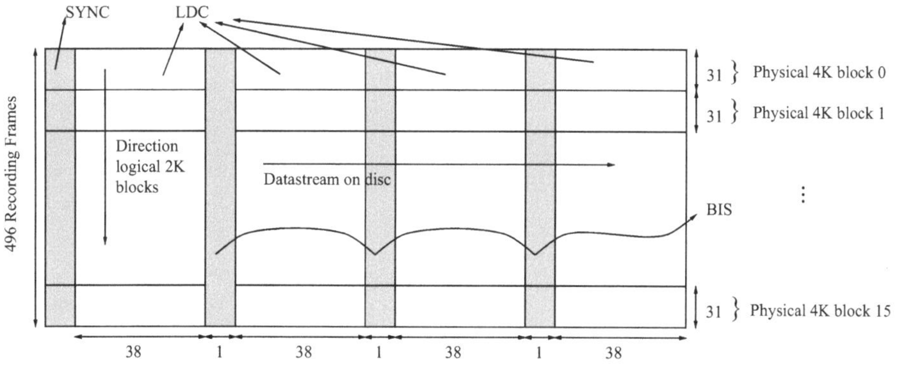

Fig. 2. ECC structure of 64 K physical cluster with LDC and BIS columns.

6.3.4.3 Data organization and data access

The protection is over physical clusters of 64K user data, which are organized in 16 physical 4K blocks. Each 4 K block is again subdivided into 31 recording frames (see Sect. 6.3.6.3). To obtain the user data of one logical 2K block we only need to decode the BIS having all address information together with the corresponding 10 RS code words in the LDC. This gives quick access to logical 2K blocks since the 64K LDC code does not have to be fully decoded.

6.3.4.4 Parameters of the picket LDC + BIS code and comparison with a conventional product code

In DVD, a product code is used for error correction.[10] The horizontal code is intended for correcting random errors and for indicating the location of burst errors. The vertical code uses erasure decoding to correct these bursts. The picket code does not have a horizontal code, all the redundancy is put in the vertically oriented LDC and BIS codes.

In the picket code, the BIS and the synchronisation patterns are used for indicating the location of bursts (see Sect. 6.3.4.2). An errors-and-erasures decoder of the LDC corrects these bursts together with random errors. Thus, compared to the vertical code in a product code, the picket code has approximately twice as many parity bytes in its vertically oriented composite codes.

Parameters	DVD	DVR
ECC rate (fraction user data bytes/ECC bytes)	0.866	0.852
Cluster size	32 kB	64 kB
Logical sector size and data	2064 bytes: — 2048 user data; — 4 EDC bytes; — 12 bytes for address, copyright management, spare	2074.5 bytes: — 2048 user data in LDC; — 4 EDC bytes in LDC; — 22.5 bytes in BIS for address, copyright management, spare
Code construction	product code	long distance code + burst indicator (Picket)
Code parameters	RS[182, 172, 11] × RS[208, 192, 17]	304 × RS[248, 216, 33] + 24 × RS[62, 30, 33]
Maximum correctable burst length (MCBL)	2912 ECC bytes	9920 ECC bytes (17.3 mm)
Number of correctable bursts of 100 ECC bytes	8–29	32–99 bursts of 175 μm
Number of correctable bursts of 200 ECC bytes	5–14	32–49 bursts of 349 μm
Number of correctable bursts of 300 ECC bytes	5–9	16–33 bursts of 524 μm
Number of correctable bursts of 600 ECC bytes	3–4	10–16 bursts of 1047 μm

Table II. Parameters and comparison of the ECC schemes of DVD (product code) and DVR (LDC + BIS Picket code).

In Table II we compare the ECC schemes of DVD and DVR. In DVD the cluster size is 32 kB, while we use a cluster size of 64 kB for our code, which again leads to a doubling of the number of parity bytes. We use this extra redundancy, together with the redundancy provided by the picket construction (described above), for increasing the interleaving length as well as the minimum distance of the vertical code. This improves the burst error capacity with a factor of 3 to 4. This is demonstrated in Table II, where burst errors of various lengths are considered. For example, when no random errors are present, between 16 and 33 bursts of 300 ECC bytes (corresponding to 524μm along a track) can be corrected in DVR, whereas the DVD-ECC can only correct between 5 and 9 bursts of 300 ECC bytes. Also the maximum correctable burst length (MCBL) in DVR is more than three times the MCBL in DVD: 9920 vs 2912 ECC bytes. Both codes have comparable ECC rates (the difference is mainly in available space for address, control, copyright management and spare area), and both codes are able to adequately correct the amount of random errors for their applications.

6.3.4.5 *Performance analysis using experimental data*

The performance of our code is illustrated by the following example, taken from our analysis presented in detail in ref. 11. On a dust sprayed disc exposed to an office environment, we find a raw byte error rate of 4×10^{-3}. The errors include many long and short burst errors. After analysis of the error patterns on this disc, a model can be made allowing us to study the error rate dependence on error classes and, *e.g.*, error density, thus generalizing the specific measurements on this disc and allowing us to determine the error rate after error correction. It is then found that our BIS code retrieves the address information and burst indication very reliably: the error rate is below 10^{-25}. The LDC-code powerfully corrects the raw byte error rate to 1.5×10^{-18} using erasure correction of the erasures flagged by the BIS-code. This has to be compared with an error rate of 5.7×10^{-7} which would have resulted after error correction with the DVD ECC.

Since the error rate after correction is very low, it is not experimentally feasible to measure the error rate after error correction when using this powerful ECC on real discs. We here demonstrate the power of the use of the BIS data as pickets for erasure flagging by comparing the number of parities used in the LDC to correct all errors. On a standard disc, we added 2–4 bursts in the information layer of 300μm length (2250 channel bits) each per ECC block. When just using the LDC, 17% of the error correction capacity was required to correct all errors. With the use of the BIS data, 72% of all errors was flagged as an erasure, all corresponding to the burst errors that were added by us, and the required correction capacity was reduced to 11%. If all burst errors were completely flagged as erasures, this could have been reduced to 8.5% (half of 17%). The difference is explained largely on the basis of the parts of the bursts before and after, respectively the first and last BIS byte related to the burst: only when all bursts start and stop exactly at a BIS position, the reduction of the number of required parity symbols by the exact factor of two is achieved. This example shows that the use of BIS for erasure flagging can result in a significant reduction in the required error correction capacity.

6.3.5 Channel modulation

The channel modulation schemes for CD-ROM and DVD-ROM were optimized for the maximum efficiency (i.e. high user capacity) within the constraints given by the modulation transfer function, i.e. the optical resolution limit. For rewritable phase-change recording however, a noise factor is introduced: when overwriting old data, differences in optical absorption and thermal properties between the old amorphous and crystalline areas result in a distortion of the newly written data. This yields variations on the effective mark position, showing up as additional jitter. We have designed our channel modulation scheme in such a way that the peculiarities of rewritable phase-change media are taken into account.

We developed a new (d=1, k=7) RLL code. The (1, 7) constraint means that we use runlengths of 2T up to 8T, with T being the channel period. The rate of this code is 2/3 (DVD's 8/16 modulation, also called EFMplus,[12] is a (2, 10) RLL code with a rate of 1/2). Using our code, the channel bit length is increased at the same data bit length compared to 8/16 modulation. This gives a larger timing tolerance, hence lower jitter (see Fig. 3) and longer recording time. Our code is named after its two new characteristic additional features: (1, 7) RLL Parity-Preserve, Prohibit Repeated Minimum Transition Runlength code, abbreviated as 17PP. We describe these features below.

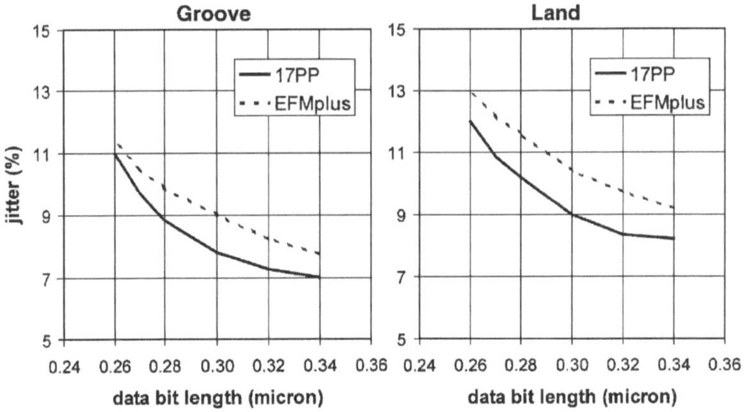

Fig. 3. Jitter comparison vs density for 17PP (d=1, k=7, rate 2/3) vs EFMplus (8/16 modulation; d=2, k=10, rate 1/2). The data is measured at λ=640 nm, NA=0.60, on a Land/Groove substrate (DVD conditions) after 10 times overwriting: a) in groove, b) on land.

6.3.5.1 *Parity preserve*

Our code has the parity-preserve property,[13] which means that the number of '1'-s in the data bit pattern before the channel encoder and in the corresponding modulated bit pattern after the channel encoder are both even, or both odd. For example, in our code the odd-parity data bit pattern '01' modulates into '010' and '10' into '001', and the even-parity '11' into '101' (or '000'). Using this property, one can efficiently obtain, and guarantee, a low DC-content of the recorded signal, thus allowing high-pass filtering of the playback signal, which makes the bit detection largely insensitive to signal level variations by *e.g.* dust and scratches, thus giving a highly reliable playback. The DC control is performed using insertion of DC-control bits in the data bit stream before the channel encoder, in contrast to alternative merging bit schemes where the so-called merging bits are inserted in the channel bit stream. We thus reduce the overhead for DC-control from 5.8% in conventional (1, 7) RLL with merging bits to 2.2% in our 17PP code, at the same DC-control block length.

This new DC-control mechanism is illustrated using the following example. Consider the data bit pattern 'P1 10 01 10 01' to be encoded with the modulation table, where P denotes a DC control bit. When 'P'='1', this data bit pattern would encode into '101 001 010 001 010', which translates in a bit pattern '001 110 011 110 011' after NRZI conversion: this has a digital sum value (DSV) of +3. With 'P'='0', it encodes into '010 001 010 001 010', which translates in '100 001 100 001 100' after NRZI conversion and has a DSV of -5. Thus we can choose between a positive and a negative digital sum value for the resulting bit sequence, and by the proper choice we can keep the low frequency content of the resulting modulation bit stream small.

A comparison of the power spectral densities between our code (using one DC control bit for every 45 data bits) and a conventional (1, 7) RLL code with merging bits for DC control at the same overhead is shown in Fig. 4.

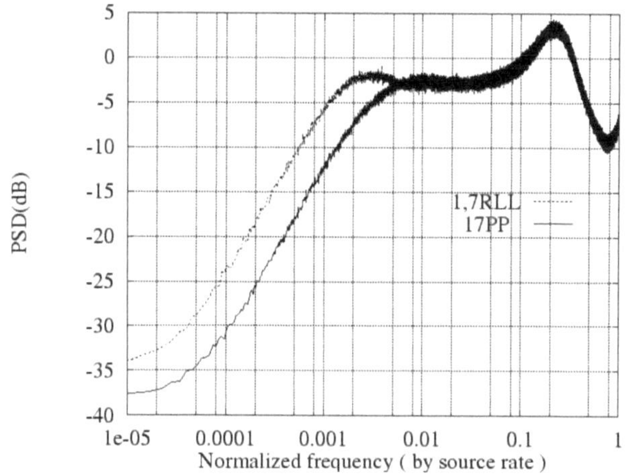

Fig. 4. Power spectral density (PSD) comparison at the same DC overhead between 17PP (1 source bit every 46 source bits) and (1, 7) RLL with merging bits (4 channel bits every 184 channel bits).

6.3.5.2 Prohibit RMTR

Our code limits the number of consecutive minimum runlengths (i.e. runs of 2T) to 6: the prohibit RMTR (repeated minimum transition runlength) property. This increases system tolerances, especially against tangential tilt as shown in Fig. 5, and hence increases the robustness of the system.

The RMTR is implemented in the modulation scheme by a careful choice of the code words and by using a substitution rule that prevents the appearance of a long sequence of the minimum runlengths. The data bit pattern '01 *11 01 11* 01' would be modulated into the channel bits '010 *101 010 101* 010', i.e. '100 *110 011 001* 100' after NRZI conversion, when the main conversion table is used. This repetition of 2T symbols is prevented by a substitution of the bits printed in italics resulting into '010 *001 000 000* 010', i.e. '100 *001 111 111* 100' after conversion to NRZI.

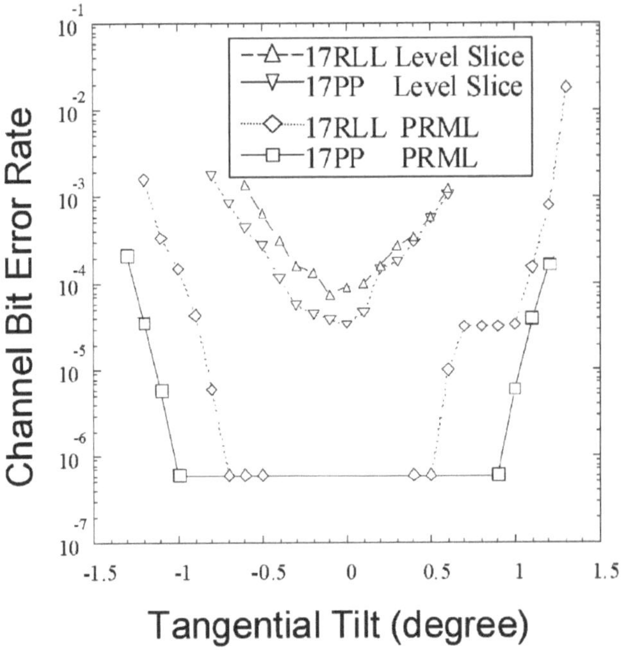

Fig. 5. Effect of the use of P-RMTR: Channel bit error rate vs tangential tilt (at a 10% increased linear density to 0.19μm/data bit, after 1000 times of overwriting) using standard slicing level detection and when using PRML detection.

6.3.6 Address format

The rigid carrier substrate contains the land/groove spiral and embossed headers. The groove forms a single spiral with a pitch of 0.90μm, with the lands in between, resulting in an effective track pitch (land-to-groove) of 0.45μm. Each track (one turn of the spiral) is divided into eight segments, shown in Fig. 6(a). Each segment starts with an embossed header area, and is followed by a wobbled groove.

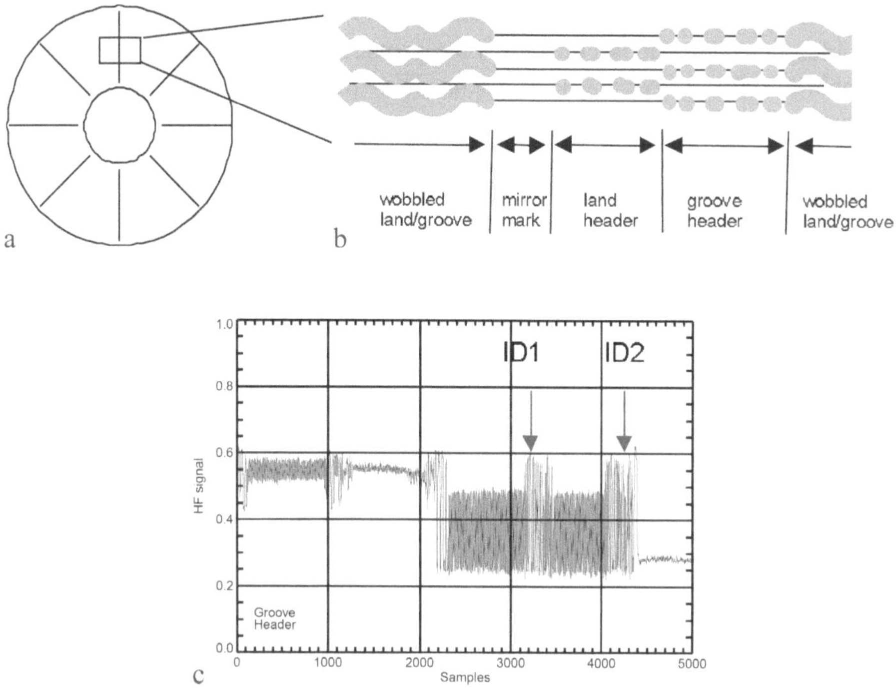

Fig. 6. Address format: a) header layout, b) header structure showing mirror mark, land header, groove header, and wobbled land/groove structure, and c) signal from groove header showing the two address fields (ID1 and ID2) and two spatially separated positions.

6.3.6.1 Wobble scheme

The wobble is used for speed control of the disc and to derivethe channel clock during recording (the channel bit length of the data is an integer fraction, 1/322, of the wobble period).

In designing a wobble scheme, two conflicting properties have to be considered in the design of the format to obtain the maximum format efficiency. On one hand, a constant linear density format allows maximum efficiency, since that gives no losses due to density variations. This implies the use of a wobble with a constant spatial frequency, i.e. a so-called CLV (constant linear velocity) wobble. On the other hand, the use of a CLV wobble cannot be applied in a land/groove system, since that would result in a wobble signal with variable amplitude on the land tracks because the wobbles in the grooves on either side have a slightly different angular frequency ('wobble beat').

Therefore, we have chosen to use a zoned CAV (constant angular velocity) wobble: the zoning of the rewritable user area is done into 99 bands of 762 tracks each (i.e. 381 groove and land tracks). Within a band, the number of wobbles per segment is constant, thus providing a single-angularfrequency wobble in both groove and land tracks. The number of wobbles per segment increases from 420 in the first track by a fixed number (6) in every next band, in such a way that the spatial wobble frequency at the start of each band is exactly the same. The wobble period is thus constant over the whole disc within ±0.8%, resulting in a practically constant linear density.

The number of wobbles added every band (6 per segment) and the size of the bands (762 tracks) are chosen for achieving the maximum efficiency. This choice is the balance point when taking into account the two sources of efficiency loss: 1) the wobble period variation, which becomes larger when the bands become larger; and 2) the loss of one track at each band boundary, i.e. the land track suffering from wobble beat due to different wobble frequencies on either side, which gives a larger efficiency loss when bands become smaller.

6.3.6.2 Header

The header has three parts: a mirror mark, a land header and a groove header (Fig. 6(b)). The mirror mark can be used as a calibration or reference field (offset control) for push pull tracking and focus. The header itself contains the track and segment numbers for addressing in the so-called identifier field (ID). In each header, this ID is repeated a second time at a physically separated position to be well protected against small dropouts, *e.g.* defects in the layer stack (Fig. 6(c)). Groove and land headers are separated in the tangential direction to prevent cross-talk between the two. The robustness of the headers is further increased by using a $d=2$ modulation code with the same channel bit length as the 17PP-encoded phase-change data, resulting in a large signal amplitude also for the shortest marks (I3) and a very wide eye opening (Fig. 7).

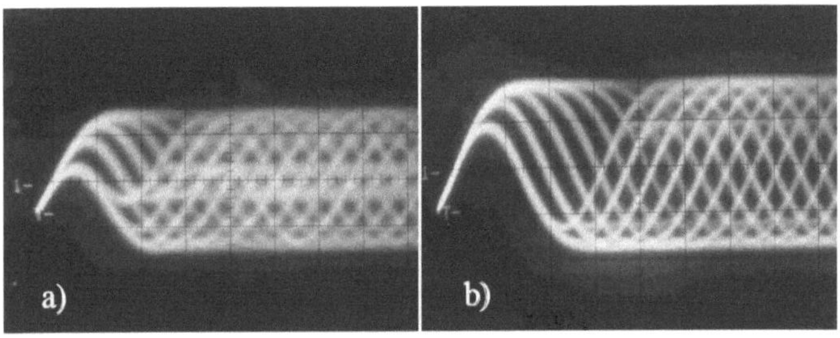

Fig. 7. Non-equalized eye patterns for a) 17PP code for (phase-change) data and b) (2, 7) RLL code for embossed header data.

This results in data-to-clock jitters below 6% in the header, and we measured an address error rate below 10^{-4}, illustrated in Fig. 8.

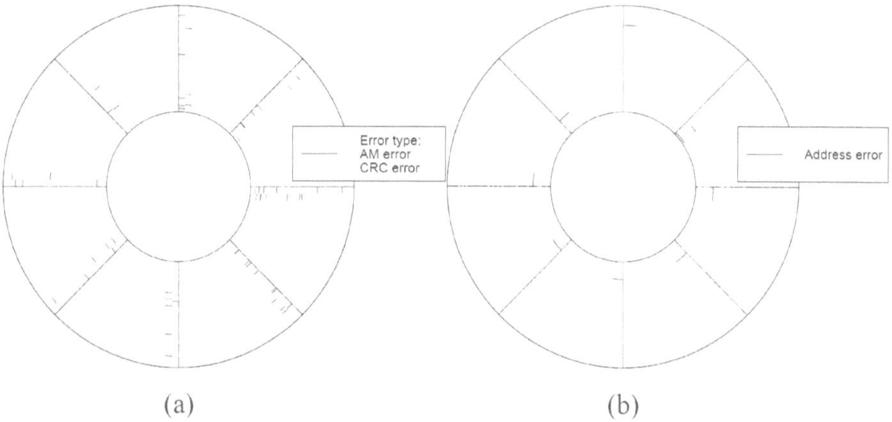

Fig. 8. Graphical presentation of header errors and the various sources of errors: an Address error occurs when none of the two address fields in a header is detected correctly. A misdetection can occur due to two reasons: the synchronization patterns of an address field, the so-called Address Marks (AM), can be missed, or the parity check symbols (CRC) can flag a detection error. Both errors are indicated for each address field of all groove headers.

6.3.6.3 Organization of data on the disc

In most optical disc systems, the physical structure of the user data and the physical structure of the address format (esp. headers) are, what one could call, synchronized. Typically, the distance between two headers is then equal to the (fixed) recording unit fragment size of, *e.g.*, 2 Kbytes. In our scheme, this is no longer the case: the distance between the headers increases every band with 6 wobble periods and thus varies from the inner diameter of the disc (where the distance is 420 wobbles, see Sect. 6.3.6.1) to the outer diameter by a factor of roughly 2.5. For maximum efficiency, the data is organized in so-called recording frames with the length of 6 wobbles, or 1932 channel bits (a SYNC, 4 times 38 LDC bytes and 3 BIS bytes, see Fig. 2), such that an integer number of these frames fits exactly in between two headers. These frames are the basic units of our recording scheme. When recording a 64k ECC cluster, equivalent to 496 frames, recording is stopped just before a header, and resumed again after the header, as shown schematically in Fig. 9. The next ECC block is written subsequently: linking between the blocks is thus not always done at a header position, but can also be done in between two headers. Of course, the start positions of all ECC blocks are known, and can be referred to by the combination of track number, segment number and wobble number.

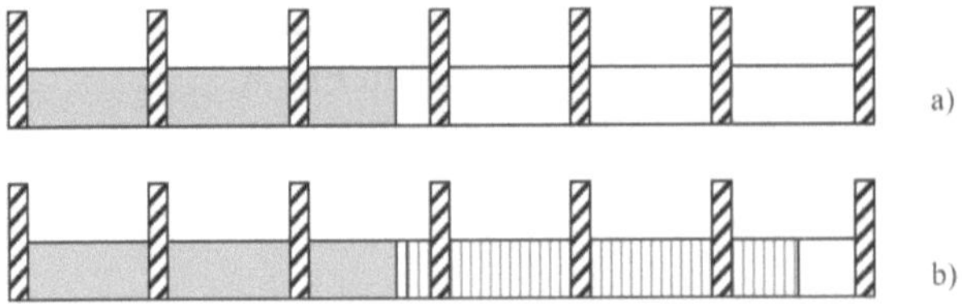

Fig. 9. Schematic representation of recording scheme: a) a first ECC block (filled gray) is written with interruptions at the header positions (diagonally hatched); b) the second ECC block is written subsequently (vertically hatched).

6.3.6.4 *Efficiency*

The combination of a fixed number of headers per revolution in a spoke-like layout, our efficient wobble-scheme with practically constant linear density and our recording scheme results in a very high efficiency for our land/groove address format: 96.6%, compared with 88% for DVD-RAM's land/groove format (which has a larger density variation over the disc and more overhead from headers)[14]. This results in a large capacity (long recording time) and highly reliable recording and playback. Moreover, the structure supports fast access.

6.3.7 System integration and evaluation

We have implemented our format on thin-cover layer[5] phase-change discs[6,7] and in an experimental optical disc drive equipped with a two-element $NA=0.85$ objective[1-3] and a red laser. The parameters are summarized in Table III.

Disc diameter	120 mm	Disc layout	Wobbled groove and land with headers
Cover layer thickness	100 ± 3 μm		
Effective track pitch	0.45 μm	Data zone division	99 ZCAV bands
Channel bit length	0.14 μm	Channel modulation	Phase-change: 17PP
Data bit length	0.21 μm		Headers: (2, 7) RLL
Total efficiency	79%	Error correction code	64 kB LDC + BIS Picket
User data capacity	9.2 Gbyte	Laser wavelength	650 nm
Channel bit rate (typ.)	62.5 Mbit/s	Numerical aperture	0.85
User data rate (typ.)	33 Mbit/s	Objective type	(rigid) dual-lens

Table III. DVR parameters.

The headers are detected highly reliably using standard slicer level detection: the data-to-clock jitter is below 6% and the header error rate is below 10^{-4} (see Sect. 6.3.6.3). The wobble is robustly detected using the high-frequent radial push-pull signal, thus providing a stable write clock that is locked to the disc. Phase-change recording using our 17PP code at a data bit length of 0.210μm and 9.2 GB disc capacity was performed with low data-to-clock jitter, less than 9%, at data rates of 33 Mbps, also after many overwrite cycles. We have recorded large streams of MPEG video data, encoded with our ECC and channel code. This data was read-back without any errors: the ECC corrected all random errors as well as the occasional burst errors (see Sect. 6.3.4.5), resulting in error-free user data (LDC) and access data (BIS).

We are currently extending and implementing the format in an experimental drive using blue-violet diode lasers (λ around 405 nm). The initial results show that a 22 GB capacity and 30–50 Mbps user data rate are feasible with our approach.

This shows that our DVR system is suitable for nextgeneration optical disc systems, after CD and DVD (Tabel IV). Moreover, we believe the main application area for such a system will be in-home consumer recording of digital video streams of both standard as well as high definition quality.

generation	first	second	third	
	CD	DVD	DVR	
	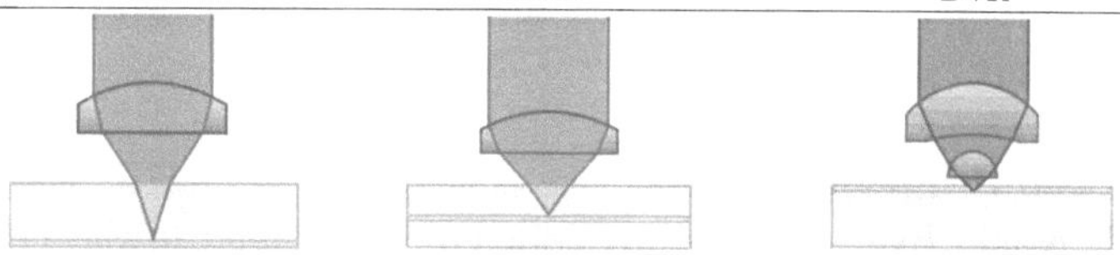			
wavelength	780 nm	650 nm	650 nm	400 nm
NA	0.45/0.50	0.60	0.85	0.85
substrate thickness	1.2 mm	0.6 mm	0.1 mm	0.1 mm
capacity (single layer)	650 MB	4.7 GB	9.2 GB	22 GB
data rate (1X)	1.2 Mbps	11 Mbps	33 Mbps	35-50 Mbps
introduced as	CD-Audio distribution	DVD-Video distribution	Digital Video recording	Digital Video recording

Table IV. Optical recording generations: CD, DVD and DVR systems.

6.3.8 Conclusions

We have presented a complete, novel format (error correction code, channel modulation code and address format) for a digital video recording system with a disc capacity of 9.2 GB and a user data rate of 33 Mbps with the use of phase-change recording with a red laser and NA=0.85 through a thin cover layer. The format is designed for optimum performance for real-time digital video recording: the address format, recording scheme and error correction scheme allow fast random access, as well as fragmented recording (e.g. for efficient use of the empty space on a partially written disc), and the efficient format combined with our 17PP channel code results in a high disc capacity (9.2 GB). This allows recording of 4 h of DVD quality video. When used in combination with our fast phase-change stacks (over 33 Mbps user data rate), it also allows dual-channel operation, e.g. writing one video programme while reading another one at MPEG2 bit rates of, say, 10 Mbps. In addition, transparent recording of HDTV formats is possible.

The DVR optical disc system parameters are summarized in Table III. The total efficiency of address format, DC-control and error correction is 79%, which is very high for a random access optical recording system. Most of the overhead is used to guarantee system robustness: a powerful and effective error correction, efficient and guaranteed DC-control, robust addressing by efficiently designed headers, and robust and reliable phase-change recording behaviour. This format allows extension to even higher capacities. With blue-violet lasers (λ around 405 nm), we will be able to obtain a capacity of 22 GB and more, which is necessary for 2 h of high definition TV recording.

Our system is well suited to be the basis of a third generation optical recording system.

Acknowledgements

We would like to thank our colleagues involved in the DVR project within Philips and Sony, in particular the teams responsible for mastering and replication, phase-change disc preparation and recording, cover layer technology, and electronical, mechanical and optical support. We acknowledge Masanobu Yamamoto, Jacques Heemskerk and Henk van Houten for their support and Masayuki Arai, Stan Baggen, Jan Bakx, Martijn Blüm, Steven Luitjens, Toshiyuki Nakagawa, Jaap Nijboer, Hiroshi Ogawa, Henk van der Put, Susumu Sensyu, Bert Stek, Hans Spruit, Ludo Tolhuizen and Kouhei Yamamoto for stimulating discussions.

References

[1] K. Yamamoto, K. Osato, I. Ichimura, F. Maeda and T. Watanabe: Jpn. J. Appl. Phys. **36** (1997) 456.

[2] K. Osato, K. Yamamoto, I. Ichimura, F. Maeda, Y. Kasami and M. Yamada: Proc. SPIE **3401** (1998) 80.

[3] Y. U. Martynov, B. H.W. Hendriks, F. Zijp, J. Aarts, J.-P. Baartman, G. van Rosmalen, J. J. H. B. Schleipen and H. van Houten: Jpn. J. Appl. Phys. **38** (1999) 1786.

[4] S. Nakamura and S. Fasol: The Blue Diode Laser (Springer Verlag, Berlin, 1997).

[5] M. M. Decré and P. H. Vromans: Jpn. J. Appl. Phys. **39** (2000) 775.

[6] Y. Kasami, Y. Kuroda, K. Seo, O. Kawakubo, S. Takagawa, M. Ono and M. Yamada: Jpn. Appl. Phys. **39** (2000) 756.

[7] B. Tieke, M. Dekker, N. Pfeffer, R. van Woudenberg, G.-F. Zhou and I. P. D. Ubbens: Jpn. J. Appl. Phys. **39** (2000) 762.

[8] I. Ichimura, F. Maeda, K. Osato, K. Yamamoto and Y. Kasami: Jpn. J. Appl. Phys. **39** (2000) 937.

[9] K. Kurokawa, M. Naito, K. Yasuda, T. Kashiwagi and O. Kawakubo: Proc. SPIE **3864** (1999) 197.

[10] DVD Specifications for Read-Only Disc: Part 1, Physical Specifications ver. 1.0.

[11] K. Yamamoto, M. Hattori and T. Narahara: Proc. SPIE **3864** (1999) 339.

[12] K. A. S. Immink: IEEE Trans. Consum. Electron. **41** (1995) 491.

[13] The parity preserve principle was first introduced by us in: J. A. H. M. Kahlman and K. A. S. Immink: U.S. Patent 5,477,222 (1995), where the principle was applied in a (d=1, k=8) RLL code.

[14] DVD Specifications for Rewritable Disc (DVD-RAM): Part 1, Physical Specifications ver. 1.0 (July 1997).

6.4 Groove-only recording under DVR conditions

M. Kuijper[1,2], I. Ubbens[3], L. Spruijt[3], J.M. ter Meulen[1,3], K. Schep[1]

[1]Philips Research Laboratories, Prof. Holstlaan 4, 5656 AA Eindhoven, The Netherlands
[2]Philips Optical Storage, Glaslaan 2, 5616 LD Eindhoven, The Netherlands
[3]Philips Optical Disc Technology Centre, Glaslaan 2, 5616 LD Eindhoven, The Netherlands

Abstract
We developed a groove-only rewritable disc for the DVR system with a blue laser diode (wavelength 405 nm). Using standard detection electronics we obtained a capacity of 23.3 GB. Higher capacities are possible with advanced detection methods. Wide system margins are obtained at 320 nm track pitch and 80 nm channel-bit length. Fast-growth materials are used for the active layer. No thermal cross-write effect is present in the central track when neighbouring tracks are repeatedly rewritten.

6.4.1 Introduction

Sony and Philips have developed an optical disc system for Digital Video Recording (DVR)[1,2]. The DVR system is based on an objective lens with a high numerical aperture (NA=0.85) and a thin cover layer with a thickness of 0.100 mm[3-6]. Using a blue-laser diode with a wavelength of 405 nm, the capacity is as high as 23.3 GB[2].

Previously, we demonstrated the feasibility of phase-change recording in a land/groove format with headers[2]. In this study, we investigated groove-only recording under DVR conditions. A groove-only format is better suited to create a compatible family of read-only, write-once, and rewritable discs (similar to the situation for DVD-ROM, DVD-R, and DVD+RW).

In addition, groove-only recording has the advantage of absence of thermal cross-write. In the beginning, quick-crystallisation materials (QCM) like $Ge_2Sb_2Te_5$ were used in DVR[7,8]. These materials are well suited for land/groove recording at low data rates since they are relatively insensitive to thermal cross-write. However, it is difficult to achieve a high data rates using QCM[7,8]. Later, fast-growth materials (FGM) were introduced in DVR making higher data rates possible. User-data rates up to 80 Mb/s have been demonstrated[9-12] and even higher data rates are expected to be feasible by using faster phase-change materials[15-17]. The introduction of FGM, however, also resulted in an increased sensitivity for thermal cross-write[10] which is in fact the limiting factor for the radial density in the present DVR system[2].

Wavelength	λ=405nm
Numerical a perture	NA=0.85
Cover layer thickness	0.100 mm
Format efficiency (with improved DVD+RW wobble)	81.7 % (including overhead for DC-control)
17PP code efficiency	66.7 %
Track pitch	320 nm (310 nm to 325 nm tested)
Channel-bit length	80 nm (80 nm to 88.8 nm tested)
User c apacity	23.3 GB

Table I. DVR on-groove conditions

In this study, we present the first results on groove-only recording using fast-growth materials under DVR conditions, see Table 1. This groove-only rewritable format creates the possibility to make a compatible family of discs and in addition reduce the sensitivity for cross-write while retaining the advantage of high data rates. Note that the format efficiency in a groove-only system can be somewhat higher compared to the land/groove DVR system due to the absence of headers and band boundaries. Note that all jitter measurements presented in this paper were obtained using standard detection electronics: a linear equaliser in combination with threshold detection.

6.4.2 Structure of the disc

The results presented here are for track pitches of 325 nm and 310 nm. The track pitch of 325 nm is derived from scaling the DVD track pitch of 740 nm from DVD (NA=0.60, λ=650 nm) to DVR conditions. Note that this track pitch is about 8% larger than the 300 nm effective data-to-data track pitch used in the DVR land/groove system[2]. The corresponding decrease in capacity can be compensated by an increase of the linear density.

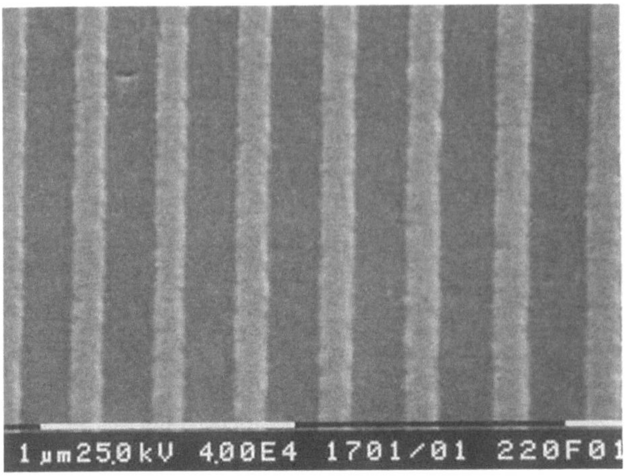

Fig. 1. SEM image of mastered groove. The LBR conditions are NA=0.90 and λ=257 nm. The trackpitch is 325 nm and the groove depth is typical 22 nm.

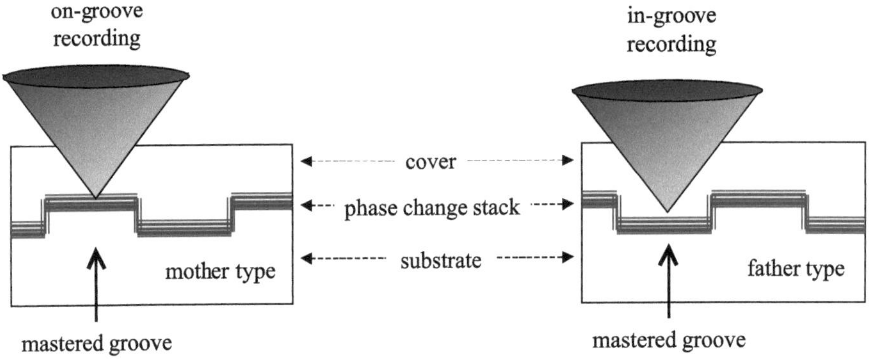

Fig. 2. With substrates moulded from a mother type stamper on-groove recording (left) is done, while with the father type substrate in-groove recording is possible (right).

The substrates were mastered using a laser-beam recorder (LBR) with a wavelength of 257 nm and a NA of 0.90, that was developed in a co-operation between Philips and Toolex. High-quality grooves are recorded, see Fig. 1. To test the influence of the groove duty cycle on the tracking signals (push-pull) and the data signal quality (jitter), we have fabricated discs with a series of groove widths by recording with different intensities of the LBR. Replication was done using both father as well as mother stampers. The mother stampers are made by growth on the father in a galvanic process (so the relief structures are opposite). The recording track is the mastered groove because of the requirement of retrieving write-clock signals and address information from a mastered wobble. For our mother stamper this implies that the mastered groove is closer to the entrance surface than the land. We therefore call discs with mother type substrates on-groove recording discs, see Fig. 2. Similarly, discs replicated using a father stamper are referred to as in-groove discs.

On the substrates, a MIPI phase-change stack was sputtered, see Fig. 3, and a 100-μm-thick cover layer applied by bonding a 75-μm-thick polycarbonate sheet with 25-μm-thick pressure-sensitive adhesive (PSA). For the stack, the FGM phase-change material developed for DVR land/groove discs was used.

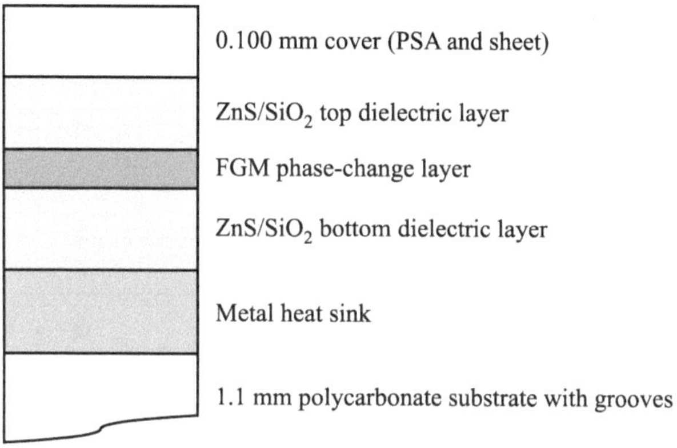

Fig. 3. Structure of the DVR recording stack.

6.4.3 On-groove *vs* in-groove recording

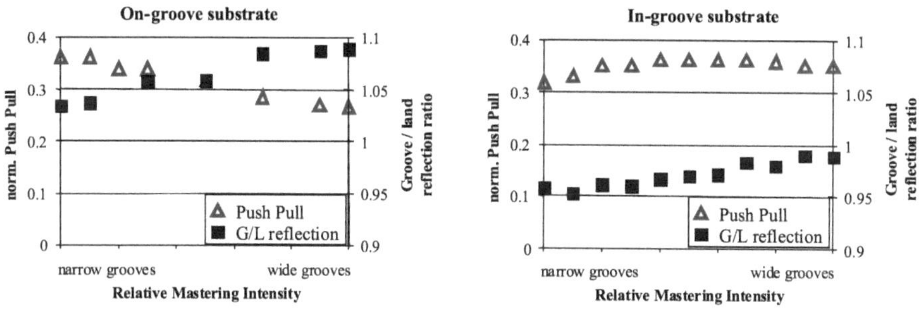

Fig. 4. The push-pull signal and groove / land reflection ratio for the on-groove substrate (left figure) and in-groove substrate (right figure) at track pitch 325 nm as function of the relative mastering intensity. The width of the groove increases from left to right in the figures.

Fig. 4 shows the push-pull signal and the groove/land reflection ratio as a function of the relative mastering intensity. The substrate does not contain the entire range of groove/land duty cycles, but is varied around a 50% groove/land duty cycle. The replicated groove/land duty factor is slightly changed by the deposited MIPI stack. On-groove grooves become wider, whereas in-groove grooves become narrower. The smallest on-groove groove/land reflection ratio is larger than one, indicating a groove width in the recording layer of larger than 50%. In that case, the push-pull signal decreases and the groove/land reflection ratio increases monotonically with wider grooves. This is in fair agreement with scalar diffraction simulations. For the in-groove case, a maximum push-pull signal is observed close to the point where the groove/land reflection ratio equals 1.0, which corresponds to a 50% duty factor of the groove width.

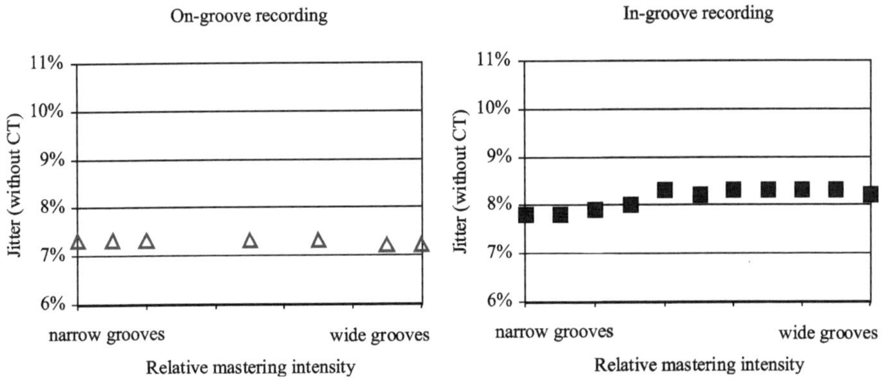

Fig. 5. Single-track jitters without cross-talk (CT) as function of the groove width for on-groove recording (left) and in-groove recording (right). Note, that the horizontal scale in the two figures is not the same.

Fig. 5 shows the jitter when recording either on-groove or in-groove at 50 Mb/s and a channel-bit length of 86.3 nm using the 17PP modulation code[1]. A significant difference is observed between the in-groove recording and on-groove recording. Their are indications that the differences arise from the optical spot quality in the recording layer[14]. Because of its superior recording behaviour, we concentrate on the on-groove recording situation for the rest of this paper.

6.4.4 On-groove recording

If we combine the results from Fig. 4 and 5 and display the relation between the tracking signal and the jitter, we observe over a wide range of groove widths only a small increase of the jitter, see Fig. 6. At the same time a wide range of push-pull levels is obtained. Thus, there is a clear trade-off between the tracking signal (push-pull) and the data quality (jitter). For robust tracking, a normalised push-pull signal of at least 0.25 is required. At track pitch of 325 nm this is always the case. However, at smaller track pitches this requirement can imply that narrower grooves closer to the 50% groove duty cycle has to be chosen, or deeper grooves with a somewhat increased jitter.

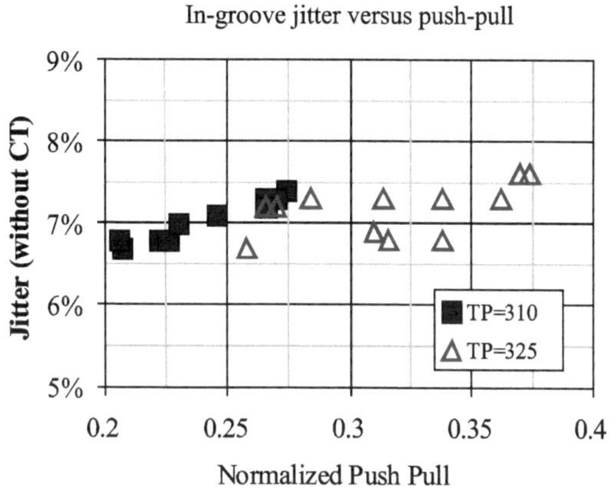

Fig. 6. On-groove jitter versus the push-pull level from different groove widths at track pitches of 325 nm and 310 nm.

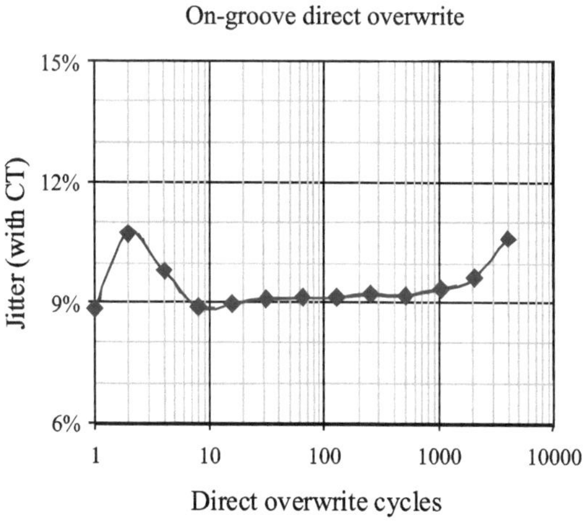

Fig. 7. Typical on-groove direct-overwrite cycles (multi-track jitter with cross-talk) at a track pitch of 325 nm and 80 nm channel-bit length.

Fig. 7 shows the typical on-groove jitter (with cross-talk) for our FGM stack as function of direct-overwrite (DOW) cycles at 36 Mb/s user data rate (66 MHz channel clock). With FGM-type phase-change stack at least 1000

DOW cycles are obtained. Although a vast amount of optical cross-talk is observed, see Fig. 8 (left), no thermal cross-write is present for the on-groove DVR disc with FGM-type phase-change recording layer. In Fig. 8 (right) the results of the cross-write experiments at a track pitch of 325 nm and a channel-bit length of 80 nm are depicted. The neighbouring tracks are 512 time overwritten. To cancel the optical cross-talk contribution, which is highly sensitive for the signal amplitude of the neighbouring tracks, we have erased the neighbouring tracks afterwards. The central track signal quality (jitter and relative modulation) has not degraded after more than 500 times overwriting neighbouring tracks. From the figure can be concluded that even with 130% overpower no cross-write effects of any importance are observed. It is believed that the metal mirror with the double barrier (groove to land to groove substrate profile) acts as a powerful heat sink and prevents the temperature in the central track to become too high to induce re-crystallisation. Also at track pitch of 310 nm we measured no cross-write effects.

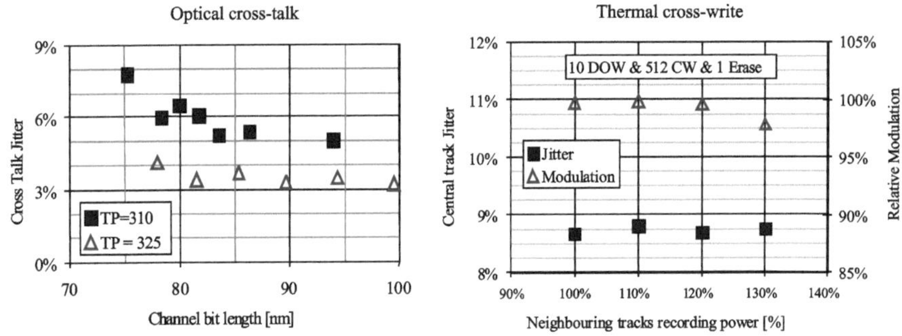

Fig. 8. On-groove optical cross-talk as function of the linear density and the track pitch (left). The cross-talk is calculated as a quadratic difference between the single-track jitter to the multi-track jitter: $\sigma_{CT}^2=\sigma_{MT}^2-\sigma_{ST}^2$. Although a vast amount of optical cross-talk is observed, no thermal cross-write is present for the on-groove DVR disc at a track pitch of 325 nm (right). Even with 30% excess recording power in the neighbouring tracks, no jitter increase is measurable (at channel-bit length of 80 nm). For the cross-write experiment the neighbouring tracks are erased after 512 DOW cycles to cancel the optical cross-talk from the measurement.

Compared to the DVR land/groove format a large amount of optical cross-talk is present for the groove-only disc, typical 4% to 6% additional jitter (by quadratic addition). Fortunately, the optical cross-talk can in principle largely be reduced by advanced signal processing[13]. The fact that no cross-write is observed allows for further optimisation of the radial versus the tangential density for other system margins. This is in contrast to the thermal cross-write that is the limiting factor on DVR land/groove discs.

In Fig. 9, the results of the variation of the tangential and the radial density are depicted. At nominal conditions a small advantage is observed from decreasing the radial density (track pitch of 325 nm instead of 310 nm) and increasing the tangential density (shorter bit length). The decrease of the jitter due to a lower tangential density is nearly cancelled by the increase of the additional optical-cross-talk jitter. For both the power as well as the two tilt margins the radial and linear densities are coupled, see Fig. 11. From power margin and tangential tilt margin point of view, a larger bit length is favourable. Also, a less critical write strategy is required and wider power margins are obtained, see Fig. 10. On the other hand, a broader radial-tilt window is possible with a larger track pitch.

From the experiments, it can be concluded that when taking all system margins into account, a user capacity of 22.5 GB identical to the land/groove DVR system[2] is feasible. However, due to the absence of thermal cross-write, a reduction of the track pitch to 320 nm is allowed. With a bitlength of 80 nm a feasible DVR groove-only rewritable disc with enough margins is obtained with a user capacity of 23.3 GB.

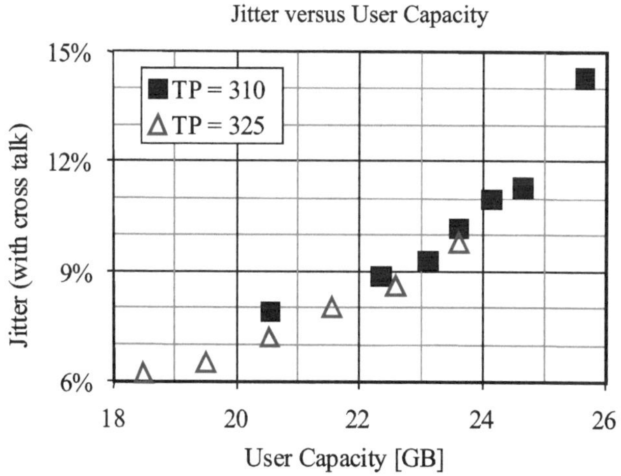

Fig. 9. On-groove jitter with cross-talk as function of the capacity. Both track pitch as bit length are varied.

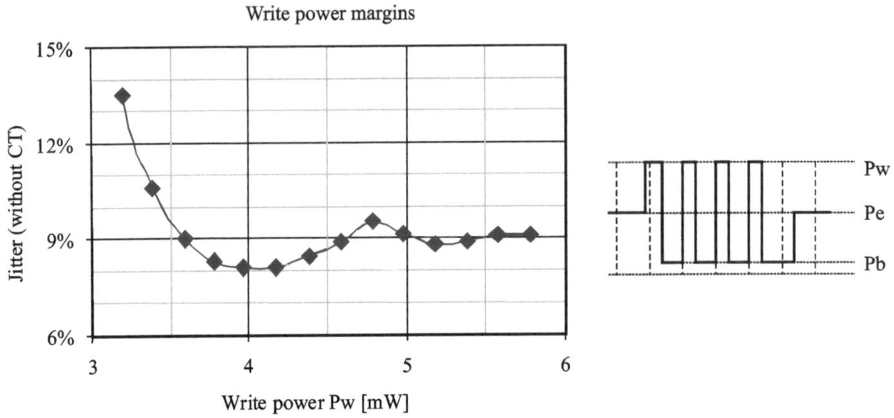

Fig. 10. Wide power margins are obtained at track pitch of 325 nm, bitlength of 80nm and at a recording speed of 36Mb/s. For the FGM type phase-change recording layer a tolerant write strategy can be used.

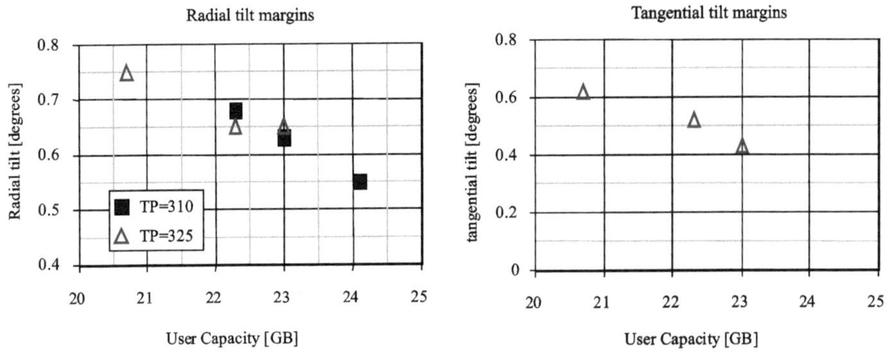

Fig. 11. Tilt margins at DVR on-groove conditions with 15% jitter crossing.

6.4.5 Wobble addressing format

The wobble addressing format for the groove-only DVR system is similar to that of DVD+RW. The wobble is used for both write-clock generation and retrieval of address information. The wobble is a predominantly single-tone sine wave that is interrupted by short modulated parts that contain the address information. The wobble has a length equal to 69 channel-bit lengths. The wobble period is chosen such that for a track pitch of 320 nm it leads to smaller wobble beat than would have been obtained using longer wobble periods. The modulation of the address bit is done using Minimum-Shift-Key (MSK) modulation of the wobble. The MSK modulation is achieved by combining wobbles of nominal frequency with wobbles of 1.5 times higher frequency. An advantage of MSK is that the phase of the wobble is continuous. An example of a MSK-modulated bit is shown in Fig. 12. Three addresses are aligned with one recording-unit block of 64 KB. This yields 6 addresses per revolution at the inner radius, which is sufficient for fast access as required for data applications.

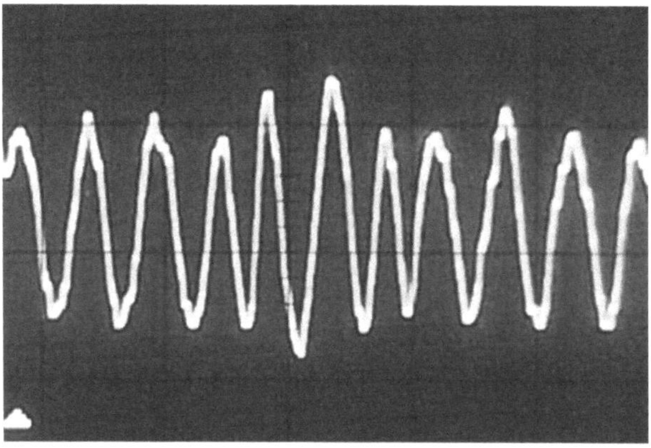

Fig. 12. Time trace of the wobble read-out signal with a MSK-bit.

6.4.6 Conclusions

We have presented first results on groove-only recording under DVR conditions, with wavelength of 405 nm, high numerical aperture of NA = 0.85, and thin cover layer of 0.1 mm. High-quality groove-only substrates are mastered with NA = 0.90 and λ=257 nm. A capacity of 23.3 GB is feasible. Wide system margins are obtained, and no thermal cross-write is present. The wobble addressing format is based on a 69T wobble and MSK modulation. Finally, we want to emphasise that the results described in this paper have been obtained using a standard detection method: a linear equaliser in combination with threshold detection. Using advanced signal processing higher capacities and / or wider margins will be obtained.

References

[1] T. Narahara et al.: Jpn. J. Appl. Phys. 39 (2000) 912.

[2] K. Schep et al.: Jpn. J. Appl. Phys. 40 (2001) 1813

[3] K. Yamamoto et al.: Jpn. J. Appl. Phys. 36 (1997) 456.

[4] K. Osato et al.: Proc. SPIE 3401 (1998) 80.

[5] Y.U. Martynov et al.: Jpn. J. Appl. Phys. 38 (1999) 1786.

[6] M.M.J. Decré and P.H.G.M. Vromans: Jpn. J. Appl. Phys. 39 (2000) 775.

[7] Y. Kasami et al.: Jpn. J. Appl. Phys. 39 (2000) 756.

[8] B. Tieke et al.: Jpn. J. Appl. Phys. 39 (2000) 762.

[9] H.J. Borg et al.: Proc. SPIE 3864 (1999) 191.

[10] M.J. Dekker et al.: Proc. SPIE 4090 (2000) 28.

[11] H.J. Borg et al.: Jpn. J. Appl. Phys. 40 (2001) 1592

[12] H. Inoue et al.: Jpn. J. Appl. Phys. 40 (2001) 1641.

[13] F. Yokogawa et al.: Jpn. J. Appl Phys 39 (2000) 819.

[14] K. Saito et al.: Postdeadline Technical Digest ODS 2001, WC9.

[15] M. Horie et al.: Technical Digest ODS 2001, 37.

[16] H. Inoue et al.: Technical Digest ODS 2001, 40.

[17] M. Shinotsuka et al.: Postdeadline Technical Digest ODS 2001, P36.

6.5 Wobble-address format of the Blu-ray Disc

Shoei Kobayashi, Shigeru Furumiya[1], Bert Stek[2], Hiromichi Ishibashi[1], Tamotsu Yamagami[1] and Kees Schep[2]

BD Development Division, AV/IT Development Group, Sony Corporation, 2-15-3 Konan, Minato-ku, Tokyo 108-6201, Japan
[1]Storage Media Systems Development Center, Matsushita Electric Industrial Co., Ltd., 1006 Kadoma, Osaka 571-8501, Japan
[2]Philips Research Laboratories, Royal Philips Electronics, Prof. Holstlaan 4, 5656 AA, Eindhoven, The Netherlands

Abstract
We explain a new address format that is adapted in the Blu-ray Disc Rewritable format. The address information is prerecorded by wobbling the groove track. It is based on a single tone carrier to obtain a stable and accurate write clock. The modulation scheme is a combination of minimum shift keying (MSK) that is strong for media noise and saw tooth wobble (STW) that is strong for wobble shift. The robustness of the format is demonstrated by measurement of the system margins.

6.5.1 Introduction

At the beginning of this year, a number of companies announced the Blu-ray Disc format. It is based on key technologies such as a blue-laser diode (λ=405 nm), an objective lens with a high numerical aperture (NA=0.85), a disc with a thin cover layer (0.1 mm thick), and phase-change recording on a groove-only substrate.[1-3] The rewritable area of the Blu-ray Disc has a fixed track pitch of 320 nm and a scalable linear density that results in a capacity of 23.3 GB or 25 GB for single layer disc and 46.6 GB or 50 GB for dual layer disc on a CD-size disc. It also has a high user-data transfer rate for recording and play-back of 36 Mbps, so it is well suited for high-definition (HD) video recording. In this paper we explain the basic concept of the wobble-address format of the Blu-ray Disc and demonstrate its robustness experimentally.

6.5.2 Requirements for the wobble-address format

The blank Blu-ray Disc has a continuous groove with a track pitch of 320 nm to generate a tracking signal. The groove is wobbled in a pre-determined manner so that it can also be used for write-clock generation, for retrieving timing and address information, and for storing disc information. The typical amplitude of the wobbling is ±10 nm.

To derive a stable and accurate write clock, the wobble is predominantly a single-tone carrier. The wobble length is chosen to reduce the effect of wobble beat on the single-tone carrier. One wobble period corresponds to exactly 69 channel bits: 69×80.0 nm $= 5.52$ μm for 23.3 GB capacity. Since the main data is written synchronized with the wobble, a geometrically shorter wobble period automatically results in a higher capacity.

For retrieving timing and address information, modulation is added to the single-tone carrier. This modulation has to be robust for different types of distortion. The first type of distortion is the almost white noise that is due to media and groove noise and due to cross-talk from the main data on the wobble signal in case of written tracks. The second type of distortion arises when the wobble phased locked loop (PLL) is in lock but there is an uncertainty in the precise wobble position with respect to the start of an address in pre-groove (ADIP) unit. This so-called wobble shift can occur, for example, after a track jump. The third type of distortion is the cross-talk from the wobble signals in the adjacent tracks. The fourth type of distortion arises from local defects including dust and scratches on the disc surface.

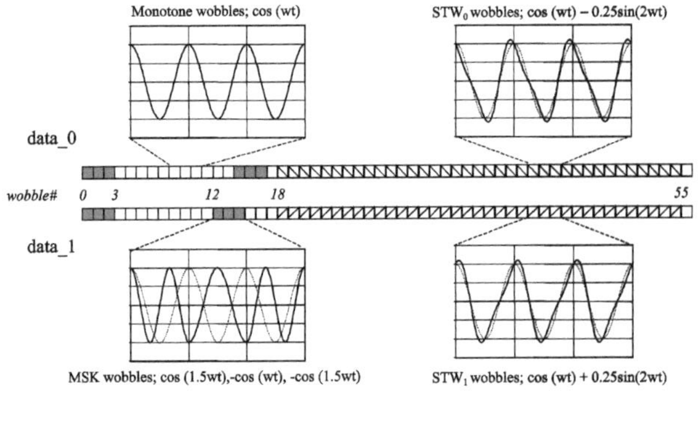

Fig. 1. Schematic representation of the ADIP units for data_0 and data_1. An ADIP unit has a length of 56 wobbles, which contain the first MSK mark for bit sync, the second MSK mark characterized by the difference of the position and 37 STWs characterized by the difference of the slope for data_0 or data_1.

6.5.3 Modulation scheme

To cope with all these distortions of different types, we have combined minimum-shift-keying (MSK) marks[1,2,4] and saw-tooth wobbles (STW)[3,5] in a single wobble-address format. Fig. 1 gives a schematic representation of the combination strategy. The wobble-address bits are stored in ADIPunits. An ADIPunit has a length of 56 wobble periods. Fig. 1 shows the ADIPunits that represent the binary data bits 'data_0' and 'data_1'. All ADIPunits start with an MSK mark, the bit sync, which can be used to identify the start of an ADIPunit. The bit sync is followed by a second MSK mark with monotone wobbles; 11 for data_0 and 9 for data_1. Note that the MSK marks are based on cosine waveforms and not on the sine waveforms presented previously.[1,2,4] This makes the low frequency component of the wobble further reduced. The 37 wobbles from positions 18 to 54 in the ADIPunits are modulated with STWs; slow falling and steep rising edge for data_0 and steep falling and slow rising edge for data_1. Note that in contrast to previous papers[3,5] only the fundamental and 2nd harmonic frequency are used in the STW, this reduces the required band width for disc mastering and limits the deterioration of other signals by the higher harmonics of the wobble. The amplitude of the 2nd harmonic in the STW equals one quarter of the fundamental tone.

A drive can use the MSK marks, or the STWs, or both to detect whether the ADIPunit is a data_0 or a data_1.

6.5.4 Detection scheme

The MSK marks and STWs can be detected using the same heterodyne circuit which consists of a carrier multiplier, an integrator and a sample and hold element as shown in Fig. 2.

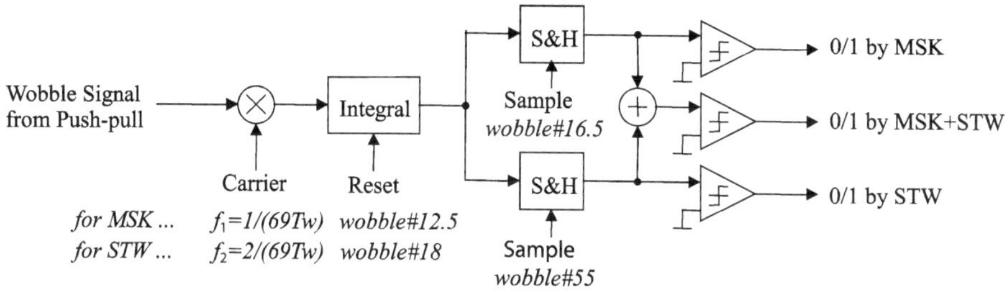

Fig. 2. An example of the heterodyne detection circuit for the MSK marks and the STWs. For the detection of the MSK marks, the carrier of $\cos(2\pi f_1 t)$ inverted at wobble#14.5 is supplied to the multiplier. For the STWs, the carrier of $\sin(2\pi f_2 t)$ compensated for phase offset by the reference STWs is supplied.

The wobble signal is multiplied by the cosine carrier of the fundamental frequency for detecting MSK in the multiplier. On the other hand it is multiplied by the sine carrier of the 2nd harmonic frequency for detecting STW. The waveform in Fig. 3 is shown to know the principle of the detection scheme. The left figure is the MSK detection and the right figure is the STWs detection. They show MSK and STWs modulation waveform, the waveform after the multiplier and the waveform after the integral in the previous circuit. It can be seen that the integral circuit integrates the output of the synchronous demodulator with better signal to noise ratio (SNR).

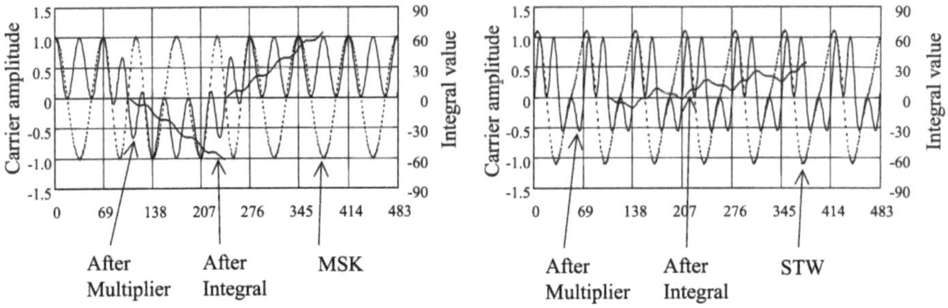

Fig. 3. The waveforms of the MSK and STW detection circuit in Fig. 2. Both show the modulation waveform, the waveform after the multiplier and the waveform after the integrator. To compare MSK and STW, both integrated periods are set to a length of 4 wobble periods.

Figure 4 shows the concept of wobble shift and the calculated SNR for detection using the circuit of Fig. 2 in case of no cross talk from adjacent tracks.

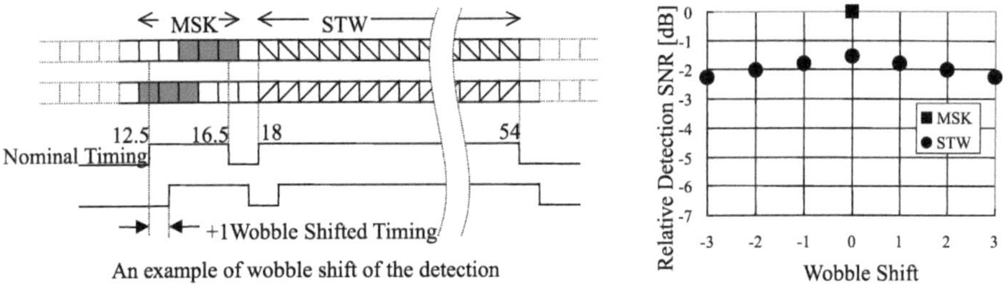

Fig. 4. The concept of wobble shift (left) and the calculated signal-to-noise ratio (right) for detection of the MSK marks and the STWs using the circuit of Fig. 2 in case of no cross talk from adjacent tracks as a function of wobble shift.

If no wobble shift occurs, the detection of MSK marks has a 1.6 dB higher SNR. In case of wobble shift, the MSK detection fails while the STW detection remains stable. Thus MSK detection is more robust for distortion by noise while STW is more robust for distortion by wobble shift. By combining MSK and STW in one address format, robustness for both noise and wobble shift is achieved. Also other advantages of MSK and STW are complementary. The STWs are spread widely over 37 wobble periods, which gives strong robustness against local defects. The MSK marks, on the other hand, are localized in only 3 wobble periods, which gives better position information for a bit sync to find the start of an ADIP unit.

The third distortion is cross talk from the wobble signals in adjacent tracks. This cross talk is quite large due to the relatively small track pitch in the Blu-ray Disc format. Two effects that arise from the cross talk can be distinguished. The first effect is due to the beat of the dominant single tone in the wobble. This beat arises due to the slightly different angular frequency of the wobbles in adjacent tracks in the constant-linear-velocity (CLV) format. The period of this wobble beat is easily calculated from the track pitch and the wobble length: $2 \times \pi \times 320\,\text{nm}/5.52\,\mu\text{m} = 2.75$ revolutions (for 23.3 GB capacity). The wobble beat gives rise to both amplitude and phase modulation of the single-tone wobble. The phase modulation causes a phase offset for the wobble PLL that thereupon degrades the wobble detection of both MSK marks and STWs. This effect is more severe for STW than for MSK because of the higher frequency of STW.

The second effect of cross talk arises according to the alignment of the ADIPunits in adjacent tracks. This effect has a period that is 56 times larger than the wobble beat, i.e. 154 tracks in case of 23.3 GB.

The effect of wobble beat can be corrected for by calibrating the phase offset using special reference ADIPunits interleaved with every five ADIP units. Because the reference ADIPunits contain 37 fixed STWs of known polarity (always equal to 'data_0'), the phase offset can be derived from them. Fig. 5 shows an example circuit compensating the phase offset between the wobble PLL and STWs. It consists of the circuit in Fig. 2 and the additional phase detector and phase adjuster. The phase offset detector detects the offset between the 2nd harmonic frequency that is play backed and the 2nd harmonic carrier that is multiplied by STWs at multiplier. The phase offset adjuster compensates the phase delay of the 2nd harmonic carrier so that the phase offset detected in the phase detector can be zero. The detection of STW becomes more stable by adopting the reference ADIP units and implementing the phase offset compensator using the reference ADIP units.

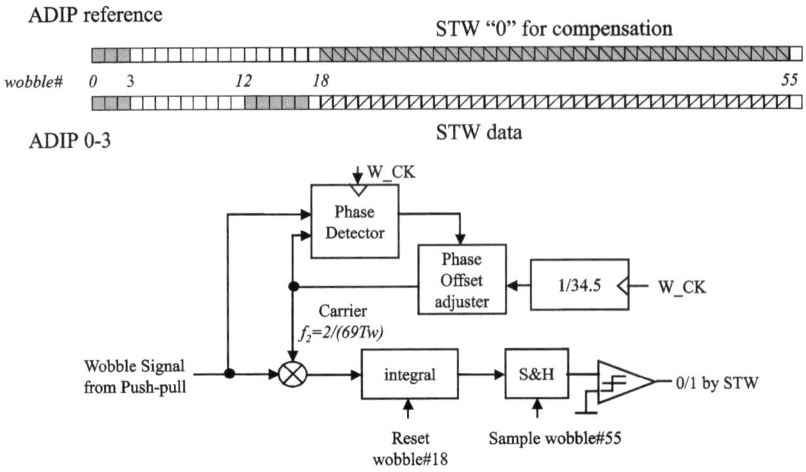

Fig. 5. The reference ADIPunit and an example circuit that compensates the phase offset for the STW using the reference ADIPunit.The circuit consists of that of Fig. 2 and an additional phase detector and phase-offset adjuster.

6.5.5 Measurements of eye-patterns and margins

The effect of cross talk on the detection of MSK and STW is shown in the left picture in Fig. 6. However, this effect of the cross talk does not close the eye pattern. Furthermore, by adding the integral values of MSK and STW, the resulting hybrid detection gives an open eye pattern shown in the right picture in Fig. 6.

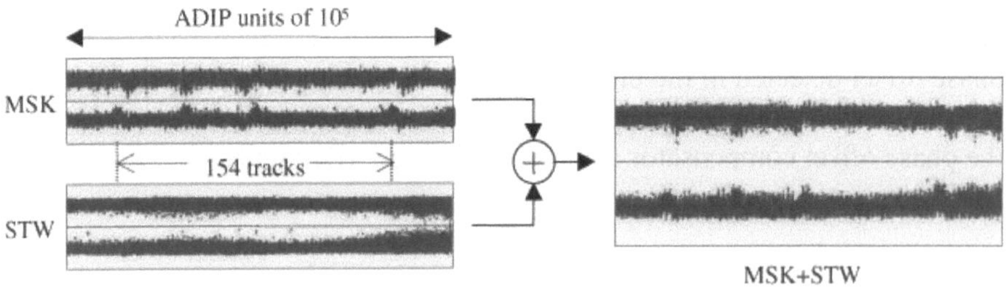

Fig. 6. The eye patterns on detection level for data_0 and data_1 for MSK-only, STW-only, and hybrid MSK + STW detection. Pictures correspond to the inputs of the three comparators in Fig. 2.

In Fig. 7 we show the margins for MSK-only, STW-only, and the hybrid MSK + STW wobble detections for both radial tilt and defocus in case of written tracks. These margins include all effects of noise of writing, wobble shift, defects, and cross talk from adjacent tracks. It is clear that the margins for the ADIP detection are larger than the margins for the main data. The bit error rate for the ADIPis far below 10^{-4} for the hybrid MSK + STW detection.

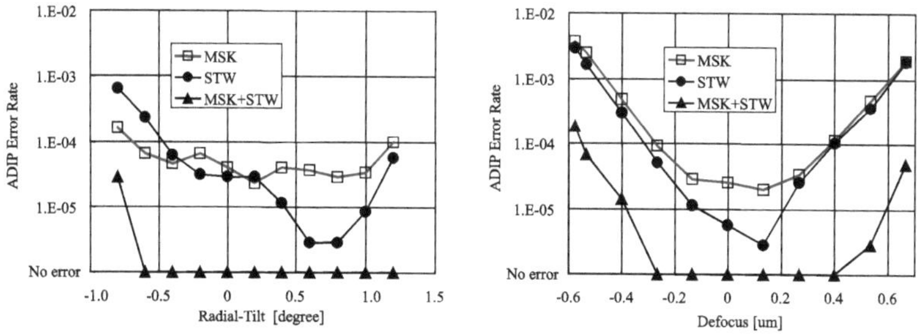

Fig. 7. Error rates of ADIPunits of a sample disc as function of radial tilt and defocus. The measurements are done with ADIPunits of 3.5 x 10^5 sample points for written tracks.

6.5.6 ADIP format

The ADIP format is summarized here. One wobble length is exactly 69 channel clocks of main data. One ADIP unit is 56 wobbles. By combining 83 consecutive ADIP units, an ADIP word is formed. Besides the ADIP units for data_0 and data_1 and for reference STWs, also ADIP units for synchronization are defined. Each ADIP word contains one address and also some additional bits to store auxiliary information such as disc information. The addresses and auxiliary data are protected by an error correction code (ECC) based on Informed Decoding.[6]

The main data format of the Blu-ray Disc is identical to the main data format that was described earlier in the references.[7,8] A recording frame in the main data format has a length of 1932 channel bits, so the length of one ADIP unit is identical to the length of two recording frames: $56 \times 69 = 2 \times 1932$. One recording unit block (RUB) in the main data contains 498 recording frames, so exactly 3 ADIP words fit into one RUB.

6.5.7 Conclusion

We explained the newly-developed wobble-address format of the Blu-ray Disc and demonstrated its robustness by margin measurements. A stable and precise write clock can be generated from the predominantly single-tone signal from the wobble. The retrieval of addresses is robust for various noise sources because of the combination of both MSK marks and STWs in the modulation scheme.

References

[1] I. Ichimura, S. Masuhara, J. Nakano, Y. Kasami, K. Yasuda, O.Kawakubo and K. Osato: Proc. SPIE **4342** (2002) 168.

[2] M. Kuijper, I. P. Ubbens, L. Spruijt, J. M. ter Meulen and K. Schep: Proc. SPIE **4342** (2002) 178.

[3] S. Furumiya, J. Minamino, H. Miyashita, A. Nakamura, M. Shouji, T. Ishida and H. Ishibashi: Proc. SPIE **4342** (2002) 186.

[4] H. Ogawa: Techn. Dig. ISOM 2001, p. 6.

[5] J. Minamino, M. Nakao, N. Kimura, M. Shouji, S. Furumiya, H. Ishibashi and E. Ohno: Jpn. J. Appl. Phys. **41** (2002) 1741.

[6] M. Blüm, L. Tolhuizen and S. Baggen: Jpn. J. Appl. Phys. **41** (2002) 1785.

[7] T. Narahara, S. Kobayashi, M. Hattori, Y. Shimpuku, G. J. van den Enden, J. A. H. M. Kahlman, M. van Dijk and R. van Woudenberg: Jpn. J. Appl. Phys. **39** (2000) 912.

[8] K. Schep, B. Stek, R. van Woudenberg, M. Blüm, S. Kobayashi, T. Narahara, T. Yamagami and H. Ogawa: Jpn. J. Appl. Phys. **40** (2001) 1813.

6.6 Liquid immersion deep-UV optical disc mastering for Blu-ray Disc Read-Only Memory

Jaap H. M. Neijzen, Erwin R. Meinders and Helmar van Santen

Philips Research Laboratories, Prof. Holstlaan 4, 5656 AA, Eindhoven, The Netherlands

Abstract
The liquid immersion mastering technique has been successfully applied to the mastering of read-only memory (ROM) discs for the Blu-ray disc (BD) system. Replicated discs with a density corresponding to 25 GB in a single layer on a 12 cm disc showed a bottom jitter of less than 5%. Results concerning process latitude and disc uniformity are presented. A full-format 25 GB ROM disc containing over 2 h of high-definition video content has been mastered according to the BD target specification. The results obtained for a reduced channel bit length show the potential of liquid immersion mastering for densities beyond 31 GB per layer.

6.6.1 Introduction

The main advantages of optical discs such as compact discs (CDs) and digital versatile discs (DVDs), are the ease and low cost of read-only memory (ROM) mass reproduction facilitated by the replication process. In the third generation system, that is the Blu-ray disc (BD)[1] system, the ROM capacities specified for a single layer of a 12 cm disc are 23.3 GB, 25 GB and the reserved capacity of 27 GB. We have successfully used far-field deep-UV mastering equipment with a wavelength (λ) of 257 nm and a numerical aperture (NA) of 0.9 in recording densities up to 23.3 GB per layer. Higher densities, however, are increasingly difficult to master and a further reduction in the size of the writing spot of the recorder is needed to keep sufficient process margins. One possibility is to consider e-beam recording,[2,3] but this requires considerable investments in mastering equipment. Another promising technique is phase transition mastering,[4] but this requires optimized phase-transition materials and specific write strategies. To maintain the advantages of conventional optical disc mastering and to further reduce the size of the writing spot, either the wavelength (λ) of the light has to be decreased or the NA of the writing objective has to be increased. Because suitable continuous wave lasers with $\lambda < 257$ nm are not yet available, the NA of the optical system has to exceed one. This can be achieved by solid immersion[5] or liquid immersion.[6–8] Liquid immersion has the advantage of a larger flying height than the flying height of several tens of nanometers required in solid immersion. In this paper, the feasibility of the deep-UV liquid immersion mastering of 25 GB ROM discs is successfully demonstrated. In addition, promising results for higher densities are presented.

6.6.2 Liquid immersion mastering concept and implementation

In liquid immersion microscopy, the immersion liquid is applied between a steady lens and a steady object. The adhesive forces of the liquid keep the object immersed. When the object moves with respect to the lens, however, the breakdown of immersion may occur, either by pulling the liquid away from the lens or by pulling air underneath the objective. The key issue in applying liquid immersion in dynamic systems such as a mastering machine therefore is to maintain a stable liquid film between the stationary lens and the moving substrate. This may be achieved by immersing the whole system in a liquid. This solution, however, is likely to suffer from vibrations, unacceptable in mastering equipment. Therefore, a better immersion concept has been developed.

Liquid immersion is achieved by locally maintaining a water film between the front-lens element of the objective and the rotating photoresist-coated disc (see Fig. 1).

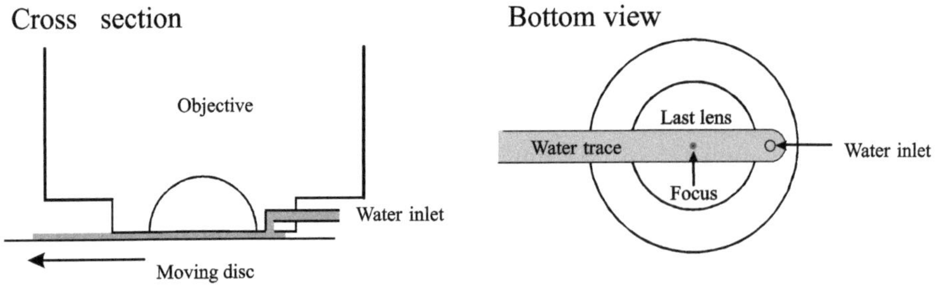

Fig. 1. Schematics of the liquid immersion concept.

Water is a natural choice for the immersion liquid as it is transparent for deep-UV light and compatible with novolac resist processing. The practical problem that water immersion objectives are as yet not commercially available for 257 nm has been solved by supplementing a commercially available far-field lens (NA=0.9, λ=257 nm) with an additional almost hemispherical lens element and a water supply system. In this way, the far-field objective is transformed into a water immersion lens. This results in a diffraction-limited optical spot corresponding to an NA slightly above 1.2. Water is continuously supplied through a hole just upstream of the immersion lens.

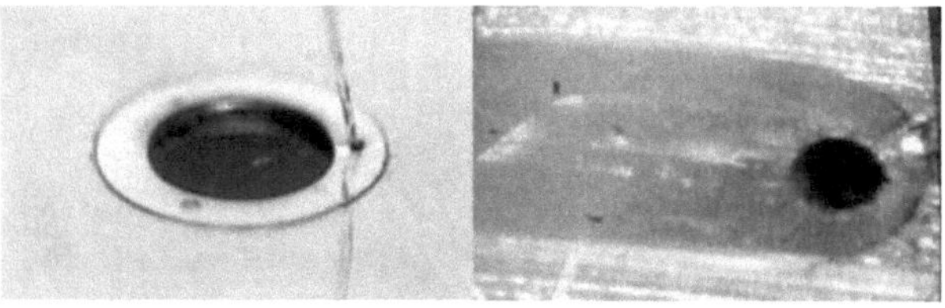

Fig. 2. Photographs of the designed liquid immersion lens head; (a) Liquid immersion water supply shown for a free-standing lens and (b) the water trace for a lens in focus (seen from below through the transparent substrate).

Figure 2(a) shows a photograph of a tiny water jet coming from the water outlet, next to the lens. If the lens approaches the substrate, the rotating disc pulls the water under the lens resulting in a stable narrow trace of water between the lens and the resist layer. The photograph in Fig. 2(b) shows an image of such a water trace made from below through the transparent substrate. Water pressure is kept sufficiently high to avoid gas inclusion. When the lens is in focus, the water trace is typically 7 μm thick and 200 μm wide at writing velocities up to 5 m/s. These dimensions of the water trace lead to a minimal liquid consumption and limit the force exerted by water on the lens. Thus, the focus actuation of the lens is not hampered by the presence of water. The successful actuation of an objective lens against a water film was demonstrated previously.[6] To avoid possible vibrations that might be caused by the water trace hitting the lens after one revolution of the substrate, a separate water removal device has been added downstream of the lens, as indicated in Fig. 3. It consists of an air bearing floating on the resist layer with an additional suction opening on the front side. This combination effectively removes the water trace from the substrate. Apart from the modifications of the water immersion lens and the addition of the water removal device, the implementation of liquid immersion mastering on a conventional deep-UV mastering machine is simple. The same substrate, resist and processing can be used as in conventional deep-UV mastering. The initial problems with the nonuniformity of the written tracks[7,8] have been solved by carefully preventing the contamination of the immersion liquid. In the next section, the feasibility of the liquid immersion technique for the mastering of ROM discs of the BD generation is demonstrated. In Sect. 6.6.4 liquid immersion mastering experiments are reported for data densities beyond that of the BD generation.

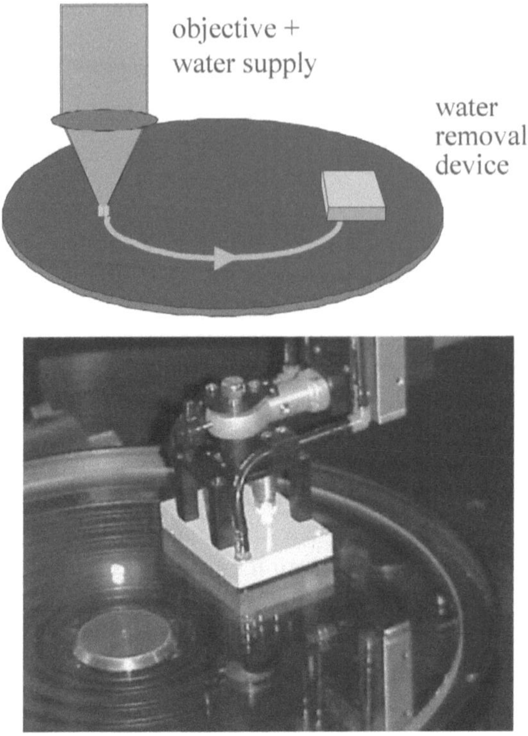

Fig. 3. Schematic drawing (upper part) of the relative position on the substrate of the lens, the water trace and the water removal air bearing device. The lower part shows a photograph of the water removal device, with the water suction opening in the front corner.

6.6.3 25 GB Blu-ray Disc ROM

To investigate the potential of the liquid immersion concept for the mastering of BD ROM discs, experiments have been carried out using an 80 nm-thick novolac resist layer. The exposure and resist process settings have been optimized and random data patterns were written according to the expected BD format. Most of the experiments were carried out for a density of 25 GB per 12 cm disc, which corresponds to a track pitch (TP) of 320 nm and a channel bit length (CBL) of 74.5 nm using a 17PP modulation code. After the development of the resist, a thin Ni layer is deposited on the substrate by sputtering. On top of the sputtered layer, a substantial Ni layer thickness is grown by galvanic metal deposition, so that the Ni layer can be separated from the resist. In this way, the pit structure is transferred to a metal father stamper having tiny bumps at the position of the exposed resist areas. The stamper is subsequently used to produce discs by glass-2p replication for making test samples or by injection moulding for the mass production of ROM discs. Fig. 4 shows an atomic force microscopy image of a data pattern on a father stamper.

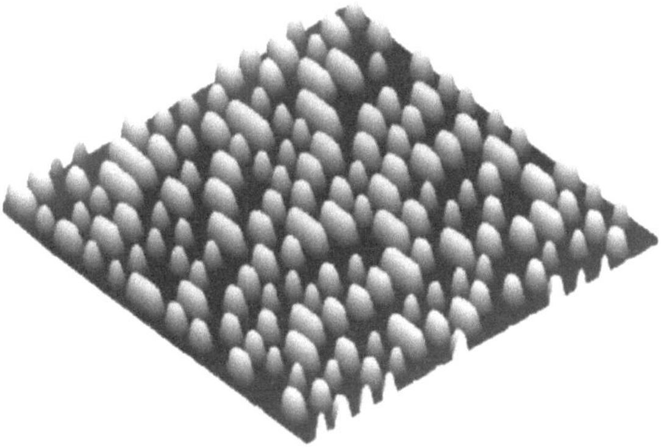

Fig. 4. Atomic force microscopy image of a data pattern on a 25 GB father stamper (TP=320 nm, CBL=74.5 nm). The size of the scanned area is 5 μm x 5 μm. The measured height of the bumps is 80 nm, which equals the original resist layer thickness.

All marks, including the smallest T2 marks with a nominal length of 149 nm, are well defined and have been written over the full depth of the original resist layer. The average width of the smallest almost circular pits is about 120-130 nm, and somewhat larger for longer symbols. This width is equal to the expected full width at half maximum value ($0.6\lambda/NA = 125$ nm) of the Airy pattern of the focussed laser. Thus, the length of the shortest effects, about 150 nm, is not limited by the laser spot size, which indicates that the resolution is sufficient for writing this density. A typical scanning-electron-microscopy (SEM) image of a replicated glass-2p disc is shown in Fig. 5. It shows the size

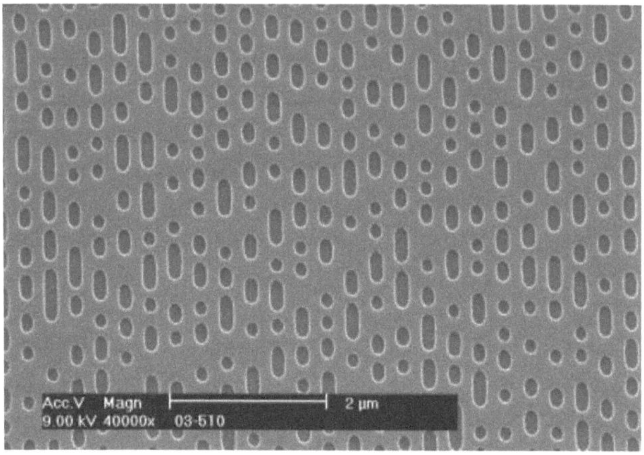

Fig.5. SEM of a replicated 25 GB disc (TP=320 nm, CBL=74.5 nm)

and reproducibility of the replicated pits. Replicated discs (both glass-2p and injection molded) with a cover layer of 100 mm have been evaluated using a BD-ROM test player (Pulstec, $\lambda=405$ nm, NA=0.85). Fig. 6 shows an example of an eye pattern obtained from a 25 GB disc using a limit equalizer.[9] Limit equalizer settings as prescribed by the BD format were used. The corresponding time-interval histogram shows a typical pit length distribution of the various symbols in the 17PP code plotted on a semi-logarithmic scale.

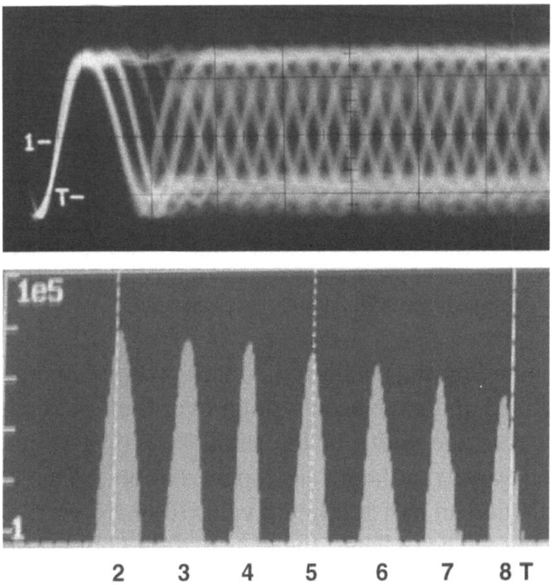

Fig. 6. Limit equalizer eye pattern for 25 GB density (TP=320 nm, CBL=74.5 nm) and corresponding time interval histogram plotted on a semi-logarithmic scale.

The figure shows perfectly separated narrow distributions for different symbols. The maximum allowable jitter in the tentative BD standard is 6.5%. Bottom jitter values below 5% have been measured whereas typical jitter values are approximately 5.5%. For the 25 GB ROM discs, we have not used any write strategy to obtain this low jitter. Also the measured normalized push-pull (NPP) and asymmetry of the signal are within the bounds prescribed by the tentative BD format (NPP > 0.1 and -0.05 < asymmetry < 0.15). In order to give an impression of the reproducibility and process latitude, the relationship between the measured asymmetry and limit equalizer jitter is plotted in Fig. 7 for a large number of discs and exposure conditions. The asymmetry increases monotonically, almost linearly, with exposure dose. Therefore, the horizontal axis can also be considered as the exposure dose axis. The data shown in Fig. 7 were obtained under slightly varying exposure settings (i.e. focus) and process conditions (i.e. resist thickness and substrate preparation).

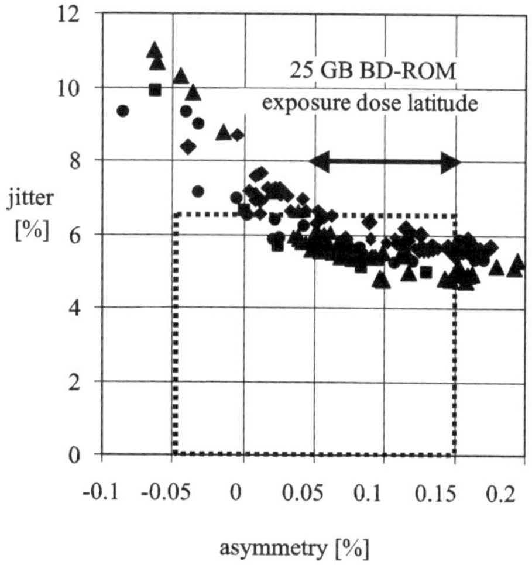

Fig. 7. Limit equalizer jitter versus asymmetry, indicating the exposure dose latitude. Different symbols refer to different measurement series.

Different symbols refer to different measurement series. The scatter on the data will be significantly smaller than that shown in this figure if the exposure and processing conditions are better preserved. The acceptable ranges of jitter and asymmetry values are indicated in Fig. 7 by the dotted box. There is a certain range of asymmetry values for which the jitter values are within the specification. This range corresponds to an exposure dose latitude of 15%, which makes it easy to reproducibly fulfill the jitter/asymmetry requirements of the BD format.

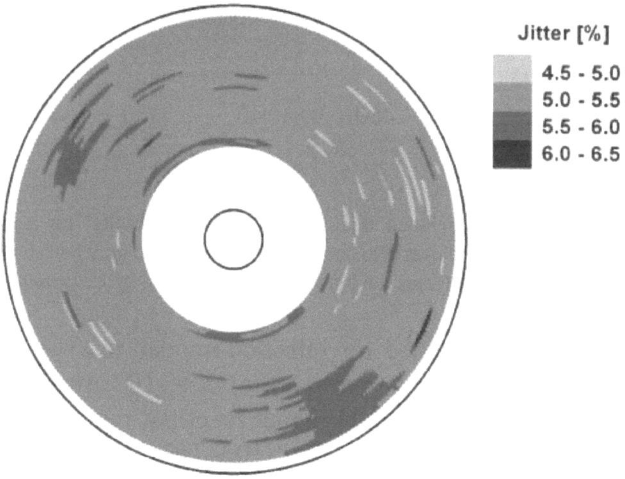

Fig.8. Jitter map of a full format 25 GB BD ROM disc.

The jitter map for a fully written disc, shown in Fig. 8, illustrates the disc uniformity that can be realized. The main part of the surface area has jitter values between 5% and 5.5%. Jitter was measured as a function of radial position at fixed azimuths. An optimum sample size of 10,000 data points was selected from jitter measurements versus the number of samples, ensuring a sufficiently large sample size to correctly measure the statistics and at the same time allowing for non-overlapping data sections. The total area of this disc is within the BD specifications with respect to jitter, asymmetry and NPP.

A number of the test discs were written with a format generator and high-definition video content was encoded according to the BD standard. The quality of readout of the data is represented by the symbol error rate (SER), also specified in the BD standard to ensure reliable readout. In Fig. 9, SER measurements, based on limit equalizer detection, are shown for a glass-2p replicated test disc as a function of disc radius.

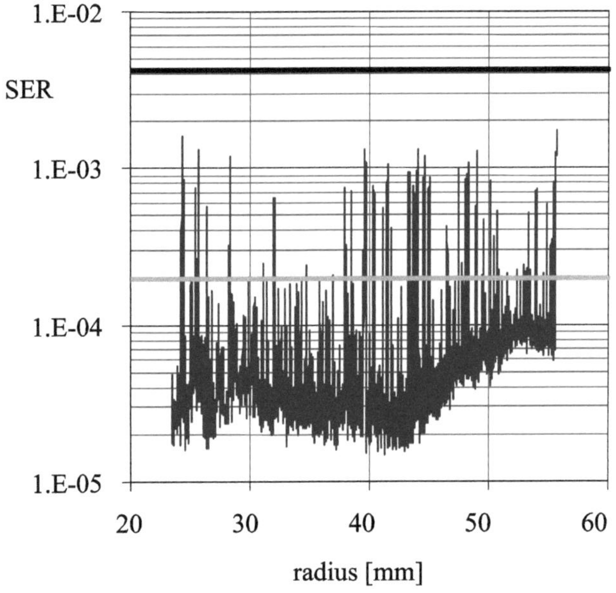

Fig. 9. SER results (limit equalizer detection; non-Viterbi results) for a 25 GB disc as a function of disc radius, averaged over 100 consecutive blocks. The gray horizontal line indicates the bottom SER target value, whereas the black line indicates the system error correction limit.

Similar results are obtained for injection-molded test discs. The bottom error rate is well below the target value of 2×10^{-4}. However, a number of localized SER spikes exceed the target level but remain below the error correction limit of the system, i.e. 4.2×10^{-3}. The occurrence of these burst errors is most likely due to spot defects on the disc. These defects are not necessarily caused by the mastering step, but may also be due to disc preparation steps such as stamper making, replication or cover layer application. We are currently investigating the origin of these defects. However, even in these burst regions, the SER is at a level that can be handled by the error correction system. This has allowed us to successfully play back the video content of the mastered 25 GB BD-ROM disc. The replicated injection-molded disc has been read out on an experimental Blu-ray ROM test player (λ=405 nm and NA=0.85) and the original data consisting of more than 2 h of high-definition video content was successfully recovered.

6.6.4 Mastering of ROM discs with a capacity beyond 25 GB

A number of experiments were carried out to explore the possibility of mastering capacities beyond 25 GB with the liquid immersion mastering technology. The track pitch was kept constant in these experiments at 320 nm. By varying channel bit length in discrete steps, we have made comparative measurements over a capacity range from 23.3 to 31 GB. A SEM image of a replicated 31 GB disc (TP=320 nm, CBL=60 nm) is shown in Fig. 10.

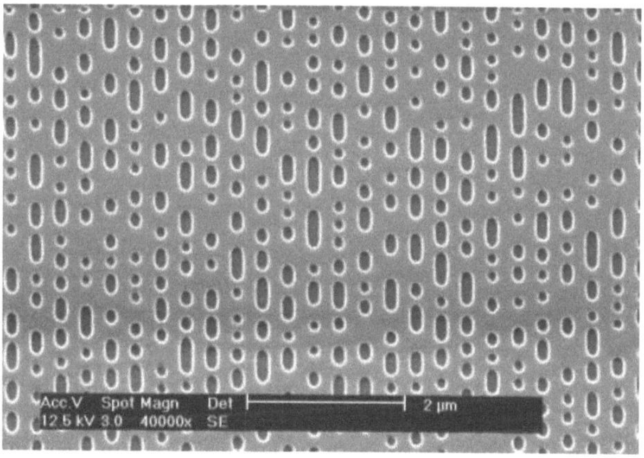

Fig. 10. SEM measurement on a replicated 31 GB disc (TP=320 nm, CBL=60 nm).

Also at this density, T2 pits with a nominal length of 120 nm are well defined and have been exposed down to the bottom of the original 80 nm resist layer. Fig. 11 shows the resulting jitter increasing with increasing capacity (decreasing CBL), using the limit equalizer settings as prescribed by the BD format up to 27 GB. A bottom jitter of 6.8% was measured for 27 GB (TP=320 nm and CBL=69 nm) using a simple write strategy. This jitter is already significantly higher than that for 25 GB, which could be obtained without a write strategy. For capacities higher than 27 GB, jitter increases markedly. The question arises whether this strong increase marks the limit of the application range of liquid immersion mastering or is it a sign of an insufficient resolution of the readout system. Are the highest frequencies in the signal for capacities above 27 GB so close to the cutoff frequency of the blue readout channel that the parameter jitter, based on threshold detection, is no longer a suitable figure of merit?

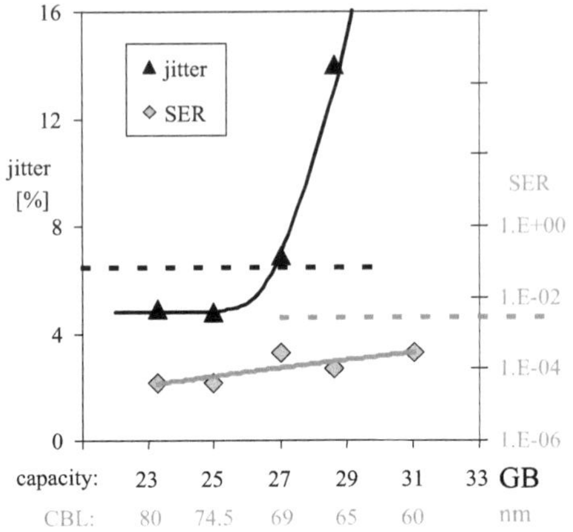

Fig. 11. Limit equalizer jitter and SER (Viterbi) at TP=320 nm increase with increasing density (decreasing CBL). The dotted black and grey line indicate the maximum allowable jitter and the system error correction limit, respectively.

This question is answered by the SER results based on Viterbi detection,[10,11] which are also shown in Fig. 11. The SER results measured at a zero tilt show only a gradual increase with increasing density. The SER stays below the system error correction limit of 4.2×10^{-3} up to the highest density measured. The SER tilt margins presented in Fig. 12 are even more relevant than the bottom SER results. Due to the adaptive equalization used, the tangential tilt margin found is even larger in this case than the radial tilt margin. The radial tilt margin can be increased further by cross talk cancellation techniques. The margins of $\pm 0.7°$ shown for the 31 GB capacity are just slightly narrower than the corresponding 25 GB margins. This behavior justifies the conclusion that the liquid immersion technique is capable of mastering these high densities. Thus, the marked increase in jitter above 27 GB proves that jitter is not a suitable figure of merit in this high-density region, and that Viterbi detection becomes inevitable. The measured tilt margins illustrate the potential to master and read back densities beyond 25 GB per layer. Even a further increase in density beyond 31 GB seems possible if the track pitch of 320 nm is reduced as well. Work on the Viterbi detection of high-data-capacity ROM discs is ongoing.[12]

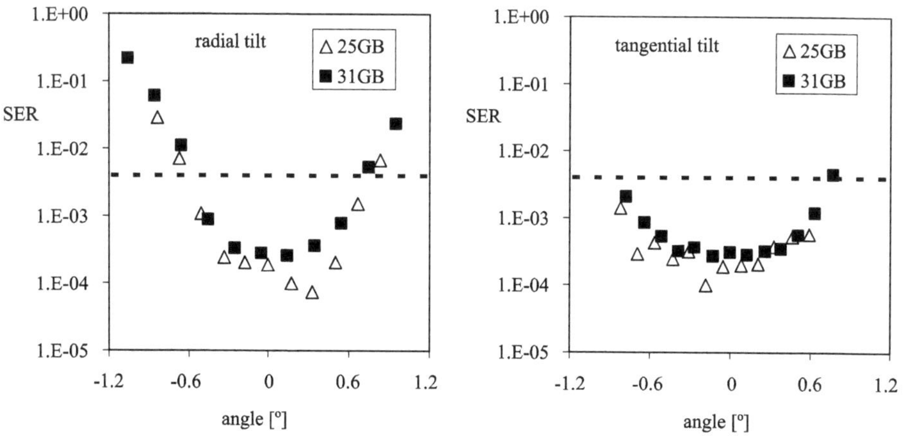

Fig. 12. Radial and tangential SER tilt margins, using Viterbi detection, at TP=320 nm for densities corresponding to 25 GB and 31 GB. The dotted lines correspond to a system error correction limit of 4.2 x 10^{-3}.

6.6.5 Concluding remarks

The results discussed in this paper show that liquid immersion mastering has developed into an attractive technology for the mass production of BD generation ROM discs. The available process margins are sufficient to write uniform full-format 25 GB discs within the specification of the Blu-ray disc format. The potential of liquid immersion mastering for higher densities is convincingly demonstrated by the results of this study for a data capacity of 31 GB.

Acknowledgements

The authors would like to acknowledge the staff of the Philips Optical Disc Technology Centre (ODTC) for essential support in making and evaluating the replicated discs. Many colleagues at Philips Research and Philips Optical Storage contributed to this work by the generation of formatted video data, by recorder measurements and data analysis using Viterbi detection. Their indispensable contribution is gratefully acknowledged.

References

[1] M. Kuijper, I. Ubbens, L. Spruijt, J. M. ter Meulen and K. Schep: Proc. SPIE **4342** (2001) 178.

[2] T. Iida, M. Katsumura, T. Imai and F. Yokogawa: Optical Data Storage 2001 Tech. Dig., p. 253.

[3] M. Takeda, M. Furuki, M. Yamamoto, Y. Aki, H. Kawase, M. Koizumi, S. Takashima, T. Miyokawa and N. Date: Int. Symp. Optical Memory and Optical Data Storage 2002 Tech. Dig., p. 18.

[4] K. Osato, S. Kai, Y. Takemoto, T. Nakao, K. Nakagawa, A. Kouchiyama and K. Aratani: Optical Data Storage 2003 Tech. Dig., p. 52.

[5] S. Imanishi, M. Takeda, M. Yamamoto, N. Mukai, K. Takagi and T. Kono: Int. Symp. Optical Memory and Optical Data Storage 2002 Tech. Dig., p. 296.

[6] H. van Santen and W. R. Koppers: Optical Data Storage 2001 Tech. Dig., p. 250.

[7] J. Neijzen and H. van Santen: Int. Symp. Optical Memory and Optical Data Storage 2002 Tech. Dig., p. 299.

[8] H. van Santen and J. Neijzen: Jpn. J. Appl. Phys. **42** (2003) 1110.

[9] S. Miyanabe, H. Kuribayashi and K. Yamamoto: Jpn. J. Appl. Phys. **38** (1999) 1715.

[10] A. J. Viterbi: IEEE Trans. Inf. Theory **IT-13** (1967) 260.

[11] R. Otte and W. Coene: IEEE Trans. Cons. Electr. **46** (2000) 1018.

[12] A. Padiy, B. Yin, C. Verschuren, R. Vlutters, T. Jansen and J. Lee: Optical Data Storage 2004 Tech. Dig., p. 34.

Index

MIX
Papier aus verantwortungsvollen Quellen
Paper from responsible sources
FSC® C105338

If you have any concerns about our products,
you can contact us on
ProductSafety@springernature.com

In case Publisher is established outside the EU,
the EU authorized representative is:
Springer Nature Customer Service Center GmbH
Europaplatz 3, 69115 Heidelberg, Germany

Printed by Libri Plureos GmbH
in Hamburg, Germany